In this book, which arose from an MSRI research workshop cosponspored by the Clay Mathematical Institute, leading experts give an overview of several areas of dynamical systems that have recently experienced substantial progress.

In symplectic geometry, a fast-growing field having its roots in classical mechanics, Cieliebak, Hofer, Latschev and Schlenk give a definitive survey of quantitative techniques and symplectic capacities, which have become a central research tool. Fisher's survey on local rigidity of group actions is a broad and up-to-date account of a flourishing subject built on the fact that for actions of noncyclic groups, topological conjugacy commonly implies smooth conjugacy.

Other articles by Eigen, Feres, Kochergin, Krieger, Navarro, Pinto, Prasad, Rand and Robinson cover subjects in hyperbolic, parabolic and symbolic dynamics as well as ergodic theory. Among the specific areas of interest are random walks and billiards, diffeomorphisms and flows on surfaces, amenability and tilings.

The articles are complemented by a fifty-page commented problem list, compiled by the editor with the help of numerous specialists. Several sections of this list focus on problems beyond the areas covered in the surveys, and all are sure to inspire and guide further research.

Mathematical Sciences Research Institute
Publications

54

Dynamics, Ergodic Theory, and Geometry
Dedicated to Anatole Katok

Mathematical Sciences Research Institute Publications

Volumes 1–4, 6–8 and 10–27 are published by Springer-Verlag

Dynamics, Ergodic Theory, and Geometry

Dedicated to Anatole Katok

Edited by

Boris Hasselblatt

Tufts University

CAMBRIDGE
UNIVERSITY PRESS

Boris Hasselblatt, Professor
Department of Mathematics, Tufts University
Medford, MA 02155-5597
Boris.Hasselblatt@Tufts.edu

Silvio Levy (*Series Editor*)
Mathematical Sciences Research Institute
17 Gauss Way, Berkeley, CA 94720
levy@msri.org

The Mathematical Sciences Research Institute wishes to acknowledge support by
the National Science Foundation and the Pacific Journal of Mathematics for the
publication of this series.

CAMBRIDGE UNIVERSITY PRESS
Cambridge, New York, Melbourne, Madrid, Cape Town,
Singapore, São Paulo, Delhi, Tokyo, Mexico City

Cambridge University Press
32 Avenue of the Americas, New York, NY 10013-2473, USA

www.cambridge.org
Information on this title: www.cambridge.org/9780521175418

© Mathematical Sciences Research Institute 2007

First published 2007
First paperback edition 2011

A catalog record for this publication is available from the British Library

Library of Congress Cataloging in Publication data

Dynamics, ergodic theory, and geometry / edited by Boris Hasselblatt.
     p.   cm. – (Mathematical Sciences Research Institute publications ; 54)
   Includes bibliographical references and index.
   ISBN 978-0-521-87541-7 (hardback)
     1. Differentiable dynamical systems. 2. Ergodic theory. 3. Geometry. I. Hasselblatt,
Boris. II. Title. III. Series.

QA614.8.D946   2007
515'.39–dc22

2007062017

ISBN 978-0-521-87541-7 Hardback
ISBN 978-0-521-17541-8 Paperback

Recent Progress in Dynamics
MSRI Publications
Volume **54**, 2007

Contents

Recent Progress in Dynamics
MSRI Publications
Volume **54**, 2007

Foreword

This volume owes its existence to the Clay Mathematics Institute / Mathematical Sciences Research Institute Workshop on "Recent Progress in Dynamics", held at the Mathematical Sciences Research Institute for a week in late September to early October 2004. This is not a proceedings volume, but most authors were participants of the workshop, and the two lead surveys reflect a good deal of what David Fisher and Helmut Hofer presented during their workshop talks.

The workshop represented a broad array of dynamical systems, not least in order to reflect the breadth of taste exhibited by Anatole Katok, whose sixtieth birthday was observed during the workshop. This was possible because we were able to invite a great number of participants from near and far, and the essential ingredient in making this possible was the generous financial support from the Clay Mathematics Institute and the Mathematical Sciences Research Institute (which is in turn supported by the National Science Foundation), as well as from the Pennsylvania State University and Tufts University. As the host of the workshop, the Mathematical Sciences Research Institute also provided administrative support for the organizers and participants. It is a pleasure to acknowledge this support.

Funding alone does not produce a successful workshop, and I want to thank my fellow workshop organizers Michael Brin, Gregory Margulis, Yakov Pesin, Peter Sarnak, Klaus Schmidt, Ralf Spatzier and Robert Zimmer. Foremost among these was Yakov Pesin, whose involvement was constant and most valuable. And a successful workshop does not by itself lead to written works of interest, so I wish to thank the authors of the articles in this volume for their contributions. Thanks also go to Silvio Levy for his smooth handling of the entire production process, and to Kathleen Hasselblatt for her support.

The aim of the workshop and this volume is to impact the development of dynamical systems, and to that end we paid some attention to making it possible for younger participants to attend the workshop, and the surveys in this volume may attract young mathematicians to those subject areas. The Mathematical Sciences Research Institute adds to the impact of the event by maintaining streaming video of the lectures given at the workshop, which enables

everyone to view these lectures (see http://www.msri.org/calendar/workshops/
WorkshopInfo/267/show_workshop). Finally, a problem list in this volume is
hoped to inspire research into subjects that posed challenges at the time of the
workshop. I hope that the readers will find this volume interesting, useful and
inspiring.

I am writing these lines on the sixty-second birthday of Anatole Katok and
wish to dedicate this volume to him.

Boris Hasselblatt
Somerville (MA), August 2006

Recent Progress in Dynamics
MSRI Publications
Volume **54**, 2007

Quantitative symplectic geometry

KAI CIELIEBAK, HELMUT HOFER,
JANKO LATSCHEV, AND FELIX SCHLENK

Dedicated to Anatole Katok on the occasion of his sixtieth birthday

A symplectic manifold (M, ω) is a smooth manifold M endowed with a nondegenerate and closed 2-form ω. By Darboux's Theorem such a manifold looks locally like an open set in some $\mathbb{R}^{2n} \cong \mathbb{C}^n$ with the standard symplectic form

$$\omega_0 = \sum_{j=1}^{n} dx_j \wedge dy_j, \qquad (0\text{--}1)$$

and so symplectic manifolds have no local invariants. This is in sharp contrast to Riemannian manifolds, for which the Riemannian metric admits various curvature invariants. Symplectic manifolds do however admit many global numerical invariants, and prominent among them are the so-called symplectic capacities.

Symplectic capacities were introduced in 1990 by I. Ekeland and H. Hofer [18; 19] (although the first capacity was in fact constructed by M. Gromov [39]). Since then, lots of new capacities have been defined [16; 29; 31; 43; 48; 58; 59; 88; 97] and they were further studied in [1; 2; 8; 9; 25; 20; 27; 30; 34; 36; 37; 40; 41; 42; 45; 47; 49; 51; 55; 56; 57; 60; 61; 62; 63; 65; 71; 72; 73; 86; 87; 89; 90; 92; 95; 96]. Surveys on symplectic capacities are [44; 49; 54; 66; 95]. Different capacities are defined in different ways, and so relations between capacities often lead to surprising relations between different aspects of symplectic geometry and Hamiltonian dynamics. This is illustrated in Section 2, where we discuss some examples of symplectic capacities and describe a few consequences of their existence. In Section 3 we present an attempt to

Cieliebak's research was partially supported by the DFG grant Ci 45/2-1. Hofer's research was partially supported by the NSF Grant DMS-0102298. Latschev held a position financed by the DFG grant Mo 843/2-1. Schlenk held a position financed by the DFG grant Schw 892/2-1.

better understand the space of all symplectic capacities, and discuss some further general properties of symplectic capacities. In Section 4, we describe several new relations between certain symplectic capacities on ellipsoids and polydiscs. Throughout the discussion we mention many open problems.

As illustrated below, many of the quantitative aspects of symplectic geometry can be formulated in terms of symplectic capacities. Of course there are other numerical invariants of symplectic manifolds which could be included in a discussion of quantitative symplectic geometry, such as the invariants derived from Hofer's bi-invariant metric on the group of Hamiltonian diffeomorphisms, [43; 79; 82], or Gromov–Witten invariants. Their relation to symplectic capacities is not well understood, and we will not discuss them here.

We start out with a brief description of some relations of symplectic geometry to neighboring fields.

1. Symplectic geometry and its neighbors

Symplectic geometry is a rather new and vigorously developing mathematical discipline. The "symplectic explosion" is described in [21]. Examples of symplectic manifolds are open subsets of $(\mathbb{R}^{2n}, \omega_0)$, the torus $\mathbb{R}^{2n}/\mathbb{Z}^{2n}$ endowed with the induced symplectic form, surfaces equipped with an area form, Kähler manifolds like complex projective space $\mathbb{CP}^n$ endowed with their Kähler form, and cotangent bundles with their canonical symplectic form. Many more examples are obtained by taking products and through more elaborate constructions, such as the symplectic blow-up operation. A diffeomorphism φ on a symplectic manifold (M, ω) is called *symplectic* or a *symplectomorphism* if $\varphi^*\omega = \omega$.

A fascinating feature of symplectic geometry is that it lies at the crossroad of many other mathematical disciplines. In this section we mention a few examples of such interactions.

Hamiltonian dynamics. Symplectic geometry originated in Hamiltonian dynamics, which originated in celestial mechanics. A time-dependent Hamiltonian function on a symplectic manifold (M, ω) is a smooth function $H: \mathbb{R} \times M \to \mathbb{R}$. Since ω is nondegenerate, the equation

$$\omega(X_H, \cdot) = dH(\cdot)$$

defines a time-dependent smooth vector field X_H on M. Under suitable assumption on H, this vector field generates a family of diffeomorphisms φ_H^t called the *Hamiltonian flow* of H. As is easy to see, each map φ_H^t is symplectic. A *Hamiltonian diffeomorphism* φ on M is a diffeomorphism of the form φ_H^1.

Symplectic geometry is the geometry underlying Hamiltonian systems. It turns out that this geometric approach to Hamiltonian systems is very fruitful. Explicit examples are discussed in Section 2 below.

Volume geometry. A volume form Ω on a manifold M is a top-dimensional nowhere vanishing differential form, and a diffeomorphism φ of M is *volume preserving* if $\varphi^*\Omega = \Omega$. Ergodic theory studies the properties of volume preserving mappings. Its findings apply to symplectic mappings. Indeed, since a symplectic form ω is nondegenerate, ω^n is a volume form, which is preserved under symplectomorphisms. In dimension 2 a symplectic form is just a volume form, so that a symplectic mapping is just a volume preserving mapping. In dimensions $2n \geq 4$, however, symplectic mappings are much more special. A geometric example for this is Gromov's Nonsqueezing Theorem stated in Section 2.2 and a dynamical example is the (partly solved) Arnol'd conjecture stating that Hamiltonian diffeomorphisms of closed symplectic manifolds have at least as many fixed points as smooth functions have critical points. For another link between ergodic theory and symplectic geometry see [81].

Contact geometry. Contact geometry originated in geometrical optics. A contact manifold (P,α) is a $(2n-1)$-dimensional manifold P endowed with a 1-form α such that $\alpha \wedge (d\alpha)^{n-1}$ is a volume form on P. The vector field X on P defined by $d\alpha(X, \cdot) = 0$ and $\alpha(X) = 1$ generates the so-called Reeb flow. The restriction of a time-independent Hamiltonian system to an energy surface can sometimes be realized as the Reeb flow on a contact manifold. Contact manifolds also arise naturally as boundaries of symplectic manifolds. One can study a contact manifold (P, α) by symplectic means by looking at its symplectization $\left(P \times \mathbb{R}, d(e^t\alpha)\right)$, see e.g. [46; 22].

Algebraic geometry. A special class of symplectic manifolds are Kähler manifolds. Such manifolds (and, more generally, complex manifolds) can be studied by looking at holomorphic curves in them. M. Gromov [39] observed that some of the tools used in the Kähler context can be adapted for the study of symplectic manifolds. One part of his pioneering work has grown into what is now called Gromov–Witten theory, see e.g. [70] for an introduction.

Many other techniques and constructions from complex geometry are useful in symplectic geometry. For example, there is a symplectic version of blowing-up, which is intimately related to the symplectic packing problem, see [64; 68] and 4.1.2 below. Another example is Donaldson's construction of symplectic submanifolds [17]. Conversely, symplectic techniques proved useful for studying problems in algebraic geometry such as Nagata's conjecture [5; 6; 68] and degenerations of algebraic varieties [7].

Riemannian and spectral geometry. Recall that the differentiable structure of a smooth manifold M gives rise to a canonical symplectic form on its cotangent bundle T^*M. Giving a Riemannian metric g on M is equivalent to prescribing its unit cosphere bundle $S_g^*M \subset T^*M$, and the restriction of the canonical 1-form from T^*M gives S^*M the structure of a contact manifold. The Reeb flow on S_g^*M is the geodesic flow (free particle motion).

In a somewhat different direction, each symplectic form ω on some manifold M distinguishes the class of Riemannian metrics which are of the form $\omega(J\cdot,\cdot)$ for some almost complex structure J.

These (and other) connections between symplectic and Riemannian geometry are by no means completely explored, and we believe there is still plenty to be discovered here. Here are some examples of known results relating Riemannian and symplectic aspects of geometry.

Lagrangian submanifolds. A middle-dimensional submanifold L of (M, ω) is called *Lagrangian* if ω vanishes on TL.

(i) Volume. Endow complex projective space $\mathbb{CP}^n$ with the usual Kähler metric and the usual Kähler form. The volume of submanifolds is taken with respect to this Riemannian metric. According to a result of Givental–Kleiner–Oh, the standard $\mathbb{RP}^n$ in $\mathbb{CP}^n$ has minimal volume among all its Hamiltonian deformations [74]. A partial result for the Clifford torus in $\mathbb{CP}^n$ can be found in [38]. The torus $S^1 \times S^1 \subset S^2 \times S^2$ formed by the equators is also volume minimizing among its Hamiltonian deformations, [50]. If L is a closed Lagrangian submanifold of $(\mathbb{R}^{2n}, \omega_0)$, there exists according to [98] a constant C depending on L such that

$$\mathrm{vol}\,(\varphi_H(L)) \geq C \quad \text{for all Hamiltonian deformations of } L. \tag{1--1}$$

(ii) Mean curvature. The mean curvature form of a Lagrangian submanifold L in a Kähler–Einstein manifold can be expressed through symplectic invariants of L, see [15].

The first eigenvalue of the Laplacian. Symplectic methods can be used to estimate the first eigenvalue of the Laplace operator on functions for certain Riemannian manifolds [80].

Short billiard trajectories. Consider a bounded domain $U \subset \mathbb{R}^n$ with smooth boundary. There exists a periodic billiard trajectory on $\overline{U}$ of length l with

$$l^n \leq C_n \mathrm{vol}(U) \tag{1--2}$$

where C_n is an explicit constant depending only on n, see [98; 30].

2. Examples of symplectic capacities

In this section we give the formal definition of symplectic capacities, and discuss a number of examples along with sample applications.

2.1. Definition. Denote by $Symp^{2n}$ the category of all symplectic manifolds of dimension $2n$, with symplectic embeddings as morphisms. A *symplectic category* is a subcategory $\mathscr{C}$ of $Symp^{2n}$ such that $(M, \omega) \in \mathscr{C}$ implies $(M, \alpha\omega) \in \mathscr{C}$ for all $\alpha > 0$.

CONVENTION. *We will use the symbol $\hookrightarrow$ to denote symplectic embeddings and $\rightarrow$ to denote morphisms in the category $\mathscr{C}$ (which may be more restrictive).*

Let $B^{2n}(r^2)$ be the open ball of radius r in $\mathbb{R}^{2n}$ and $Z^{2n}(r^2) = B^2(r^2) \times \mathbb{R}^{2n-2}$ the open cylinder (the reason for this notation will become apparent below). Unless stated otherwise, open subsets of $\mathbb{R}^{2n}$ are always equipped with the canonical symplectic form $\omega_0 = \sum_{j=1}^{n} dy_j \wedge dx_j$. We will suppress the dimension $2n$ when it is clear from the context and abbreviate

$$B := B^{2n}(1), \qquad Z := Z^{2n}(1).$$

Now let $\mathscr{C} \subset Symp^{2n}$ be a symplectic category containing the ball B and the cylinder Z. A *symplectic capacity* on $\mathscr{C}$ is a covariant functor c from $\mathscr{C}$ to the category $([0, \infty], \leq)$ (with $a \leq b$ as morphisms) satisfying

(MONOTONICITY): $c(M, \omega) \leq c(M', \omega')$ if there exists a morphism $(M, \omega) \rightarrow (M', \omega')$;

(CONFORMALITY): $c(M, \alpha\omega) = \alpha \, c(M, \omega)$ for $\alpha > 0$;

(NONTRIVIALITY): $0 < c(B)$ and $c(Z) < \infty$.

Note that the (Monotonicity) axiom just states the functoriality of c. A symplectic capacity is said to be *normalized* if

(NORMALIZATION): $c(B) = 1$.

As a frequent example we will use the set Op^{2n} of open subsets in $\mathbb{R}^{2n}$. We make it into a symplectic category by identifying $(U, \alpha^2\omega_0)$ with the symplectomorphic manifold $(\alpha U, \omega_0)$ for $U \subset \mathbb{R}^{2n}$ and $\alpha > 0$. We agree that the morphisms in this category shall be symplectic embeddings induced by *global* symplectomorphisms of $\mathbb{R}^{2n}$. With this identification, the (Conformality) axiom above takes the form

(CONFORMALITY)$'$: $c(\alpha U) = \alpha^2 c(U)$ for $U \in Op^{2n}$, $\alpha > 0$.

2.2. Gromov radius. In view of Darboux's Theorem one can associate with each symplectic manifold (M, ω) the numerical invariant

$$c_B(M, \omega) := \sup\{\alpha > 0 \mid B^{2n}(\alpha) \hookrightarrow (M, \omega)\}$$

called the *Gromov radius* of (M, ω), [39]. It measures the symplectic size of (M, ω) in a geometric way, and is reminiscent of the injectivity radius of a Riemannian manifold. Note that it clearly satisfies the (Monotonicity) and (Conformality) axioms for a symplectic capacity. It is equally obvious that $c_B(B) = 1$.

If M is 2-dimensional and connected, then $\pi c_B(M, \omega) = \int_M \omega$, i.e. c_B is proportional to the volume of M, see [89]. The following theorem from Gromov's seminal paper [39] implies that in higher dimensions the Gromov radius is an invariant very different from the volume.

NONSQUEEZING THEOREM (GROMOV, 1985). *The cylinder $Z \in Symp^{2n}$ satisfies $c_B(Z) = 1$.*

Therefore the Gromov radius is a normalized symplectic capacity on $Symp^{2n}$. Gromov originally obtained this result by studying properties of moduli spaces of pseudo-holomorphic curves in symplectic manifolds.

It is important to realize that the existence of at least one capacity c with $c(B) = c(Z)$ also *implies* the Nonsqueezing Theorem. We will see below that each of the other important techniques in symplectic geometry (such as variational methods and the global theory of generating functions) gave rise to the construction of such a capacity, and hence an independent proof of this fundamental result.

It was noted in [18] that the following result, originally established by Eliashberg and by Gromov using different methods, is also an easy consequence of the existence of a symplectic capacity.

THEOREM (ELIASHBERG, GROMOV). *The group of symplectomorphisms of a symplectic manifold (M, ω) is closed for the compact-open C^0-topology in the group of all diffeomorphisms of M.*

2.3. Symplectic capacities via Hamiltonian systems. The next four examples of symplectic capacities are constructed via Hamiltonian systems. A crucial role in the definition or the construction of these capacities is played by the action functional of classical mechanics. For simplicity, we assume that $(M, \omega) = (\mathbb{R}^{2n}, \omega_0)$. Given a Hamiltonian function $H: S^1 \times \mathbb{R}^{2n} \to \mathbb{R}$ which is periodic in the time-variable $t \in S^1 = \mathbb{R}/\mathbb{Z}$ and which generates a global flow φ_H^t, the action functional on the loop space $C^\infty(S^1, \mathbb{R}^{2n})$ is defined as

$$\mathcal{A}_H(\gamma) = \int_\gamma y\,dx - \int_0^1 H\big(t, \gamma(t)\big)dt. \tag{2-1}$$

Its critical points are exactly the 1-periodic orbits of φ_H^t. Since the action functional is neither bounded from above nor from below, critical points are saddle points. In his pioneering work [83; 84], P. Rabinowitz designed special minimax principles adapted to the hyperbolic structure of the action functional to find such critical points. We give a heuristic argument why this works. Consider the space of loops

$$E = H^{1/2}(S^1, \mathbb{R}^{2n}) = \left\{ z \in L^2\left(S^1; \mathbb{R}^{2n}\right) \,\middle|\, \sum_{k \in \mathbb{Z}} |k|\,|z_k|^2 < \infty \right\}$$

where $z = \sum_{k \in \mathbb{Z}} e^{2\pi k t J} z_k$, $z_k \in \mathbb{R}^{2n}$, is the Fourier series of z and J is the standard complex structure of $\mathbb{R}^{2n} \cong \mathbb{C}^n$. The space E is a Hilbert space with inner product

$$\langle z, w \rangle = \langle z_0, w_0 \rangle + 2\pi \sum_{k \in \mathbb{Z}} |k|\,\langle z_k, w_k \rangle,$$

and there is an orthogonal splitting $E = E^- \oplus E^0 \oplus E^+$, $z = z^- + z^0 + z^+$, into the spaces of $z \in E$ having nonzero Fourier coefficients $z_k \in \mathbb{R}^{2n}$ only for $k < 0$, $k = 0$, $k > 0$. The action functional $\mathcal{A}_H: C^\infty(S^1, \mathbb{R}^{2n}) \to \mathbb{R}$ extends to E as

$$\mathcal{A}_H(z) = \left(\tfrac{1}{2}\|z^+\|^2 - \tfrac{1}{2}\|z^-\|^2 \right) - \int_0^1 H(t, z(t))\,dt. \tag{2-2}$$

Notice now the hyperbolic structure of the first term $\mathcal{A}_0(x)$, and that the second term is of lower order. Some of the critical points $z(t) \equiv const$ of $\mathcal{A}_0$ should thus persist for $H \neq 0$.

2.3.1. Ekeland–Hofer capacities. The first construction of symplectic capacities via Hamiltonian systems was carried out by Ekeland and Hofer [18; 19]. To give the heuristics, we consider a bounded domain $U \subset \mathbb{R}^{2n}$ with smooth boundary ∂U. A *closed characteristic* γ on ∂U is an embedded circle in ∂U tangent to the characteristic line bundle

$$\mathcal{L}_U = \{(x, \xi) \in T\partial U \mid \omega_0(\xi, \eta) = 0 \text{ for all } \eta \in T_x\partial U\}.$$

If ∂U is represented as a regular energy surface $\{x \in \mathbb{R}^{2n} \mid H(x) = const\}$ of a smooth function H on $\mathbb{R}^{2n}$, then the Hamiltonian vector field X_H restricted to ∂U is a section of $\mathcal{L}_U$, and so the traces of the periodic orbits of X_H on ∂U are the closed characteristics on ∂U. The *action* of a closed characteristic γ on ∂U is defined as $\mathscr{A}(\gamma) = \left| \int_\gamma y \, dx \right|$. The set

$$\Sigma(U) = \{ k \, \mathscr{A}(\gamma) \mid k = 1, 2, \ldots; \, \gamma \text{ is a closed characteristic on } \partial U \}$$

is called the *action spectrum* of U. Now one would like to associate with U suitable elements of $\Sigma(U)$. Without further assumptions on U, however, the set $\Sigma(U)$ may be empty (see [32; 33; 35]), and there is no obvious way to achieve (Monotonicity). To salvage this naive idea, Ekeland and Hofer considered for each bounded open subset U of $\mathbb{R}^{2n}$ the space $\mathscr{F}(U)$ of time-independent Hamiltonian functions $H \colon \mathbb{R}^{2n} \to [0, \infty)$ satisfying

- $H \equiv 0$ on some open neighbourhood of $\overline{U}$, and
- $H(z) = a|z|^2$ for $|z|$ large, where $a > \pi$, $a \notin \mathbb{N}\pi$.

Notice that the circle S^1 acts on the Hilbert space E by time-shift $x(t) \mapsto x(t + \theta)$ for $\theta \in S^1 = \mathbb{R}/\mathbb{Z}$. The special form of $H \in \mathscr{F}(U)$ guarantees that for each $k \in \mathbb{N}$ the equivariant minimax value

$$c_{H,k} := \inf \left\{ \sup_{\gamma \in \xi} \mathscr{A}_H(\gamma) \mid \xi \subset E \text{ is } S^1\text{-equivariant and ind}(\xi) \geq k \right\}$$

is a critical value of the action functional (2–2). Here, ind(ξ) denotes a suitable Fadell–Rabinowitz index [26; 19] of the intersection $\xi \cap S^+$ of ξ with the unit sphere $S^+ \subset E^+$. The *k-th Ekeland–Hofer capacity* c_k^{EH} on the symplectic category Op^{2n} is now defined as

$$c_k^{\text{EH}}(U) := \inf \left\{ c_{H,k} \mid H \in \mathscr{F}(U) \right\}$$

if $U \subset \mathbb{R}^{2n}$ is bounded and as

$$c_k^{\text{EH}}(U) := \sup \left\{ c_k^{\text{EH}}(V) \mid V \subset U \text{ bounded} \right\}$$

in general. It turns out that these numbers are indeed symplectic capacities. Moreover, they realize the naive idea of picking out suitable elements of $\Sigma(U)$ for many U: A bounded open subset U of $\mathbb{R}^{2n}$ is said to be of *restricted contact type* if its boundary ∂U is smooth and if there exists a vector field v on $\mathbb{R}^{2n}$ which is transverse to ∂U and whose Lie derivative satisfies $L_v \omega_0 = \omega_0$. Examples are bounded star-shaped domains with smooth boundary.

PROPOSITION (EKELAND AND HOFER, 1990). *If U is of restricted contact type, then $c_k^{\text{EH}}(U) \in \Sigma(U)$ for each $k \in \mathbb{N}$.*

Since the index appearing in the definition of $c_{H,k}$ is monotone, it is immediate from the definition that $c_1^{EH} \leq c_2^{EH} \leq c_3^{EH} \leq \ldots$ form an increasing sequence. Their values on the ball and cylinder are

$$c_k^{EH}(B) = \left[\frac{k+n-1}{n}\right]\pi \quad \text{and} \quad c_k^{EH}(Z) = k\pi,$$

where $[x]$ denotes the largest integer $\leq x$. Hence the existence of c_1^{EH} gives an independent proof of Gromov's Nonsqueezing Theorem. Using the capacity c_n^{EH}, Ekeland and Hofer [19] also proved the following nonsqueezing result.

THEOREM (EKELAND AND HOFER, 1990). *The cube*

$$P = B^2(1) \times \ldots \times B^2(1) \subset \mathbb{C}^n$$

can be symplectically embedded into the ball $B^{2n}(r^2)$ *if and only if* $r^2 \geq n$.

Other illustrations of the use of Ekeland–Hofer capacities in studying embedding problems for ellipsoids and polydiscs appear in Section 4.

2.3.2. Hofer–Zehnder capacity. (See [48; 49].) Given a symplectic manifold (M, ω) we consider the class $\mathcal{S}(M)$ of simple Hamiltonian functions $H \colon M \to [0, \infty)$ characterized by the following properties:

- $H = 0$ near the (possibly empty) boundary of M;
- The critical values of H are 0 and max H.

Such a function is called *admissible* if the flow φ_H^t of H has no nonconstant periodic orbits with period $T \leq 1$.

The *Hofer–Zehnder capacity* c_{HZ} on $Symp^{2n}$ is defined as

$$c_{HZ}(M) := \sup\{\max H \mid H \in \mathcal{S}(M) \text{ is admissible}\}$$

It measures the symplectic size of M in a dynamical way. Easily constructed examples yield the inequality $c_{HZ}(B) \geq \pi$. In [48; 49], Hofer and Zehnder applied a minimax technique to the action functional (2–2) to show that $c_{HZ}(Z) \leq \pi$, so

$$c_{HZ}(B) = c_{HZ}(Z) = \pi,$$

providing another independent proof of the Nonsqueezing Theorem. Moreover, for every symplectic manifold (M, ω) the inequality $\pi c_B(M) \leq c_{HZ}(M)$ holds.

The importance of understanding the Hofer–Zehnder capacity comes from the following result proved in [48; 49].

THEOREM (HOFER AND ZEHNDER, 1990). *Let* $H \colon (M, \omega) \to \mathbb{R}$ *be a proper autonomous Hamiltonian. If* $c_{HZ}(M) < \infty$, *then for almost every* $c \in H(M)$ *the energy level* $H^{-1}(c)$ *carries a periodic orbit.*

Variants of the Hofer–Zehnder capacity which can be used to detect periodic orbits in a prescribed homotopy class where considered in [59; 88].

2.3.3. Displacement energy (See [43; 55].) Next, let us measure the symplectic size of a subset by looking at how much energy is needed to displace it from itself. Fix a symplectic manifold (M, ω). Given a compactly supported Hamiltonian $H : [0, 1] \times M \to \mathbb{R}$, set

$$\|H\| := \int_0^1 \left(\sup_{x \in M} H(t, x) - \inf_{x \in M} H(t, x) \right) dt.$$

The *energy* of a compactly supported Hamiltonian diffeomorphism φ is

$$E(\varphi) := \inf \left\{ \|H\| \mid \varphi = \varphi_H^1 \right\}.$$

The *displacement energy* of a subset A of M is now defined as

$$e(A, M) := \inf \{ E(\varphi) \mid \varphi(A) \cap A = \varnothing \}$$

if A is compact and as

$$e(A, M) := \sup \{ e(K, M) \mid K \subset A \text{ is compact} \}$$

for a general subset A of M.

Now consider the special case $(M, \omega) = (\mathbb{R}^{2n}, \omega_0)$. Simple explicit examples show $e(Z, \mathbb{R}^{2n}) \leq \pi$. In [43], H. Hofer designed a minimax principle for the action functional (2–2) to show that $e(B, \mathbb{R}^{2n}) \geq \pi$, so that

$$e(B, \mathbb{R}^{2n}) = e(Z, \mathbb{R}^{2n}) = \pi.$$

It follows that $e(\cdot, \mathbb{R}^{2n})$ is a symplectic capacity on the symplectic category Op^{2n} of open subsets of $\mathbb{R}^{2n}$.

One important feature of the displacement energy is the inequality

$$c_{\mathrm{HZ}}(U) \leq e(U, M) \tag{2–3}$$

holding for open subsets of many (and possibly all) symplectic manifolds, including $(\mathbb{R}^{2n}, \omega_0)$. Indeed, this inequality and the Hofer–Zehnder Theorem imply existence of periodic orbits on almost every energy surface of any Hamiltonian with support in U provided only that U is displaceable in M. The proof of this inequality uses the spectral capacities introduced in Section 2.3.4 below.

As an application, consider a closed Lagrangian submanifold L of $(\mathbb{R}^{2n}, \omega_0)$. Viterbo [98] used an elementary geometric construction to show that

$$e \left(L, \mathbb{R}^{2n} \right) \leq C_n \left(\mathrm{vol}(L) \right)^{2/n}$$

for an explicit constant C_n. By a result of Chekanov [12], $e\left(L, \mathbb{R}^{2n}\right) > 0$. Since $e\left(\varphi_H(L), \mathbb{R}^{2n}\right) = e\left(L, \mathbb{R}^{2n}\right)$ for every Hamiltonian diffeomorphism of L, we obtain Viterbo's inequality (1–1).

2.3.4. Spectral capacities. (See [31; 45; 49; 75; 76; 77; 86; 97].) For simplicity, we assume again $(M, \omega) = (\mathbb{R}^{2n}, \omega_0)$. Denote by $\mathcal{H}$ the space of compactly supported Hamiltonian functions $H: S^1 \times \mathbb{R}^{2n} \to \mathbb{R}$. An *action selector* σ selects for each $H \in \mathcal{H}$ the action $\sigma(H) = \mathcal{A}_H(\gamma)$ of a "topologically visible" 1-periodic orbit γ of φ_H^t in a suitable way. Such *action selectors* were constructed by Viterbo [97], who applied minimax to generating functions, and by Hofer and Zehnder [45; 49], who applied minimax directly to the action functional (2–2). An outline of their constructions can be found in [30].

Given an action selector σ for $(\mathbb{R}^{2n}, \omega_0)$, one defines the *spectral capacity* c_σ on the symplectic category Op^{2n} by

$$c_\sigma(U) := \sup\{\sigma(H) \mid H \text{ is supported in } S^1 \times U\}.$$

It follows from the defining properties of an action selector (not given here) that $c_{\mathrm{HZ}}(U) \le c_\sigma(U)$ for any spectral capacity c_σ. Elementary considerations also imply $c_\sigma(U) \le e(U, \mathbb{R}^{2n})$, see [30; 45; 49; 97]. In this way one in particular obtains the important inequality (2–3) for $M = \mathbb{R}^{2n}$.

Here is another application of action selectors:

THEOREM (VITERBO, 1992). *Every nonidentical compactly supported Hamiltonian diffeomorphism of $\left(\mathbb{R}^{2n}, \omega_0\right)$ has infinitely many nontrivial periodic points.*

Moreover, the existence of an action selector is an important ingredient in Viterbo's proof of the estimate (1–2) for billiard trajectories.

Using the Floer homology of (M, ω) filtered by the action functional, an action selector can be constructed for many (and conceivably for all) symplectic manifolds (M, ω), [31; 75; 76; 77; 86]. This existence result implies the energy-capacity inequality (2–3) for arbitrary open subsets U of such (M, ω), which has many applications [87].

2.4. Lagrangian capacity. In [16] a capacity is defined on the category of $2n$-dimensional symplectic manifolds (M, ω) with $\pi_1(M) = \pi_2(M) = 0$ (with symplectic embeddings as morphisms) as follows. The *minimal symplectic area* of a Lagrangian submanifold $L \subset M$ is

$$A_{\min}(L) := \inf\left\{\int_\sigma \omega \,\Big|\, \sigma \in \pi_2(M, L), \int_\sigma \omega > 0\right\} \in [0, \infty].$$

The *Lagrangian capacity* of (M, ω) is defined as

$$c_L(M, \omega) := \sup \{A_{\min}(L) \mid L \subset M \text{ is an embedded Lagrangian torus}\}.$$

Its values on the ball and cylinder are

$$c_L(B) = \pi/n, \qquad c_L(Z) = \pi.$$

As the cube $P = B^2(1) \times \ldots \times B^2(1)$ contains the standard Clifford torus $T^n \subset \mathbb{C}^n$, and is contained in the cylinder Z, it follows that $c_L(P) = \pi$. Together with $c_L(B) = \pi/n$ this gives an alternative proof of the nonsqueezing result of Ekeland and Hofer mentioned in Section 2.3.1. There are also applications of the Lagrangian capacity to Arnold's chord conjecture and to Lagrangian (non)embedding results into uniruled symplectic manifolds [16].

3. General properties and relations between symplectic capacities

In this section we study general properties of and relations between symplectic capacities. We begin by introducing some more notation. Define the *ellipsoids* and *polydiscs*

$$E(a) := E(a_1, \ldots, a_n) := \left\{ z \in \mathbb{C}^n \;\middle|\; \frac{|z_1|^2}{a_1} + \ldots + \frac{|z_n|^2}{a_n} < 1 \right\}$$

$$P(a) := P(a_1, \ldots, a_n) := B^2(a_1) \times \ldots \times B^2(a_n)$$

for $0 < a_1 \leq \ldots \leq a_n \leq \infty$. Note that in this notation the ball, cube and cylinder are $B = E(1, \ldots, 1)$, $P = P(1, \ldots, 1)$ and $Z = E(1, \infty, \ldots, \infty) = P(1, \infty, \ldots, \infty)$.

Besides $Symp^{2n}$ and Op^{2n}, two symplectic categories that will frequently play a role below are

Ell^{2n}: the category of ellipsoids in $\mathbb{R}^{2n}$, with symplectic embeddings induced by global symplectomorphisms of $\mathbb{R}^{2n}$ as morphisms,

Pol^{2n}: the category of polydiscs in $\mathbb{R}^{2n}$, with symplectic embeddings induced by global symplectomorphisms of $\mathbb{R}^{2n}$ as morphisms.

3.1. Generalized symplectic capacities. From the point of view of this work, it is convenient to have a more flexible notion of symplectic capacities, whose axioms were originally designed to explicitly exclude such invariants as the volume. We thus define a *generalized symplectic capacity* on a symplectic category $\mathscr{C}$ as a covariant functor c from $\mathscr{C}$ to the category $([0, \infty], \leq)$ satisfying only the (Monotonicity) and (Conformality) axioms of Section 2.1.

Now examples such as the *volume capacity* on $Symp^{2n}$ are included into the discussion. It is defined as

$$c_{\text{vol}}(M, \omega) := \left(\frac{\text{vol}(M, \omega)}{\text{vol}(B)} \right)^{1/n},$$

where $\text{vol}(M, \omega) := \int_M \omega^n / n!$ is the symplectic volume. For $n \geq 2$ we have $c_{\text{vol}}(B) = 1$ and $c_{\text{vol}}(Z) = \infty$, so c_{vol} is a normalized generalized capacity but not a capacity. Many more examples appear below.

3.2. Embedding capacities.

Let $\mathscr{C}$ be a symplectic category. Every object (X, Ω) of $\mathscr{C}$ induces two generalized symplectic capacities on $\mathscr{C}$,

$$c_{(X,\Omega)}(M, \omega) := \sup \{\alpha > 0 \mid (X, \alpha\Omega) \to (M, \omega)\},$$
$$c^{(X,\Omega)}(M, \omega) := \inf \{\alpha > 0 \mid (M, \omega) \to (X, \alpha\Omega)\},$$

Here the supremum and infimum over the empty set are set to 0 and ∞, respectively. Note that

$$c_{(X,\Omega)}(M, \omega) = \left(c^{(M,\omega)}(X, \Omega)\right)^{-1}. \tag{3-1}$$

EXAMPLE 1. Suppose that $(X, \alpha\Omega) \to (X, \Omega)$ for some $\alpha > 1$. Then

$$c_{(X,\Omega)}(X, \Omega) = \infty \quad \text{and} \quad c^{(X,\Omega)}(X, \Omega) = 0,$$

so

$$c_{(X,\Omega)}(M, \omega) = \begin{cases} \infty & \text{if } (X, \beta\Omega) \to (M, \omega) \text{ for some } \beta > 0, \\ 0 & \text{if } (X, \beta\Omega) \to (M, \omega) \text{ for no } \beta > 0, \end{cases}$$

$$c^{(X,\Omega)}(M, \omega) = \begin{cases} 0 & \text{if } (M, \omega) \to (X, \beta\Omega) \text{ for some } \beta > 0, \\ \infty & \text{if } (M, \omega) \to (X, \beta\Omega) \text{ for no } \beta > 0. \end{cases}$$

The following fact follows directly from the definitions.

FACT 1. *Suppose that there exists no morphism $(X, \alpha\Omega) \to (X, \Omega)$ for any $\alpha > 1$. Then $c_{(X,\Omega)}(X, \Omega) = c^{(X,\Omega)}(X, \Omega) = 1$, and for every generalized capacity c with $0 < c(X, \Omega) < \infty$,*

$$c_{(X,\Omega)}(M, \omega) \leq \frac{c(M, \omega)}{c(X, \Omega)} \leq c^{(X,\Omega)}(M, \omega) \qquad \text{for all } (M, \omega) \in \mathscr{C}.$$

In other words, $c_{(X,\Omega)}$ (resp. $c^{(X,\Omega)}$) is the minimal (resp. maximal) generalized capacity c with $c(X, \Omega) = 1$.

Important examples on $Symp^{2n}$ arise from the ball $B = B^{2n}(1)$ and cylinder $Z = Z^{2n}(1)$. By Gromov's Nonsqueezing Theorem and volume reasons we have for $n \geq 2$:

$$c_B(Z) = 1, \qquad c^Z(B) = 1, \qquad c^B(Z) = \infty, \qquad c_Z(B) = 0.$$

In particular, for every normalized symplectic capacity c,

$$c_B(M, \omega) \leq c(M, \omega) \leq c(Z) \, c^Z(M, \omega) \qquad \text{for all } (M, \omega) \in Symp^{2n}. \quad (3\text{–}2)$$

Recall that the capacity c_B is the Gromov radius defined in Section 2.2. The capacities c_B and c^Z are not comparable on Op^{2n}: Example 3 below shows that for every $k \in \mathbb{N}$ there is a bounded star-shaped domain U_k of $\mathbb{R}^{2n}$ such that

$$c_B(U_k) \leq 2^{-k} \qquad and \qquad c^Z(U_k) \geq \pi k^2,$$

see also [42].

We now turn to the question which capacities can be represented as *embedding capacities* $c_{(X, \Omega)}$ or $c^{(X, \Omega)}$.

EXAMPLE 2. Consider the subcategory $\mathscr{C} \subset Op^{2n}$ of connected open sets. Then every generalized capacity c on $\mathscr{C}$ can be represented as the capacity $c^{(X, \Omega)}$ of embeddings into a (possibly uncountable) union (X, Ω) of objects in $\mathscr{C}$.

For this, just define (X, Ω) as the disjoint union of all (X_ι, Ω_ι) in the category $\mathscr{C}$ with $c(X_\iota, \Omega_\iota) = 0$ or $c(X_\iota, \Omega_\iota) = 1$.

PROBLEM 1. *Which (generalized) capacities can be represented as $c^{(X, \Omega)}$ for a connected symplectic manifold (X, Ω)?*

PROBLEM 2. *Which (generalized) capacities can be represented as the capacity $c_{(X, \Omega)}$ of embeddings from a symplectic manifold (X, Ω)?*

EXAMPLE 3. Embedding capacities give rise to some curious generalized capacities. For example, consider the capacity c^Y of embeddings into the symplectic manifold $Y := \amalg_{k \in \mathbb{N}} B^{2n}(k^2)$. It only takes values 0 and ∞, with $c^Y(M, \omega) = 0$ if and only if (M, ω) embeds symplectically into Y; see Example 1. If M is connected, $\mathrm{vol}(M, \omega) = \infty$ implies $c^Y(M, \omega) = \infty$. On the other hand, for every $\varepsilon > 0$ there exists an open subset $U \subset \mathbb{R}^{2n}$, diffeomorphic to a ball, with $\mathrm{vol}(U) < \varepsilon$ and $c^Y(U) = \infty$. To see this, consider for $k \in \mathbb{N}$ an open neighbourhood U_k of volume $< 2^{-k} \varepsilon$ of the linear cone over the Lagrangian torus $\partial B^2(k^2) \times \ldots \times \partial B^2(k^2)$. The Lagrangian capacity of U_k clearly satisfies $c_L(U_k) \geq \pi k^2$. The open set $U := \cup_{k \in \mathbb{N}} U_k$ satisfies $\mathrm{vol}(U) < \varepsilon$ and $c_L(U) = \infty$, hence U does not embed symplectically into any ball. By appropriate choice of the U_k we can arrange that U is diffeomorphic to a ball; see [86, Proposition A.3]. $\diamond$

Special embedding spaces. Given an arbitrary pair of symplectic manifolds (X, Ω) and (M, ω), it is a difficult problem to determine or even estimate $c_{(X,\Omega)}(M, \omega)$ and $c^{(X,\Omega)}(M, \omega)$. We thus consider two special cases.

1. *Embeddings of skinny ellipsoids.* Assume (M, ω) is an ellipsoid $E(a, \ldots, a, 1)$ with $0 < a \leq 1$, and (X, Ω) is connected and has finite volume. Upper bounds for the function

$$e^{(X,\Omega)}(a) = c^{(X,\Omega)}\left(E(a, \ldots, a, 1)\right), \quad a \in (0, 1],$$

are obtained from symplectic embedding results of ellipsoids into (X, Ω), and lower bounds are obtained from computing other (generalized) capacities and using Fact 1. In particular, the volume capacity yields

$$\frac{\left(e^{(X,\Omega)}(a)\right)^n}{a^{n-1}} \geq \frac{\mathrm{vol}(B)}{\mathrm{vol}(X, \Omega)}.$$

The only known general symplectic embedding results for ellipsoids are obtained via multiple symplectic folding. The following result is part of Theorem 3 in [86], which in our setting reads

FACT 2. *Assume that (X, Ω) is a connected $2n$-dimensional symplectic manifold of finite volume. Then*

$$\lim_{a \to 0} \frac{\left(e^{(X,\Omega)}(a)\right)^n}{a^{n-1}} = \frac{\mathrm{vol}(B)}{\mathrm{vol}(X, \Omega)}.$$

For a restricted class of symplectic manifolds, Fact 2 can be somewhat improved. The following result is part of Theorem 6.25 of [86].

FACT 3. *Assume that X is a bounded domain in $\left(\mathbb{R}^{2n}, \omega_0\right)$ with piecewise smooth boundary or that (X, Ω) is a compact connected $2n$-dimensional symplectic manifold. If $n \leq 3$, there exists a constant $C > 0$ depending only on (X, Ω) such that*

$$\frac{\left(e^{(X,\Omega)}(a)\right)^n}{a^{n-1}} \leq \frac{\mathrm{vol}(B)}{\mathrm{vol}(X, \Omega)\left(1 - Ca^{1/n}\right)} \quad \text{for all } a < \frac{1}{C^n}.$$

These results have their analogues for polydiscs $P(a, \ldots, a, 1)$. The analogue of Fact 3 is known in all dimensions.

2. *Packing capacities.* Given an object (X, Ω) of $\mathscr{C}$ and $k \in \mathbb{N}$, we denote by $\coprod_k (X, \Omega)$ the disjoint union of k copies of (X, Ω) and define

$$c_{(X,\Omega;k)}(M, \omega) := \sup\left\{\alpha > 0 \,\middle|\, \coprod_k (X, \alpha\Omega) \hookrightarrow (M, \omega)\right\}.$$

If $\mathrm{vol}(X, \Omega)$ is finite, we see as in Fact 1 that

$$c_{(X,\Omega;k)}(M,\omega) \leq \frac{1}{c_{\mathrm{vol}}\left(\coprod_k(X,\Omega)\right)} c_{\mathrm{vol}}(M,\omega). \qquad (3\text{--}3)$$

We say that (M,ω) admits a *full k-packing* by (X,Ω) if equality holds in (3–3). For $k_1, \ldots, k_n \in \mathbb{N}$ a full $k_1 \cdots k_n$-packing of $B^{2n}(1)$ by $E\left(\frac{1}{k_1}, \ldots, \frac{1}{k_n}\right)$ is given in [94]. E. Ophstein recently showed in [78] that for every closed symplectic manifold (M,ω) with $[\omega] \in H^2(M; \mathbb{Q})$ there exists a full 1-packing by some ellipsoid. Full k-packings by balls and obstructions to full k-packings by balls are studied in [3; 4; 39; 53; 63; 68; 86; 94].

Assume that also $\mathrm{vol}(M,\omega)$ is finite. Studying the capacity $c_{(X,\Omega;k)}(M,\omega)$ is equivalent to studying the *packing number*

$$p_{(X,\Omega;k)}(M,\omega) = \sup_\alpha \frac{\mathrm{vol}\left(\left(\coprod_k(X, \alpha\Omega)\right)\right)}{\mathrm{vol}(M,\omega)}$$

where the supremum is taken over all α for which $\coprod_k (X, \alpha\Omega)$ symplectically embeds into (M,ω). Clearly, $p_{(X,\Omega;k)}(M,\omega) \leq 1$, and equality holds if and only if equality holds in (3–3). Results in [68] together with the above-mentioned full packings of a ball by ellipsoids from [94] imply

FACT 4. *If X is an ellipsoid or a polydisc, then*

$$p_{(X,k)}(M,\omega) \to 1 \ as \ k \to \infty$$

for every symplectic manifold (M,ω) of finite volume.

Note that if the conclusion of Fact 4 holds for X and Y, then it also holds for $X \times Y$.

PROBLEM 3. *For which bounded convex subsets X of $\mathbb{R}^{2n}$ is the conclusion of Fact 4 true?*

In [68] and [3; 4], the packing numbers $p_{(X,k)}(M)$ are computed for $X = B^4$ and $M = B^4$ or $\mathbb{C}P^2$. Moreover, the following fact is shown in [3; 4]:

FACT 5. *If $X = B^4$, then for every closed connected symplectic 4-manifold (M,ω) with $[\omega] \in H^2(M; \mathbb{Q})$ there exists $k_0(M,\omega)$ such that*

$$p_{(X,k)}(M,\omega) = 1 \ for \ all \ k \geq k_0(M,\omega).$$

PROBLEM 4. *For which bounded convex subsets X of $\mathbb{R}^{2n}$ and which connected symplectic manifolds (M,ω) of finite volume is the conclusion of Fact 5 true?*

3.3. Operations on capacities. We say that a function $f \colon [0, \infty]^n \to [0, \infty]$ is *homogeneous* and *monotone* if

$$f(\alpha x_1, \ldots, \alpha x_n) = \alpha f(x_1, \ldots, x_n) \qquad \text{for all } \alpha > 0,$$

$$f(x_1, \ldots, x_i, \ldots, x_n) \le f(x_1, \ldots, y_i, \ldots, x_n) \qquad \text{for } x_i \le y_i.$$

If f is homogeneous and monotone and $c_1, \ldots, c_n$ are generalized capacities, then $f(c_1, \ldots, c_n)$ is again a generalized capacity. If in addition

$$0 < f(1, \ldots, 1) < \infty$$

and $c_1, \ldots, c_n$ are capacities, then $f(c_1, \ldots, c_n)$ is a capacity. Compositions and pointwise limits of homogeneous monotone functions are again homogeneous and monotone. Examples include $\max(x_1, \ldots, x_n)$, $\min(x_1, \ldots, x_n)$, and the weighted (arithmetic, geometric, harmonic) means

$$\lambda_1 x_1 + \ldots + \lambda_n x_n, \qquad x_1^{\lambda_1} \cdots x_n^{\lambda_n}, \qquad \frac{1}{\frac{\lambda_1}{x_1} + \ldots + \frac{\lambda_n}{x_n}},$$

with $\lambda_1, \ldots, \lambda_n \ge 0$, $\lambda_1 + \ldots + \lambda_n = 1$.

There is also a natural notion of convergence of capacities. We say that a sequence c_n of generalized capacities on $\mathscr{C}$ *converges pointwise* to a generalized capacity c if $c_n(M, \omega) \to c(M, \omega)$ for every $(M, \omega) \in \mathscr{C}$.

These operations yield lots of dependencies between capacities, and it is natural to look for generating systems. In a very general form, this can be formulated as follows.

PROBLEM 5. *For a given symplectic category $\mathscr{C}$, find a minimal generating system $\mathscr{G}$ for the (generalized) symplectic capacities on $\mathscr{C}$. This means that every (generalized) symplectic capacity on $\mathscr{C}$ is the pointwise limit of homogeneous monotone functions of elements in $\mathscr{G}$, and no proper subcollection of $\mathscr{G}$ has this property.*

This problem is already open for Ell^{2n} and Pol^{2n}. One may also ask for generating systems allowing fewer operations, e.g. only max and min, or only positive linear combinations. We will formulate more specific versions of this problem below. The following simple fact illustrates the use of operations on capacities.

FACT 6. *Let $\mathscr{C}$ be a symplectic category containing B (resp. P). Then every generalized capacity c on $\mathscr{C}$ with $c(B) \ne 0$ (resp. $c(P) \ne 0$) is the pointwise limit of capacities.*

Indeed, if $c(B) \ne 0$, then c is the pointwise limit as $k \to \infty$ of the capacities

$$c_k = \min(c, k\, c_B),$$

and likewise with $c(P)$, c_P instead of $c(B)$, c_B.

EXAMPLE 4. (i) The generalized capacity $c \equiv 0$ on Op^{2n} is not a pointwise limit of capacities, and so the assumption $c(B) \neq 0$ in Fact 6 cannot be omitted.

(ii) The assumption $c(B) \neq 0$ is not always necessary:

(a) Define a generalized capacity c on Op^{2n} by

$$c(U) = \begin{cases} 0 & \text{if } \mathrm{vol}(U) < \infty, \\ c_B(U) & \text{if } \mathrm{vol}(U) = \infty. \end{cases}$$

Then $c(B) = 0$ and $c(Z) = 1$, and c is the pointwise limit of the capacities

$$c_k = \max\left(c, \tfrac{1}{k}c_B\right).$$

(b) Define a generalized capacity c on Op^{2n} by

$$c(U) = \begin{cases} 0 & \text{if } c_B(U) < \infty, \\ \infty & \text{if } c_B(U) = \infty. \end{cases}$$

Then $c(B) = 0 = c(Z)$ and $c(\mathbb{R}^{2n}) = \infty$, and $c = \lim_{k \to \infty} \tfrac{1}{k}c_B$.

(iii) We do not know whether the generalized capacity $c_{\mathbb{R}^{2n}}$ on Op^{2n} is the pointwise limit of capacities.

PROBLEM 6. *Given a symplectic category $\mathscr{C}$ containing B or P and Z, characterize the generalized capacities which are pointwise limits of capacities.*

3.4. Continuity. There are several notions of continuity for capacities on open subsets of $\mathbb{R}^{2n}$, see [1; 18]. For example, consider a *smooth family of hypersurfaces* $(S_t)_{-\varepsilon < t < \varepsilon}$ in $\mathbb{R}^{2n}$, each bounding a compact subset with interior U_t. Recall that S_0 is said to be of *restricted contact type* if there exists a vector field v on $\mathbb{R}^{2n}$ which is transverse to S_0 and whose Lie derivative satisfies $L_v\omega_0 = \omega_0$. Let c be a capacity on Op^{2n}. As the flow of v is conformally symplectic, the (Conformality) axiom implies the following (see [49, p. 116]):

FACT 7. *If S_0 is of restricted contact type, the function $t \mapsto c(U_t)$ is Lipschitz continuous at 0.*

Fact 7 fails without the hypothesis of restricted contact type. For example, if S_0 possesses no closed characteristic (such S_0 exist by [32; 33; 35]), then by Theorem 3 in Section 4.2 of [49] the function $t \mapsto c_{\mathrm{HZ}}(U_t)$ is not Lipschitz continuous at 0. V. Ginzburg [34] presents an example of a smooth family of hypersurfaces (S_t) (albeit not in $\mathbb{R}^{2n}$) for which the function $t \mapsto c_{\mathrm{HZ}}(U_t)$ is not smoother than $1/2$-Hölder continuous. These considerations lead to

PROBLEM 7. *Are capacities continuous on all smooth families of domains bounded by smooth hypersurfaces?*

3.5. Convex sets. Here we restrict to the subcategory $Conv^{2n} \subset Op^{2n}$ of convex open subsets of $\mathbb{R}^{2n}$, with embeddings induced by global symplectomorphisms of $\mathbb{R}^{2n}$ as morphisms. Recall that a subset $U \subset \mathbb{R}^{2n}$ is *star-shaped* if U contains .a point p such that for every $q \in U$ the straight line between p and q belongs to U. In particular, convex domains are star-shaped.

FACT 8. (Extension after Restriction Principle [18]) *Assume that $\varphi: U \hookrightarrow \mathbb{R}^{2n}$ is a symplectic embedding of a bounded star-shaped domain $U \subset \mathbb{R}^{2n}$. Then for any compact subset K of U there exists a symplectomorphism Φ of $\mathbb{R}^{2n}$ such that $\Phi|_K = \varphi|_K$.*

This principle continues to hold for some, but not all, symplectic embeddings of unbounded star-shaped domains, see [86]. We say that a capacity c defined on a symplectic subcategory of Op^{2n} has the *exhaustion property* if

$$c(U) = \sup\{ c(V) \mid V \subset U \text{ is bounded } \}. \qquad (3\text{--}4)$$

The capacities introduced in Section 2 all have this property, but the capacity in Example 3 does not. By Fact 8, all statements about capacities defined on a subcategory of $Conv^{2n}$ and having the exhaustion property remain true if we allow all symplectic embeddings (not just those coming from global symplectomorphisms of $\mathbb{R}^{2n}$) as morphisms.

FACT 9. *Let U and V be objects in $Conv^{2n}$. Then there exists a morphism $\alpha U \to V$ for every $\alpha \in (0, 1)$ if and only if $c(U) \leq c(V)$ for all generalized capacities c on $Conv^{2n}$.*

Indeed, the necessity of the condition is obvious, and the sufficiency follows by observing that $\alpha U \to U$ for all $\alpha \in (0, 1)$ and $1 \leq c_U(U) \leq c_U(V)$. What happens for $\alpha = 1$ is not well understood, see Section 3.6 for related discussions. The next example illustrates that the conclusion of Fact 9 is wrong without the convexity assumption.

EXAMPLE 5. Consider the open annulus $A = B(4) \setminus B(1)$ in $\mathbb{R}^2$. If $\frac{3}{4} < \alpha^2 < 1$, then αA cannot be embedded into A by a global symplectomorphism. Indeed, volume considerations show that any potential such global symplectomorphism would have to map A homotopically nontrivially into itself. This would force the image of the ball $\alpha B(1)$ to cover all of $B(1)$, which is impossible for volume reasons.   ◊

Assume now that c is a normalized symplectic capacity on $Conv^{2n}$. Using John's ellipsoid, Viterbo [98] noticed that there is a constant C_n depending only on n such that

$$c^Z(U) \leq C_n c_B(U) \quad \text{for all } U \in Conv^{2n}$$

and so, in view of (3–2),

$$c_B(U) \le c(U) \le C_n c(Z) c_B(U) \quad \text{for all } U \in Conv^{2n}. \tag{3–5}$$

In fact, $C_n \le (2n)^2$ and $C_n \le 2n$ on centrally symmetric convex sets.

PROBLEM 8. *What is the optimal value of the constant C_n appearing in (3–5)? In particular, is $C_n = 1$?*

Note that $C_n = 1$ would imply uniqueness of capacities satisfying $c(B) = c(Z) = 1$ on $Conv^{2n}$. In view of Gromov's Nonsqueezing Theorem, $C_n = 1$ on Ell^{2n} and Pol^{2n}. More generally, this equality holds for all convex Reinhardt domains [42]. In particular, for these special classes of convex sets

$$\pi c_B = c_1^{\text{EH}} = c_{\text{HZ}} = e(\cdot, \mathbb{R}^{2n}) = \pi c^Z.$$

3.6. Recognition. One may ask how complete the information provided by all symplectic capacities is. Consider two objects (M, ω) and (X, Ω) of a symplectic category $\mathscr{C}$.

QUESTION 1. *Assume $c(M, \omega) \le c(X, \Omega)$ for all generalized symplectic capacities c on $\mathscr{C}$. Does it follow that $(M, \omega) \hookrightarrow (X, \Omega)$ or even that $(M, \omega) \to (X, \Omega)$?*

QUESTION 2. *Assume $c(M, \omega) = c(X, \Omega)$ for all generalized symplectic capacities c on $\mathscr{C}$. Does it follow that (M, ω) is symplectomorphic to (X, Ω) or even that $(M, \omega) \cong (X, \Omega)$ in the category $\mathscr{C}$?*

Note that if $(M, \alpha\omega) \to (M, \omega)$ for all $\alpha \in (0, 1)$ then, under the assumptions of Question 1, the argument leading to Fact 9 yields $(M, \alpha\omega) \to (X, \Omega)$ for all $\alpha \in (0, 1)$.

EXAMPLE 6. (i) Set $U = B^2(1)$ and $V = B^2(1) \backslash \{0\}$. For each $\alpha < 1$ there exists a symplectomorphism of $\mathbb{R}^2$ with $\varphi(\alpha U) \subset V$, so that monotonicity and conformality imply $c(U) = c(V)$ for all generalized capacities c on Op^2. Clearly, $U \hookrightarrow V$, but $U \nrightarrow V$, and U and V are not symplectomorphic.

(ii) Set $U = B^2(1)$ and let $V = B^2(1) \backslash \{(x, y) \mid x \ge 0, \ y = 0\}$ be the slit disc. As is well-known, U and V are symplectomorphic. Fact 8 implies $c(U) = c(V)$ for all generalized capacities c on Op^2, but clearly $U \nrightarrow V$. In dimensions $2n \ge 4$ there are bounded convex sets U and V with smooth boundary which are symplectomorphic while $U \nrightarrow V$, see [24].

(iii) Let U and V be ellipsoids in Ell^{2n}. The answer to Question 1 is unknown even for Ell^4. For $U = E(1, 4)$ and $V = B^4(2)$ we have $c(U) \le c(V)$ for all generalized capacities that can presently be computed, but it is unknown whether

$U \hookrightarrow V$, (see 4.1.2 below). By Fact 10 below the answer to Question 2 is "yes" on Ell^{2n}.

(iv) Let U and V be polydiscs in Pol^{2n}. Again, the answer to Question 1 is unknown even for Pol^4. However, in this dimension the Gromov radius together with the volume capacity determine a polydisc, so that the answer to Question 2 is "yes" on Pol^4.   ◊

PROBLEM 9. *Are two polydiscs in dimension $2n \geq 6$ with equal generalized symplectic capacities symplectomorphic?*

To conclude this section, we mention a specific example in which $c(U) = c(V)$ for all known (but possibly not for all) generalized symplectic capacities.

EXAMPLE 7. Consider the subsets

$$U = E(2, 6) \times E(3, 3, 6) \quad \text{and} \quad V = E(2, 6, 6) \times E(3, 3)$$

of $\mathbb{R}^{10}$. Then $c(U) = c(V)$ whenever $c(B) = c(Z)$ by the Nonsqueezing Theorem, the volumes agree, and $c_k^{\mathrm{EH}}(U) = c_k^{EH}(V)$ for all k by the product formula (3–8). It is unknown whether $U \hookrightarrow V$ or $V \hookrightarrow U$ or $U \rightarrow V$. Symplectic homology as constructed in [28; 93] does not help in these problems because a computation based on [29] shows that all symplectic homologies of U and V agree.

3.7. Hamiltonian representability. Consider a bounded domain $U \subset \mathbb{R}^{2n}$ with smooth boundary of restricted contact type (see Section 2.3.1 for the definition). As in 2.3.1 we consider the action spectrum

$$\Sigma(U) = \left\{ k \left| \int_\gamma y \, dx \right| \, \middle| \, k = 1, 2, \ldots; \, \gamma \text{ is a closed characteristic on } \partial U \right\}$$

of U. This set is nowhere dense in $\mathbb{R}$ (compare [49, Section 5.2]), and it is easy to see that $\Sigma(U)$ is closed and $0 \notin \Sigma(U)$. For many capacities c constructed via Hamiltonian systems, such as Ekeland–Hofer capacities c_k^{EH} and spectral capacities c_σ, one has $c(U) \in \Sigma(U)$, see [19; 41]. Moreover,

$$c_{\mathrm{HZ}}(U) = c_1^{\mathrm{EH}}(U) = \min(\Sigma(U)) \quad \text{if } U \text{ is convex.} \tag{3–6}$$

One might therefore be tempted to ask

QUESTION 3. *Is it true that $\pi c(U) \in \Sigma(U)$ for every normalized symplectic capacity c on Op^{2n} and every domain U with boundary of restricted contact type?*

The following example due to D. Hermann [42] shows that the answer to Question 3 is "no".

EXAMPLE 8. Choose any U with boundary of restricted contact type such that

$$c_B(U) < c^Z(U). \tag{3-7}$$

Examples are bounded star-shaped domains U with smooth boundary which contain the Lagrangian torus $S^1 \times \ldots \times S^1$ but have small volume: According to [91], $c^Z(U) \geq 1$, while $c_B(U)$ is as small as we like. Now notice that for each $t \in [0, 1]$,

$$c_t = (1-t)c_B + tc^Z$$

is a normalized symplectic capacity on Op^{2n}. By (3–7), the interval

$$\{c_t(U) \mid t \in [0, 1]\} = [c_B(U), c^Z(U)]$$

has positive measure and hence cannot lie in the nowhere dense set $\Sigma(U)$.   ◊

D. Hermann also pointed out that the argument in Example 8 together with (3–6) implies that the question "$C_n = 1$?" posed in Problem 8 is equivalent to Question 3 for convex sets.

3.8. Products. Consider a family of symplectic categories $\mathscr{C}^{2n}$ in all dimensions $2n$ such that

$$(M, \omega) \in \mathscr{C}^{2m}, \quad (N, \sigma) \in \mathscr{C}^{2n} \implies (M \times N, \omega \oplus \sigma) \in \mathscr{C}^{2(m+n)}.$$

We say that a collection $c \colon \coprod_{n=1}^\infty \mathscr{C}^{2n} \to [0, \infty]$ of generalized capacities has the *product property* if

$$c(M \times N, \omega \oplus \sigma) = \min\{c(M, \omega), c(N, \sigma)\}$$

for all $(M, \omega) \in \mathscr{C}^{2m}$, $(N, \sigma) \in \mathscr{C}^{2n}$. If $\mathbb{R}^2 \in \mathscr{C}^2$ and $c(\mathbb{R}^2) = \infty$, the product property implies the *stability property*

$$c(M \times \mathbb{R}^2, \omega \oplus \omega_0) = c(M, \omega)$$

for all $(M, \omega) \in \mathscr{C}^{2m}$.

EXAMPLE 9. (i) Let Σ_g be a closed surface of genus g endowed with an area form ω. Then

$$c_B\left(\Sigma_g \times \mathbb{R}^2, \omega \oplus \omega_0\right) = \begin{cases} c_B\left(\Sigma_g, \omega\right) = \frac{1}{\pi}\omega\left(\Sigma_g\right) & \text{if } g = 0, \\ \infty & \text{if } g \geq 1. \end{cases}$$

While the result for $g = 0$ follows from Gromov's Nonsqueezing Theorem, the result for $g \geq 1$ belongs to Polterovich [69, Exercise 12.4] and Jiang [52]. Since c_B is the smallest normalized symplectic capacity on $Symp^{2n}$, we find that no collection c of symplectic capacities defined on the family $\coprod_{n=1}^\infty Symp^{2n}$ with $c\left(\Sigma_g, \omega\right) < \infty$ for some $g \geq 1$ has the product or stability property.

(ii) On the family of polydiscs $\coprod_{n=1}^{\infty} Pol^{2n}$, the Gromov radius, the Lagrangian capacity and the unnormalized Ekeland–Hofer capacities c_k^{EH} all have the product property (see Section 4.2). The volume capacity is not stable.

(iii) Let $U \in Op^{2m}$ and $V \in Op^{2n}$ have smooth boundary of restricted contact type (see Section 3.4 for the definition). The formula

$$c_k^{EH}(U \times V) = \min_{i+j=k} \left(c_i^{EH}(U) + c_j^{EH}(V) \right), \qquad (3\text{–}8)$$

in which we set $c_0^{EH} \equiv 0$, was conjectured by Floer and Hofer [95] and has been proved by Chekanov [13] as an application of his equivariant Floer homology. Consider the collection of sets $U_1 \times \ldots \times U_l$, where each $U_i \in Op^{2n_i}$ has smooth boundary of restricted contact type, and $\sum_{i=1}^{l} n_i = n$. We denote by RCT^{2n} the corresponding category with symplectic embeddings induced by global symplectomorphisms of $\mathbb{R}^{2n}$ as morphisms. If v_i are vector fields on $\mathbb{R}^{2n_i}$ with $L_{v_i}\omega_0 = \omega_0$, then $L_{v_1+\ldots+v_l}\omega_0 = \omega_0$ on $\mathbb{R}^{2n}$. Elements of RCT^{2n} can therefore be exhausted by elements of RCT^{2n} with smooth boundary of restricted contact type. This and the exhaustion property (3–4) of the c_k^{EH} shows that (3–8) holds for all $U \in RCT^{2m}$ and $V \in RCT^{2n}$, implying in particular that Ekeland–Hofer capacities are stable on $RCT := \coprod_{n=1}^{\infty} RCT^{2n}$. Moreover, (3–8) yields that

$$c_k^{EH}(U \times V) \leq \min \left(c_k^{EH}(U), c_k^{EH}(V) \right),$$

and it shows that c_1^{EH} on RCT has the product property. Using (3–8) together with an induction over the number of factors and $c_2^{EH}(E(a_1, \ldots, a_n)) \leq 2a_1$ we also see that c_2^{EH} has the product property on products of ellipsoids. For $k \geq 3$, however, the Ekeland–Hofer capacities c_k^{EH} on RCT do not have the product property. As an example, for $U = B^4(4)$ and $V = E(3, 8)$ we have

$$c_3^{EH}(U \times V) = 7 < 8 = \min \left(c_3^{EH}(U), c_3^{EH}(V) \right).$$

PROBLEM 10. *Characterize the collections of (generalized) capacities on polydiscs that have the product (resp. stability) property.*

Next consider a collection c of generalized capacities on open subsets Op^{2n}. In general, it will not be stable. However, we can stabilize c to obtain stable generalized capacities $c^{\pm} \colon \coprod_{n=1}^{\infty} Op^{2n} \to [0, \infty]$,

$$c^+(U) := \limsup_{k \to \infty} c(U \times \mathbb{R}^{2k}), \qquad c^-(U) := \liminf_{k \to \infty} c(U \times \mathbb{R}^{2k}).$$

Notice that $c(U) = c^+(U) = c^-(U)$ for all $U \in \coprod_{n=1}^{\infty} Op^{2n}$ if and only if c is stable. If c consists of capacities and there exist constants $a, A > 0$ such that

$$a \leq c\left(B^{2n}(1) \right) \leq c\left(Z^{2n}(1) \right) \leq A \qquad \text{for all } n \in \mathbb{N},$$

then $c^{\pm}$ are collections of capacities. Thus there exist plenty of stable capacities on Op^{2n}. However, we have

PROBLEM 11. *Decide stability of specific collections of capacities on $Conv^{2n}$ or Op^{2n}, e.g.: Gromov radius, Ekeland–Hofer capacity, Lagrangian capacity, and the embedding capacity c_P of the unit cube.*

PROBLEM 12. *Does there exist a collection of capacities on $\coprod_{n=1}^{\infty} Conv^{2n}$ or $\coprod_{n=1}^{\infty} Op^{2n}$ with the product property?*

3.9. Higher order capacities? Following [44], we briefly discuss the concept of higher order capacities. Consider a symplectic category $\mathscr{C} \subset Symp^{2n}$ containing Ell^{2n} and fix $d \in \{1, \ldots, n\}$. A *symplectic d-capacity* on $\mathscr{C}$ is a generalized capacity satisfying

(d-NONTRIVIALITY): $0 < c(B)$ and

$$\begin{cases} c\big(B^{2d}(1) \times \mathbb{R}^{2(n-d)}\big) < \infty, \\ c\big(B^{2(d-1)}(1) \times \mathbb{R}^{2(n-d+1)}\big) = \infty. \end{cases}$$

For $d = 1$ we recover the definition of a symplectic capacity, and for $d = n$ the volume capacity c_{vol} is a symplectic n-capacity.

PROBLEM 13. *Does there exist a symplectic d-capacity on a symplectic category $\mathscr{C}$ containing Ell^{2n} for some $d \in \{2, \ldots, n-1\}$?*

Problem 13 on $Symp^{2n}$ is equivalent to the following symplectic embedding problem.

PROBLEM 14. *Does there exist a symplectic embedding*

$$B^{2(d-1)}(1) \times \mathbb{R}^{2(n-d+1)} \hookrightarrow B^{2d}(R) \times \mathbb{R}^{2(n-d)} \tag{3–9}$$

for some $R < \infty$ and $d \in \{2, \ldots, n-1\}$?

Indeed, the existence of such an embedding would imply that no symplectic d-capacity can exist on $Symp^{2n}$. Conversely, if no such embedding exists, then the embedding capacity $c^{Z_{2d}}$ into $Z_{2d} = B^{2d}(1) \times \mathbb{R}^{2(n-d)}$ would be an example of a d-capacity on $Symp^{2n}$. The Ekeland–Hofer capacity c_d^{EH} shows that R is at least 2 if a symplectic embedding (3–9) exists. The known symplectic embedding techniques are not designed to effectively use the unbounded factor of the target space in (3–9). E.g., multiple symplectic folding only shows that there exists a function $f:[1,\infty) \to \mathbb{R}$ with $f(a) < \sqrt{2a} + 2$ such that for each $a \geq 1$ there exists a symplectic embedding

$$B^2(1) \times B^2(a) \times \mathbb{R}^2 \hookrightarrow B^4(f(a)) \times \mathbb{R}^2$$

of the form $\varphi \times id_2$, see [86, Section 4.3.2].

4. Ellipsoids and polydiscs

In this section we investigate generalized capacities on the categories of ellipsoids Ell^{2n} and polydiscs Pol^{2n} in more detail. All (generalized) capacities c in this section are defined on some symplectic subcategory of Op^{2n} containing at least one of the above categories and are assumed to have the exhaustion property (3–4).

4.1. Ellipsoids.

4.1.1. Arbitrary dimension. We first describe the values of the capacities introduced in Section 2 on ellipsoids.

The values of the Gromov radius c_B on ellipsoids are

$$c_B\big(E(a_1,\dots,a_n)\big) = \min\{a_1,\dots,a_n\}.$$

More generally, monotonicity implies that this formula holds for all symplectic capacities c on Op^{2n} with $c(B) = c(Z) = 1$ and hence also for $\frac{1}{\pi}c_1^{EH}$, $\frac{1}{\pi}c_{HZ}$, $\frac{1}{\pi}e(\cdot,\mathbb{R}^{2n})$ and c^Z.

The values of the Ekeland–Hofer capacities on the ellipsoid $E(a_1,\dots,a_n)$ can be described as follows [19]. Write the numbers $m\,a_i\pi$, $m \in \mathbb{N}$, $1 \le i \le n$, in increasing order as $d_1 \le d_2 \le \dots$, with repetitions if a number occurs several times. Then

$$c_k^{EH}\big(E(a_1,\dots,a_n)\big) = d_k.$$

The values of the Lagrangian capacity on ellipsoids are presently not known. In [16], Cieliebak and Mohnke make the following conjecture:

CONJECTURE 1.

$$c_L\big(E(a_1,\dots,a_n)\big) = \frac{\pi}{1/a_1 + \dots + 1/a_n}.$$

Since $\mathrm{vol}\big(E(a_1,\dots,a_n)\big) = a_1 \cdots a_n \mathrm{vol}(B)$, the values of the volume capacity on ellipsoids are

$$c_{\mathrm{vol}}\big(E(a_1,\dots,a_n)\big) = (a_1 \cdots a_n)^{1/n}.$$

In view of conformality and the exhaustion property, a (generalized) capacity on Ell^{2n} is determined by its values on the ellipsoids $E(a_1,\dots,a_n)$ with $0 < a_1 \le \dots \le a_n = 1$. So we can view each (generalized) capacity c on ellipsoids as a function

$$c(a_1,\dots,a_{n-1}) := c\left(E(a_1,\dots,a_{n-1},1)\right)$$

on the set $\{0 < a_1 \le \dots \le a_{n-1} \le 1\}$. By Fact 7, this function is continuous. This identification with functions yields a notion of *uniform convergence* for capacities on Ell^{2n}.

For what follows, it is useful to have normalized versions of the Ekeland–Hofer capacities, so in dimension $2n$ we define

$$\bar{c}_k := \frac{c_k^{\mathrm{EH}}}{[\frac{k+n-1}{n}]\pi}.$$

PROPOSITION 1. *As $k \to \infty$, for every $n \geq 2$ the normalized Ekeland–Hofer capacities $\bar{c}_k$ converge uniformly on Ell^{2n} to the normalized symplectic capacity c_∞ given by*

$$c_\infty\left(E(a_1, \ldots, a_n)\right) = \frac{n}{1/a_1 + \ldots + 1/a_n}.$$

REMARK. Note that Conjecture 1 asserts that c_∞ agrees with the normalized Lagrangian capacity $\bar{c}_L = nc_L/\pi$ on Ell^{2n}.

PROOF OF PROPOSITION 1. Fix $\varepsilon > 0$. We need to show that $|\bar{c}_k(a) - c_\infty(a)| \leq \varepsilon$ for every vector $a = (a_1, \ldots, a_n)$ with $0 < a_1 \leq a_2 \leq \ldots \leq a_n = 1$ and all sufficiently large k. Abbreviate $\delta = \varepsilon/n$.

Case 1. $a_1 \leq \delta$. Then

$$c_k^{\mathrm{EH}}(a) \leq k\delta\pi, \qquad \bar{c}_k(a) \leq n\delta, \qquad c_\infty(a) \leq n\delta$$

from which we conclude $|\bar{c}_k(a) - c_\infty(a)| \leq n\delta = \varepsilon$ for all $k \geq 1$.

Case 2. $a_1 > \delta$. Let $k \geq 2\frac{n-1}{\delta} + 2$. For the unique integer l with

$$\pi l\, a_n \leq c_k^{\mathrm{EH}}(a) < \pi(l+1)a_n$$

we then have $l \geq 2$. In the increasing sequence of the numbers $m a_i$ ($m \in \mathbb{N}$, $1 \leq i \leq n$), the first $[l\, a_n/a_i]$ multiples of a_i occur no later than $l\, a_n$. By the description of the Ekeland–Hofer capacities on ellipsoids given above, this yields the estimates

$$\frac{(l-1)a_n}{a_1} + \ldots + \frac{(l-1)a_n}{a_n} \leq k \leq \frac{(l+1)a_n}{a_1} + \ldots + \frac{(l+1)a_n}{a_n}.$$

With $\gamma := a_n/a_1 + \ldots + a_n/a_n$ this becomes

$$(l-1)\gamma \leq k \leq (l+1)\gamma.$$

Using $\gamma \geq n$, we derive the inequalities

$$\left[\frac{k+n-1}{n}\right] \leq \frac{k}{n} + 1 \leq \frac{(l+1)\gamma + n}{n} \leq \frac{(l+2)\gamma}{n},$$

$$\left[\frac{k+n-1}{n}\right] \geq \frac{k}{n} \geq \frac{(l-1)\gamma}{n}.$$

With the definition of $\bar{c}_k$ and the estimate above for c_k^{EH}, we find

$$\frac{n\,l\,a_n}{(l+2)\gamma} \leq \bar{c}_k(a) = \frac{c_k^{\mathrm{EH}}(a)}{\left[\frac{k+n-1}{n}\right]\pi} \leq \frac{n(l+1)a_n}{(l-1)\gamma}.$$

Since $c_\infty(a) = n\,a_n/\gamma$, this becomes

$$\frac{l}{l+2}c_\infty(a) \leq \bar{c}_k(a) \leq \frac{l+1}{l-1}c_\infty(a),$$

which in turn implies

$$|\bar{c}_k(a) - c_\infty(a)| \leq \frac{2c_\infty(a)}{l-1}.$$

Since $a_1 > \delta$ we have

$$\gamma \leq \frac{n}{\delta}, \qquad l+1 \geq \frac{k}{\gamma} \geq \frac{k\delta}{n},$$

from which we conclude

$$|\bar{c}_k(a) - c_\infty(a)| \leq \frac{2}{l-1} \leq \frac{2n}{k\delta - 2n} \leq \varepsilon$$

for k sufficiently large. $\qquad\qquad\qquad\qquad\qquad\qquad\qquad\qquad\square$

We turn to the question whether Ekeland–Hofer capacities generate the space of all capacities on ellipsoids by suitable operations. First note some easy facts.

FACT 10. *An ellipsoid $E \subset \mathbb{R}^{2n}$ is uniquely determined by its Ekeland–Hofer capacities $c_1^{\mathrm{EH}}(E), c_2^{\mathrm{EH}}(E), \ldots$..*

Indeed, if $E(a)$ and $E(b)$ are two ellipsoids with $a_i = b_i$ for $i < k$ and $a_k < b_k$, then the multiplicity of a_k in the sequence of Ekeland–Hofer capacities is one higher for $E(a)$ than for $E(b)$, so not all Ekeland–Hofer capacities agree.

FACT 11. *For every $k \in \mathbb{N}$ there exist ellipsoids E and E' with $c_i^{\mathrm{EH}}(E) = c_i^{\mathrm{EH}}(E')$ for $i < k$ and $c_k^{\mathrm{EH}}(E) \neq c_k^{\mathrm{EH}}(E')$.*

For example, we can take $E = E(a)$ and $E' = E(b)$ with $a_1 = b_1 = 1$, $a_2 = k - 1/2$, $b_2 = k + 1/2$, and $a_i = b_i = 2k$ for $i \geq 3$. So formally, every generalized capacity on ellipsoids is a function of the Ekeland–Hofer capacities, and the Ekeland–Hofer capacities are functionally independent. However, Ekeland–Hofer capacities do not form a generating system for symplectic capacities on Ell^{2n} (see Example 10 below), and on bounded ellipsoids each finite set of Ekeland–Hofer capacities is determined by the (infinitely many) other Ekeland–Hofer capacities:

LEMMA 1. *Let $d_1 \leq d_2 \leq \ldots$ be an increasing sequence of real numbers obtained from the sequence $c_1^{\mathrm{EH}}(E) \leq c_2^{\mathrm{EH}}(E) \leq \ldots$ of Ekeland–Hofer capacities of a bounded ellipsoid $E \in Ell^{2n}$ by removing at most N_0 numbers. Then E can be recovered uniquely.*

PROOF. We first consider the special case in which $E = E(a_1, \ldots, a_n)$ is such that $a_i/a_j \in \mathbb{Q}$ for all i, j. In this case, the sequence $d_1 \leq d_2 \leq \ldots$ contains infinitely many blocks of n consecutive equal numbers. We traverse the sequence until we have found $N_0 + 1$ such blocks, for each block $d_k = d_{k+1} = \ldots = d_{k+n-1}$ recording the number $g_k := d_{k+n} - d_k$. The minimum of the g_k for the $N_0 + 1$ first blocks equals a_1. After deleting each occurring positive integer multiple of a_1 once from the sequence $d_1 \leq d_2 \leq \ldots$, we can repeat the same procedure to determine a_2, and so on.

In general, we do not know whether or not $a_i/a_j \in \mathbb{Q}$ for all i, j. To reduce to the previous case, we split the sequence $d_1 \leq d_2 \leq \ldots$ into (at most n) subsequences of numbers with rational quotients. More precisely we traverse the sequence, grouping the d_i into increasing subsequences $s_1, s_2, \ldots$, where each new number is added to the first subsequence s_j whose members are rational multiples of it. Furthermore, in this process we record for each sequence s_j the maximal length l_j of a block of consecutive equal numbers seen so far. We stop when

(i) the sum of the l_j equals n, and

(ii) each subsequence s_j contains at least $N_0 + 1$ blocks of l_j consecutive equal numbers.

Now the previously described procedure in the case that $a_i/a_j \in \mathbb{Q}$ for all i, j can be applied for each subsequence s_j separately, where l_j replaces n in the above argument. □

REMARK. If the volume of E is known, one does not need to know N_0 in Fact 1. The proof of this is left to the interested reader. ◇

The set of Ekeland–Hofer capacities does *not* form a generating system for symplectic capacities on Ell^{2n}. Indeed, the volume capacity c_{vol} is not the pointwise limit of homogeneous monotone functions of Ekeland–Hofer capacities:

EXAMPLE 10. Consider the ellipsoids $E = E(1, \ldots, 1, 3^n + 1)$ and $F = E(3, \ldots, 3)$ in Ell^{2n}. As is easy to see,

$$c_k^{\mathrm{EH}}(E) < c_k^{\mathrm{EH}}(F) \qquad \text{for all } k. \tag{4–1}$$

Assume that f_i is a sequence of homogeneous monotone functions of Ekeland–Hofer capacities which converge pointwise to c_{vol}. By (4–1) and the monotonicity of the f_i we would find that $c_{vol}(E) \leq c_{vol}(F)$. This is not true.

PROBLEM 15. *Do the Ekeland–Hofer capacities together with the volume capacity form a generating system for symplectic capacities on Ell^{2n}?*

If the answer to this problem is "yes", this is a very difficult problem as Lemma 2 below illustrates.

4.1.2. Ellipsoids in dimension 4.

A generalized capacity on ellipsoids in dimension 4 is represented by a function $c(a) := c\big(E(a, 1)\big)$ of a single real variable $0 < a \leq 1$. This function has the following two properties.

(i) The function $c(a)$ is nondecreasing.
(ii) The function $c(a)/a$ is nonincreasing.

The first property follows directly from the (Monotonicity) axiom. The second property follows from (Monotonicity) and (Conformality): For $a \leq b$, $E(b, 1) \subset E\left(\frac{b}{a}a, \frac{b}{a}\right)$, hence $c(b) \leq \frac{b}{a}c(a)$. Note that property (ii) is equivalent to the estimate

$$\frac{c(b) - c(a)}{b - a} \leq \frac{c(a)}{a} \tag{4–2}$$

for $0 < a < b$, so the function $c(a)$ is Lipschitz continuous at all $a > 0$. We will restrict our attention to *normalized* (generalized) capacities, so the function c also satisfies

(iii) $c(1) = 1$.

An ellipsoid $E(a_1, \ldots, a_n)$ embeds into $E(b_1, \ldots, b_n)$ by a *linear* symplectic embedding only if $a_i \leq b_i$ for all i, see [49]. Hence for normalized capacities on the category *LinEll*^{4} of ellipsoids with *linear* embeddings as morphisms, properties (i), (ii) and (iii) are the only restrictions on the function $c(a)$. On *Ell*^{4}, nonlinear symplectic embeddings ("folding") yield additional constraints which are still not completely known; see [86] for the presently known results.

By Fact 1, the embedding capacities c_B and c^B are the smallest, resp. largest, normalized capacities on ellipsoids. By Gromov's Nonsqueezing Theorem, $c_B(a) = \bar{c}_1(a) = a$. The function $c^B(a)$ is not completely known. Fact 1 applied to $\bar{c}_2$ yields

$$c^B(a) = 1 \text{ if } a \in \left[\tfrac{1}{2}, 1\right] \quad \text{and} \quad c^B(a) \geq 2a \text{ if } a \in \left(0, \tfrac{1}{2}\right],$$

and Fact 1 applied to c_{vol} yields $c^B(a) \geq \sqrt{a}$. Folding constructions provide upper bounds for $c^B(a)$. Lagrangian folding [94] yields $c^B(a) \leq l(a)$ where

$$
l(a) = \begin{cases} (k+1)a & \text{for } \dfrac{1}{k(k+1)} \leq a \leq \dfrac{1}{(k-1)(k+1)} \\[2ex] \dfrac{1}{k} & \text{for } \dfrac{1}{k(k+2)} \leq a \leq \dfrac{1}{k(k+1)} \end{cases}
$$

and multiple symplectic folding [86] yields $c^B(a) \leq s(a)$ where the function $s(a)$ is as shown in Figure 1. While symplectically folding once yields $c^B(a) \leq a + 1/2$ for $a \in (0, 1/2]$, the function $s(a)$ is obtained by symplectically folding "infinitely many times", and it is known that

$$
\liminf_{\varepsilon \to 0^+} \frac{c^B\left(\frac{1}{2}\right) - c^B\left(\frac{1}{2} - \varepsilon\right)}{\varepsilon} \geq \frac{8}{7}.
$$

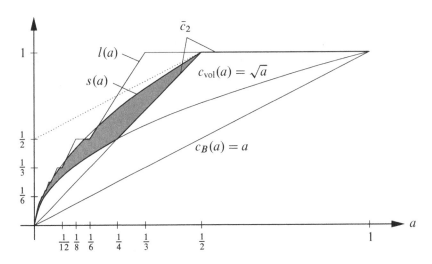

Figure 1. Lower and upper bounds for $c^B(a)$.

Let us come back to Problem 15.

LEMMA 2. *If the Ekeland–Hofer capacities and the volume capacity form a generating system for symplectic capacities on Ell^{2n}, then $c^B\left(\frac{1}{4}\right) = \frac{1}{2}$.*

We recall that $c^B\left(\frac{1}{4}\right) = \frac{1}{2}$ means that the ellipsoid $E(1, 4)$ symplectically embeds into $B^4(2 + \varepsilon)$ for every $\varepsilon > 0$.

PROOF OF LEMMA 2. We can assume that all capacities are normalized. By assumption, there exists a sequence f_i of homogeneous and monotone functions in the $\bar{c}_k$ and in c_{vol} forming normalized capacities which pointwise converge to

c^B. As is easy to see, $\bar{c}_k\left(E\left(\frac{1}{4},1\right)\right) \le \bar{c}_k\left(B^4\left(\frac{1}{2}\right)\right)$ for all k, and $c_{\text{vol}}\left(E\left(\frac{1}{4},1\right)\right) = c_{\text{vol}}\left(B^4\left(\frac{1}{2}\right)\right)$. Since the f_i are monotone and converge in particular at $E\left(\frac{1}{4},1\right)$ and $B^4\left(\frac{1}{2}\right)$ to c^B, we conclude that $c^B\left(\frac{1}{4}\right) = c^B\left(E\left(\frac{1}{4},1\right)\right) \le c^B\left(B^4\left(\frac{1}{2}\right)\right) = \frac{1}{2}$, which proves Lemma 2. $\qquad\square$

In view of Lemma 2, the following problem is a special case of Problem 15.

PROBLEM 16. *Is it true that* $c^B\left(\frac{1}{4}\right) = \frac{1}{2}$?

The best upper bound for $c^B\left(\frac{1}{4}\right)$ presently known is $s\left(\frac{1}{4}\right) \approx 0.6729$. Answering Problem 16 in the affirmative means to construct for each $\varepsilon > 0$ a symplectic embedding $E\left(\frac{1}{4},1\right) \to B^4\left(\frac{1}{2}+\varepsilon\right)$. We do not believe that such embeddings can be constructed "by hand". A strategy for studying symplectic embeddings of 4-dimensional ellipsoids by algebrogeometric tools is proposed in [6].

Our next goal is to represent the (normalized) Ekeland–Hofer capacities as embedding capacities. First we need some preparations.

From the above discussion of c^B it is clear that capacities and folding also yield bounds for the functions $c^{E(1,b)}$ and $c_{E(1,b)}$. We content ourselves with noting

LEMMA 3. *Let* $N \in \mathbb{N}$ *be given. Then for* $N \le b \le N+1$ *we have*

$$c^{E(1,b)}(a) = \begin{cases} 1/b & \text{for} \quad 1/(N+1) \le a \le 1/b, \\ a & \text{for} \quad 1/b \le a \le 1 \end{cases} \qquad (4\text{--}3)$$

and

$$c_{E(1,b)}(a) = \begin{cases} a & \text{for} \quad 0 < a \le 1/b, \\ 1/b & \text{for} \quad 1/b \le a \le 1/N, \end{cases} \qquad (4\text{--}4)$$

see Figure 2.

REMARK. Note that (4–4) completely describes $c_{E(1,b)}$ on the whole interval $(0,1]$ for $1 \le b \le 2$.

PROOF. As both formulas are proved similarly, we only prove (4–3). The first Ekeland–Hofer capacity gives the lower bound $c^{E(1,b)}(a) \ge a$ for all $a \in (0,1]$. Note that for $a \ge 1/b$ this bound is achieved by the standard embedding, so that the second claim follows.

For $1/(N+1) \le a \le 1/N$ we have $\bar{c}_{N+1}(E(a,1)) = 1$ and $\bar{c}_{N+1}(E(1,b)) = b$. By Fact 1 we see that $c^{E(1,b)} \ge 1/b$ on this interval, and this bound is again achieved by the standard embedding. This completes the proof of (4–3). $\qquad\square$

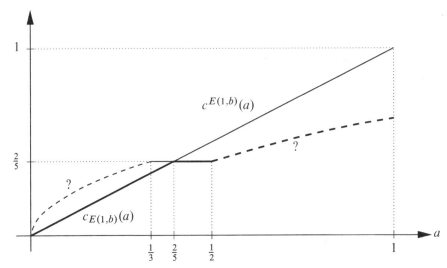

Figure 2. The functions $c^{E(1,b)}(a)$ and $c_{E(1,b)}(a)$ for $b = \frac{5}{2}$.

REMARK. Consider the functions

$$e^b(a) := c^{E(1,b)}(a), \quad a \in (0, 1], \, b \ge 1.$$

Notice that $e^1 = c^B$. By Gromov's Nonsqueezing Theorem and monotonicity,

$$a = c_B(a) = c^Z(a) \le e^b(a) \le c^B(a), \quad a \in (0, 1], \, b \ge 1.$$

Since $e^b(a) = \left(c_{E(a,1)}\big(E(1, b)\big)\right)^{-1}$ by equation (3–1), we see that for each $a \in (0, 1]$ the function $b \mapsto e^b(a)$ is monotone decreasing and continuous. By (4–3), it satisfies $e^b(a) = a$ for $a \ge 1/b$. In particular, we see that the family of graphs $\{\text{graph}(e^b) \mid 1 \le b < \infty\}$ fills the whole region between the graphs of c_B and c^B; see Figure 1. ◇

The normalized Ekeland–Hofer capacities are represented by piecewise linear functions $\bar{c}_k(a)$. Indeed, $\bar{c}_1(a) = a$ for all $a \in (0, 1]$, and for $k \ge 2$ the following formula follows straight from the definition.

LEMMA 4. *Setting* $m := \left\lceil \frac{k+1}{2} \right\rceil$, *the function* $\bar{c}_k : (0, 1] \to (0, 1]$ *is given by*

$$\bar{c}_k(a) = \begin{cases} \dfrac{k+1-i}{m} \cdot a & \text{for } \dfrac{i-1}{k+1-i} \le a \le \dfrac{i}{k+1-i}, \\[2mm] \dfrac{i}{m} & \text{for } \dfrac{i}{k+1-i} \le a \le \dfrac{i}{k-i}. \end{cases} \tag{4-5}$$

Here i takes integer values between 1 and m.

Figure 3 shows the first six of the $\bar{c}_k$ and their limit function c_∞ according to Proposition 1.

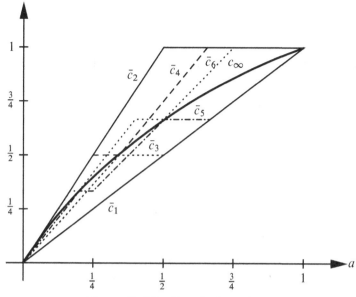

Figure 3. The first six $\bar{c}_k$ and c_∞.

In dimension 4, the uniform convergence $\bar{c}_k \to c_\infty$ is very transparent, as can be seen in Figure 3. One readily checks that $\bar{c}_k - c_\infty \geq 0$ if k is even, in which case $\|\bar{c}_k - c_\infty\| = \frac{1}{k+1}$, and that $\bar{c}_k - c_\infty \leq 0$ if $k = 2m - 1$ is odd, in which case $\|\bar{c}_k - c_\infty\| = \frac{m-1}{mk}$ if $k \geq 3$. Note that the sequences of the even (resp. odd) $\bar{c}_k$ are almost, but not quite, decreasing (resp. increasing). We still have

COROLLARY 1. *For all $r, s \in \mathbb{N}$, we have*

$$\bar{c}_{2rs} \leq \bar{c}_{2r}.$$

This will be a consequence of the following characterization of Ekeland–Hofer capacities.

LEMMA 5. *Fix $k \in \mathbb{N}$ and denote by $[a_l, b_l]$ the interval on which $\bar{c}_k$ has the value $l/\left[\frac{k+1}{2}\right]$. Then*

(a) $\bar{c}_k \leq c$ *for every capacity c such that $\bar{c}_k(a_l) \leq c(a_l)$ for all $l = 1, 2, \ldots, \left[\frac{k+1}{2}\right]$.*
(b) $\bar{c}_k \geq c$ *for every capacity c such that $\bar{c}_k(b_l) \geq c(b_l)$ for all $l = 1, 2, \ldots, \left[\frac{k}{2}\right]$*
and

$$\lim_{a \to 0} \frac{c(a)}{a} \leq \frac{k}{\left[\frac{k+1}{2}\right]}.$$

PROOF. Formula (4–2) and Lemma 4 show that where a normalized Ekeland–Hofer capacity grows, it grows with maximal slope. In particular, going left from the left end point a_l of a plateau a normalized Ekeland–Hofer capacity drops with the fastest possible rate until it reaches the level of the next lower plateau and then stays there, showing the minimality. Similarly, going right from the right end point b_l of some plateau a normalized Ekeland–Hofer capacity grows with the fastest possible rate until it reaches the next higher level, showing the maximality. □

PROOF OF COROLLARY 1. The right end points of plateaus for $\bar{c}_{2r}$ are given by $b_i = \frac{i}{2r-i}$. Thus we compute

$$\bar{c}_{2r}\left(\frac{i}{2r-i}\right) = \frac{i}{r} = \frac{is}{rs} = \bar{c}_{2rs}\left(\frac{is}{2rs-is}\right) = \bar{c}_{2rs}\left(\frac{i}{2r-i}\right)$$

and the claim follows from the characterization of $\bar{c}_{2r}$ by maximality. □

Lemma 3 and the piecewise linearity of the $\bar{c}_k$ suggest that they may be representable as embedding capacities into a disjoint union of finitely many ellipsoids. This is indeed the case.

PROPOSITION 2. *The normalized Ekeland–Hofer capacity $\bar{c}_k$ on Ell^4 is the capacity c^{X_k} of embeddings into the disjoint union of ellipsoids*

$$X_k = Z\left(\frac{m}{k}\right) \sqcup \coprod_{j=1}^{\left[\frac{k}{2}\right]} E\left(\frac{m}{k-j}, \frac{m}{j}\right),$$

where $m = \left[\frac{k+1}{2}\right]$.

PROOF. The proposition clearly holds for $k = 1$. We thus fix $k \geq 2$. Recall from Lemma 4 that $\bar{c}_k$ has $\left[\frac{k}{2}\right]$ plateaus, the j-th of which has height $\frac{j}{m}$ and starts at $a_j := j/(k+1-j)$ and ends at $b_j := j/(k-j)$. The j-th ellipsoid in Proposition 2 is found as follows: In view of (4–3) we first select an ellipsoid $E(1, b)$ so that the point $\frac{1}{b}$ corresponds to b_j. This ellipsoid is then rescaled to achieve the correct height $\frac{j}{m}$ of the plateau (note that by conformality, $\alpha c^{E(\alpha,\alpha b)} = c^{E(1,b)}$ for $\alpha > 0$). We obtain the candidate ellipsoid

$$E_j = E\left(\frac{m}{k-j}, \frac{m}{j}\right).$$

The slope of $\bar{c}_k$ following its j-th plateau and the slope of c^{E_j} after its plateau both equal $\frac{k-j}{m}$. The cylinder is added to achieve the correct behavior near $a = 0$. We are thus left with showing that for each $1 \leq j \leq \left[\frac{k}{2}\right]$,

$$\bar{c}_k(a) \leq c^{E_j}(a) \quad \text{for all } a \in (0, 1].$$

According to Lemma 5 (a) it suffices to show that for each $1 \leq j \leq \left[\frac{k}{2}\right]$ and each $1 \leq l \leq \left[\frac{k}{2}\right]$ we have

$$\bar{c}_k(a_l) = \frac{l}{m} \leq c^{E_j}(a_l). \tag{4-6}$$

For $l > j$, the estimate (4–6) follows from the fact that $\bar{c}_k = c^{E_j}$ near b_j and from the argument given in the proof of Lemma 5 (a), and for $l = j$ the estimate (4–6) follows from (4–3) of Lemma 3 by a direct computation. We will deal with the other cases

$$1 \leq l < j \leq \left[\frac{k}{2}\right]$$

by estimating $c^{E_j}(a_l)$ from below, using Fact 1 with $c = c_{\mathrm{vol}}$ and $c = \bar{c}_2$.
    Fix j and recall that $c_{\mathrm{vol}}(E(x, y)) = \sqrt{xy}$, so

$$c^{E_j}(a_l) \geq \frac{c_{\mathrm{vol}}(E(a_l, 1))}{c_{\mathrm{vol}}\left(E\left(\frac{m}{k-j}, \frac{m}{j}\right)\right)} = \sqrt{\frac{lj(k-j)}{(k+1-l)m^2}} = \frac{l}{m} \cdot \sqrt{\frac{j(k-j)}{(k+1-l)l}}$$

gives the desired estimate (4–6) if $j(k-j) \geq -l^2 + (k+1)l$. Computing the roots $l_{\pm}$ of this quadratic inequality in l, we find that this is the case if

$$l \leq l_- = \frac{1}{2}\left(k + 1 - \sqrt{1 + 2k + (k - 2j)^2}\right).$$

Computing the normalized second Ekeland–Hofer capacity under the assumption that $a_l \leq \frac{1}{2}$, we find that $\bar{c}_2(E(a_l, 1)) = 2a_l = \frac{2l}{k+1-l}$ and $\bar{c}_2(E_j) \leq \frac{m}{j}$, so

$$c^{E_j}(a_l) \geq \frac{\bar{c}_2(E(a_l, 1))}{\bar{c}_2\left(E\left(\frac{m}{k-j}, \frac{m}{j}\right)\right)} \geq \frac{2l}{k+1-l} \cdot \frac{j}{m} = \frac{l}{m} \cdot \frac{2j}{k+1-l},$$

which gives the required estimate (4–6) if

$$l \geq k + 1 - 2j.$$

Note that for $\frac{1}{2} \leq a_l \leq 1$ we have $\bar{c}_2(E(a_l, 1)) = 1$ and hence

$$\frac{\bar{c}_2(E(a_l, 1))}{\bar{c}_2\left(E\left(\frac{m}{k-j}, \frac{m}{j}\right)\right)} \geq \frac{j}{m} > \frac{l}{m}$$

trivially, because we only consider $l < j$.

So combining the results from the two capacities, we find that the desired estimate (4–6) holds provided either $l \leq l_- = \frac{1}{2}\left(k + 1 - \sqrt{1 + 2k + (k - 2j)^2}\right)$ or $l \geq k + 1 - 2j$. As we only consider $l < j$, it suffices to verify that

$$\min(j - 1, k + 1 - 2j) \leq \frac{1}{2}\left(k + 1 - \sqrt{1 + 2k + (k - 2j)^2}\right)$$

for all positive integers j and k satisfying $1 \leq j \leq \left[\frac{k}{2}\right]$. This indeed follows from another straightforward computation, completing the proof of Proposition 2. □

Using the results above, we find a presentation of the normalized capacity $c_\infty = \lim_{k \to \infty} \bar{c}_k$ on Ell^4 as embedding capacity into a countable disjoint union of ellipsoids. Indeed, the space X_{4r} appearing in the statement of Proposition 2 is obtained from X_{2r} by adding r more ellipsoids. Combined with Proposition 1 this yields the presentation

$$c_\infty = c^X \quad \text{on } Ell^4,$$

where $X = \bigsqcup_{r=1}^{\infty} X_{2r}$ is a disjoint union of countably many ellipsoids. Together with Conjecture 1, the following conjecture suggests a much more efficient presentation of c_∞ as an embedding capacity.

CONJECTURE 2. *The restriction of the normalized Lagrangian capacity $\bar{c}_L$ to Ell^4 equals the embedding capacity c^X, where X is the connected subset $B(1) \cup Z(\frac{1}{2})$ of $\mathbb{R}^4$.*

For the embedding capacities *from* ellipsoids, we have the following analogue of Proposition 2.

PROPOSITION 3. *The normalized Ekeland–Hofer capacity $\bar{c}_k$ on Ell^4 is the maximum of finitely many capacities $c_{E_{k,j}}$ of embeddings of ellipsoids $E_{k,j}$,*

$$\bar{c}_k(a) = \max\{c_{E_{k,j}}(a) \mid 1 \leq j \leq m\}, \quad a \in (0, 1],$$

where

$$E_{k,j} = E\left(\frac{m}{k + 1 - j}, \frac{m}{j}\right)$$

with $m = \left[\frac{k+1}{2}\right]$.

PROOF. The ellipsoids $E_{k,j}$ are determined using (4–4) in Lemma 3. According to Lemma 5 (b), this time it suffices to check that for all $1 \leq j \leq l \leq \left[\frac{k}{2}\right]$ the values of the corresponding capacities at the right end points $b_l = \frac{l}{k-1}$ of plateaus of $\bar{c}_k$ satisfy

$$c_{E_{k,j}}(b_l) \leq \frac{l}{m} = \bar{c}_k(b_l). \tag{4–7}$$

The case $l = j$ follows from (4–4) in Lemma 3 by a direct computation. For the remaining cases

$$1 \le j < l \le \left[\frac{k}{2}\right]$$

we use three different methods, depending on the value of j. If $j \le \frac{k-1}{3}$, then Fact 1 with $c = c_{\text{vol}}$ gives (4–7) by a computation similar to the one in the proof of Proposition 2. If $j \ge \frac{k+1}{3}$, then $a_j = \frac{j}{k+1-j} \ge \frac{1}{2}$, so that (4–4) in Lemma 3 shows that $c_{E_{k,j}}$ is constant on $[a_j, 1]$, proving (4–7) in this case. Finally, if $j = \frac{k}{3}$ and $l \ge j + 1$, then $\bar{c}_2(E_{k,j}) = \frac{2m}{k+1-j}$ and $\bar{c}_2(b_l) = 1$, so with Fact 1 we get

$$c_{E_{k,j}}(b_l) \le \frac{k+1-j}{2m},$$

which is smaller than $\frac{l}{m}$ for the values of j and l we consider here. This completes the proof of Proposition 3. □

Here is the corresponding conjecture for the normalized Lagrangian capacity.

CONJECTURE 3. *The restriction of the normalized Lagrangian capacity $\bar{c}_L$ to Ell^{2n} equals the embedding capacity $c_{P(1/n,\dots,1/n)}$ of the cube of radius $1/\sqrt{n}$.*

4.2. Polydiscs.

4.2.1. Arbitrary dimension. Again we first describe the values of the capacities in Section 2 on polydiscs.

The values of the Gromov radius c_B on polydiscs are

$$c_B\big(P(a_1,\dots,a_n)\big) = \min\{a_1,\dots,a_n\}.$$

As for ellipsoids, this also determines the values of c_1^{EH}, c_{HZ}, $e(\cdot, \mathbb{R}^{2n})$ and c^Z. According to [19], the values of Ekeland–Hofer capacities on polydiscs are

$$c_k^{\text{EH}}\big(P(a_1,\dots,a_n)\big) = k\pi \min\{a_1,\dots,a_n\}.$$

Using Chekanov's result [11] that $A_{\min}(L) \le e(L, \mathbb{R}^{2n})$ for every closed Lagrangian submanifold $L \subset \mathbb{R}^{2n}$, one finds the values of the Lagrangian capacity on polydiscs to be

$$c_L\big(P(a_1,\dots,a_n)\big) = \pi \min\{a_1,\dots,a_n\}.$$

Since $\text{vol}\big(P(a_1,\dots,a_n)\big) = a_1 \cdots a_n \cdot \pi^n$ and $\text{vol}(B^{2n}) = \frac{\pi^n}{n!}$, the values of the volume capacity on polydiscs are

$$c_{\text{vol}}\big(P(a_1,\dots,a_n)\big) = (a_1 \cdots a_n \cdot n!)^{1/n}.$$

As in the case of ellipsoids, a (generalized) capacity c on Pol^{2n} can be viewed as a function

$$c(a_1, \ldots, a_{n-1}) := c\left(P(a_1, \ldots, a_{n-1}, 1)\right)$$

on the set $\{0 < a_1 \leq \ldots \leq a_{n-1} \leq 1\}$. Directly from the definitions and the computations above we obtain the following easy analogue of Proposition 1.

PROPOSITION 4. *As $k \to \infty$, the normalized Ekeland–Hofer capacities $\bar{c}_k$ converge on Pol^{2n} uniformly to the normalized Lagrangian capacity $\bar{c}_L = nc_L/\pi$.*

Propositions 4 and 1 (together with Conjecture 1) give rise to

PROBLEM 17. *What is the largest subcategory of Op^{2n} on which the normalized Lagrangian capacity is the limit of the normalized Ekeland–Hofer capacities?*

4.2.2. Polydiscs in dimension 4.

Again, a normalized (generalized) capacity on polydiscs in dimension 4 is represented by a function $c(a) := c\left(P(a, 1)\right)$ of a single real variable $0 < a \leq 1$, which has the properties (i), (ii), (iii). Contrary to ellipsoids, these properties are not the only restrictions on a normalized capacity on 4-dimensional polydiscs even if one restricts to *linear* symplectic embeddings as morphisms. Indeed, the linear symplectomorphism

$$(z_1, z_2) \mapsto \frac{1}{\sqrt{2}}(z_1 + z_2, z_1 - z_2)$$

of $\mathbb{R}^4$ yields a symplectic embedding

$$P(a, b) \hookrightarrow P\left(\frac{a+b}{2} + \sqrt{ab}, \frac{a+b}{2} + \sqrt{ab}\right)$$

for any $a, b > 0$, which implies

FACT 12. *For any normalized capacity c on $LinPol^4$,*

$$c(a) \leq \frac{1}{2} + \frac{a}{2} + \sqrt{a}.$$

Still, we have the following easy analogues of Propositions 2 and 3.

PROPOSITION 5. *The normalized Ekeland–Hofer capacity $\bar{c}_k$ on Pol^4 is the capacity c^{Y_k}, where*

$$Y_k = Z\left(\frac{[\frac{k+1}{2}]}{k}\right),$$

as well as the capacity $c_{Y_k'}$, where

$$Y_k' = B\left(\frac{[\frac{k+1}{2}]}{k}\right).$$

COROLLARY 2. *The identity $\bar{c}_k = c^{X_k}$ of Proposition 2 extends to $Ell^4 \cup Pol^4$.*

PROOF. Note that Y_k is the first component of the space X_k of Proposition 2. It thus remains to show that for each of the ellipsoid components E_j of X_k,

$$\bar{c}_k\left(P(a,1)\right) \leq c^{E_j}\left(P(a,1)\right), \quad a \in (0,1].$$

This follows at once from the observation that for each j we have $c_k^{\mathrm{EH}}\left(E_j\right) = [\frac{k+1}{2}]\pi$, whereas $c_k^{\mathrm{EH}}\left(P(a,1)\right) = ka\pi$. □

PROBLEM 18. *Does the equality $\bar{c}_k = c^{X_k}$ hold on a larger class of open subsets of $\mathbb{R}^4$?*

References

[1] S. Bates, *Some simple continuity properties of symplectic capacities*, The Floer memorial volume, 185–193, Progr. Math. **133**, Birkhäuser, Basel, 1995.

[2] S. Bates, *A capacity representation theorem for some non-convex domains*, Math. Z. **227**, 571–581 (1998).

[3] P. Biran, *Symplectic packing in dimension 4*, Geom. Funct. Anal. **7**, 420–437 (1997).

[4] P. Biran, *A stability property of symplectic packing*, Invent. Math. **136**, 123–155 (1999).

[5] P. Biran, *Constructing new ample divisors out of old ones*, Duke Math. J. **98**, 113–135 (1999).

[6] P. Biran, *From symplectic packing to algebraic geometry and back*, European Congress of Mathematics, Vol. II (Barcelona, 2000), 507–524, Progr. Math. **202**, Birkhäuser, Basel, 2001.

[7] P. Biran, *Geometry of symplectic intersections*, Proceedings of the International Congress of Mathematicians, Vol. II (Beijing, 2002), 241–255, Higher Ed. Press, Beijing, 2002.

[8] P. Biran and K. Cieliebak, *Symplectic topology on subcritical manifolds*, Comment. Math. Helv. **76**, 712–753 (2001).

[9] P. Biran, L. Polterovich and D. Salamon, *Propagation in Hamiltonian dynamics and relative symplectic homology*, Duke Math. J. **119**, 65–118 (2003).

[10] F. Bourgeois, Ya. Eliashberg, H. Hofer, K. Wysocki and E. Zehnder, *Compactness results in symplectic field theory*, Geom. Topol. **7**, 799–888 (2003).

[11] Y. Chekanov, *Hofer's symplectic energy and Lagrangian intersections*, Contact and Symplectic Geometry, ed. C. B. Thomas, Publ. Newton Inst. **8**, 296–306, Cambridge University Press, Cambridge, 1996.

[12] Y. Chekanov, *Lagrangian intersections, symplectic energy, and areas of holomorphic curves*, Duke Math. J. **95**, 213–226 (1998).

[13] Y. Chekanov, talk on a hike on Üetliberg on a sunny day in May 2004.

[14] K. Cieliebak, A. Floer and H. Hofer, *Symplectic homology. II. A general construction*, Math. Z. **218**, 103–122 (1995).

[15] K. Cieliebak and E. Goldstein, *A note on mean curvature, Maslov class and symplectic area of Lagrangian immersions*, J. Symplectic Geom. **2**, 261–266 (2004).

[16] K. Cieliebak and K. Mohnke, *Punctured holomorphic curves and Lagrangian embeddings*, preprint, 2003.

[17] S. Donaldson, *Symplectic submanifolds and almost-complex geometry*, J. Differential Geom. **44**, 666–705 (1996).

[18] I. Ekeland and H. Hofer, *Symplectic topology and Hamiltonian dynamics*, Math. Z. **200**, 355–378 (1989).

[19] I. Ekeland and H. Hofer, *Symplectic topology and Hamiltonian dynamics II*, Math. Z. **203**, 553–567 (1990).

[20] I. Ekeland and S. Mathlouthi, *Calcul numérique de la capacité symplectique*, Progress in variational methods in Hamiltonian systems and elliptic equations (L'Aquila, 1990), 68–91, Pitman Res. Notes Math. Ser. **243**, Longman Sci. Tech., Harlow, 1992.

[21] Y. Eliashberg, *Symplectic topology in the nineties*, Symplectic geometry. Differential Geom. Appl. **9**, 59–88 (1998).

[22] Y. Eliashberg, A. Givental and H. Hofer, *Introduction to symplectic field theory*, GAFA 2000 (Tel Aviv, 1999), Geom. Funct. Anal. 2000, Special Volume, Part II, 560–673.

[23] Y. Eliashberg and M. Gromov, *Convex symplectic manifolds*, Several complex variables and complex geometry, Part 2 (Santa Cruz, CA, 1989), 135–162, Proc. Sympos. Pure Math. **52**, Part 2, American Mathematical Society, Providence, RI, 1991.

[24] Y. Eliashberg and H. Hofer, *Unseen symplectic boundaries*, Manifolds and geometry, Sympos. Math. XXXVI (Pisa, 1993) 178–189, Cambridge Univ. Press, Cambridge, 1996.

[25] Y. Eliashberg and H. Hofer, *An energy-capacity inequality for the symplectic holonomy of hypersurfaces flat at infinity*, Symplectic geometry, 95–114, London Math. Soc. Lecture Note Ser. **192**, Cambridge Univ. Press, Cambridge, 1993.

[26] E. Fadell and P. Rabinowitz, *Generalized cohomological index theories for Lie group actions with an application to bifurcation questions for Hamiltonian systems*, Invent. Math. **45**, 139–173 (1978).

[27] A. Floer, H. Hofer and C. Viterbo, *The Weinstein conjecture in $P \times \mathbb{C}^l$*, Math. Z. **203**, 469–482 (1990).

[28] A Floer, H. Hofer, *Symplectic homology. I. Open sets in $\mathbb{C}^n$*, Math. Z. **215**, 37–88 (1994).

[29] A Floer, H. Hofer and K. Wysocki, *Applications of symplectic homology. I*, Math. Z. **217**, 577–606 (1994).

[30] U. Frauenfelder, V. Ginzburg and F. Schlenk, *Energy capacity inequalities via an action selector*, Geometry, spectral theory, groups, and dynamics, 129–152, Contemp. Math. **387**, Amer. Math. Soc., Providence, RI, 2005.

[31] U. Frauenfelder and F. Schlenk, *Hamiltonian dynamics on convex symplectic manifolds*, Israel J. of Maths. **159**, 1–56 (2007).

[32] V. Ginzburg, *An embedding $S^{2n-1} \to \mathbf{R}^{2n}$, $2n - 1 \geq 7$, whose Hamiltonian flow has no periodic trajectories*, Internat. Math. Res. Notices 1995, 83–97 (1995).

[33] V. Ginzburg, *A smooth counterexample to the Hamiltonian Seifert conjecture in $\mathbf{R}^6$*, Internat. Math. Res. Notices 1997, 641–650 (1997).

[34] V. Ginzburg, *The Weinstein conjecture and theorems of nearby and almost existence*, The breadth of symplectic and Poisson geometry, 139–172, Progr. Math. **232**, Birkhäuser, Boston, MA, 2005.

[35] V. Ginzburg and B. Gürel, *A C^2-smooth counterexample to the Hamiltonian Seifert conjecture in $\mathbb{R}^4$*, Ann. of Math. (2) **158**, 953–976 (2003).

[36] V. Ginzburg and B. Gürel, *Relative Hofer–Zehnder capacity and periodic orbits in twisted cotangent bundles*, Duke Math. J. **123**, 1–47 (2004).

[37] V. Ginzburg and E. Kerman, *Periodic orbits in magnetic fields in dimensions greater than two*, Geometry and topology in dynamics (Winston-Salem, NC, 1998, and San Antonio, TX, 1999), 113–121, Contemp. Math. **246**, American Mathematical Society, Providence, RI, 1999.

[38] E. Goldstein, *Some estimates related to Oh's conjecture for the Clifford tori in $\mathbb{CP}^n$*, preprint, math.DG/0311460.

[39] M. Gromov, *Pseudo holomorphic curves in symplectic manifolds*, Invent. Math. **82**, 307–347 (1985).

[40] D. Hermann, *Holomorphic curves and Hamiltonian systems in an open set with restricted contact-type boundary*, Duke Math. J. **103**, 335–374 (2000).

[41] D. Hermann, *Inner and outer Hamiltonian capacities*, Bull. Soc. Math. France **132**, 509–541 (2004).

[42] D. Hermann, *Symplectic capacities and symplectic convexity*, Preprint, 2005.

[43] H. Hofer, *On the topological properties of symplectic maps*, Proc. Roy. Soc. Edinburgh Sect. A **115**, 25–38 (1990).

[44] H. Hofer, *Symplectic capacities*, Geometry of low-dimensional manifolds, 2 (Durham, 1989), 15–34, London Math. Soc. Lecture Note Ser. **151**, Cambridge Univ. Press, Cambridge, 1990.

[45] H. Hofer, *Estimates for the energy of a symplectic map*, Comment. Math. Helv. **68**, 48–72 (1993).

[46] H. Hofer, *Pseudoholomorphic curves in symplectizations with applications to the Weinstein conjecture in dimension three*, Invent. Math. **114**, 515–563 (1993).

[47] H. Hofer and C. Viterbo, *The Weinstein conjecture in the presence of holomorphic spheres*, Comm. Pure Appl. Math. **45**, 583–622 (1992).

[48] H. Hofer and E. Zehnder, *A new capacity for symplectic manifolds*, Analysis, et cetera, 405–427, Academic Press, Boston, MA, 1990.

[49] H. Hofer and E. Zehnder, *Symplectic Invariants and Hamiltonian Dynamics*, Birkhäuser, Basel (1994).

[50] H. Iriyeh, H. Ono and T. Sakai, *Integral geometry and Hamiltonian volume minimizing property of a totally geodesic Lagrangian torus in $S^2 \times S^2$*, Proc. Japan Acad. Ser. A Math. Sci. **79**, 167–170 (2003).

[51] M.-Y. Jiang, *Hofer–Zehnder symplectic capacity for two-dimensional manifolds*, Proc. Roy. Soc. Edinburgh Sect. A **123**, 945–950 (1993).

[52] M.-Y. Jiang, *Symplectic embeddings from* $\mathbf{R}^{2n}$ *into some manifolds*, Proc. Roy. Soc. Edinburgh Sect. A **130**, 53–61 (2000).

[53] B. Kruglikov, *A remark on symplectic packings*, Dokl. Akad. Nauk **350**, 730–734 (1996).

[54] F. Lalonde, *Energy and capacities in symplectic topology*, Geometric topology (Athens, GA, 1993), 328–374, AMS/IP Stud. Adv. Math. **2.1**, American Mathematical Society, Providence, RI, 1997.

[55] F. Lalonde and D. McDuff, *The geometry of symplectic energy*, Ann. of Math. (2) **141**, 349–371 (1995).

[56] F. Lalonde and D. McDuff, *Hofer's* L^∞*-geometry: energy and stability of Hamiltonian flows. I, II*, Invent. Math. **122**, 1–33, 35–69 (1995).

[57] F. Lalonde and C. Pestieau, *Stabilisation of symplectic inequalities and applications*, Northern California Symplectic Geometry Seminar, 63–71, AMS Transl. Ser. 2 **196**, American Mathematical Society, Providence, RI, 1999.

[58] F. Lalonde and M. Pinsonnault, *The topology of the space of symplectic balls in rational 4-manifolds*, Duke Math. J. **122**, 347–397 (2004).

[59] G. Lu, *The Weinstein conjecture on some symplectic manifolds containing the holomorphic spheres*, Kyushu J. Math. **52**, 331–351 (1998) and **54**, 181–182 (2000).

[60] G. Lu, *Symplectic capacities of toric manifolds and combinatorial inequalities*, C. R. Math. Acad. Sci. Paris **334**, 889–892 (2002).

[61] L. Macarini, *Hofer–Zehnder capacity and Hamiltonian circle actions*, Commun. Contemp. Math. **6**, 913–945 (2004).

[62] L. Macarini and F. Schlenk, *A refinement of the Hofer–Zehnder theorem on the existence of closed trajectories near a hypersurface*, Bull. London Math. Soc. **37**, 297–300 (2005).

[63] F. Maley, J. Mastrangeli, L. Traynor, *Symplectic packings in cotangent bundles of tori*, Experiment. Math. **9**, 435–455 (2000).

[64] D. McDuff, *Blowing up and symplectic embeddings in dimension 4*, Topology **30**, 409–421 (1991).

[65] D. McDuff, *Symplectic manifolds with contact type boundaries*, Invent. Math. **103**, 651–671 (1991).

[66] D. McDuff, *Symplectic topology and capacities*, Prospects in mathematics (Princeton, NJ, 1996), 69–81, American Mathematical Society, Providence, RI, 1999.

[67] D. McDuff, *Geometric variants of the Hofer norm*, J. Symplectic Geom. **1**, 197–252 (2002).

[68] D. McDuff and L. Polterovich, *Symplectic packings and algebraic geometry*, Invent. Math. **115**, 405–429 (1994).

[69] D. McDuff and D. Salamon, *Introduction to symplectic topology*, Second edition, Oxford Mathematical Monographs, The Clarendon Press, Oxford University Press, New York, 1998.

[70] D. McDuff and D. Salamon, *J-holomorphic curves and symplectic topology*, AMS Colloquium Publications **52**, American Mathematical Society, Providence, RI, 2004.

[71] D. McDuff and J. Slimowitz, *Hofer–Zehnder capacity and length minimizing Hamiltonian paths*, Geom. Topol. **5**, 799–830 (2001).

[72] D. McDuff and L. Traynor, *The 4-dimensional symplectic camel and related results*, Symplectic geometry, 169–182, London Math. Soc. Lecture Note Ser. **192**, Cambridge Univ. Press, Cambridge (1993).

[73] E. Neduv, *Prescribed minimal period problems for convex Hamiltonian systems via Hofer–Zehnder symplectic capacity*, Math. Z. **236**, 99–112 (2001).

[74] Y.-G. Oh, *Second variation and stabilities of minimal Lagrangian submanifolds in Kähler manifolds*, Invent. Math. **101**, 501–519 (1990).

[75] Y.-G. Oh, *Chain level Floer theory and Hofer's geometry of the Hamiltonian diffeomorphism group*, Asian J. Math. **6**, 579–624 (2002).

[76] Y.-G. Oh, *Construction of spectral invariants of Hamiltonian paths on closed symplectic manifolds*, The breadth of symplectic and Poisson geometry, 525–570, Progr. Math. **232**, Birkhäuser Boston, Boston, MA, 2005.

[77] Y.-G. Oh, *Spectral invariants, analysis of the Floer moduli space, and geometry of the Hamiltonian diffeomorphism group*, Duke Math. J. **130**, 199–295 (2005).

[78] E. Opshtein, *Maximal symplectic packings in $\mathcal{P}^2$*, To appear in *Compos. Math.*

[79] L. Polterovich, *Gromov's K-area and symplectic rigidity*, Geom. Funct. Anal. **6**, 726–739 (1996).

[80] L. Polterovich, *Symplectic aspects of the first eigenvalue*, J. Reine Angew. Math. **502**, 1–17 (1998).

[81] L. Polterovich, *Hamiltonian loops from the ergodic point of view*, J. Eur. Math. Soc. **1**, 87–107 (1999).

[82] L. Polterovich, *The geometry of the group of symplectic diffeomorphisms*, Lectures in Mathematics ETH Zürich, Birkhäuser, Basel, 2001.

[83] P. Rabinowitz, *Periodic solutions of Hamiltonian systems*, Comm. Pure Appl. Math. **31**, 157–184 (1978).

[84] P. Rabinowitz, *Periodic solutions of a Hamiltonian system on a prescribed energy surface*, J. Differential Equations **33**, 336–352 (1979).

[85] F. Schlenk, *Symplectic embedding of ellipsoids*, Israel J. of Math. **138**, 215–252 (2003).

[86] F. Schlenk, *Embedding problems in symplectic geometry*, De Gruyter Expositions in Mathematics **40**, Walter de Gruyter, Berlin, 2005.

[87] F. Schlenk, *Applications of Hofer's geometry to Hamiltonian dynamics*, Comment. Math. Helv. **81**, 105–121 (2006).

[88] M. Schwarz, *On the action spectrum for closed symplectically aspherical manifolds*, Pacific J. Math. **193**, 419–461 (2000).

[89] K.-F. Siburg, *Symplectic capacities in two dimensions*, Manuscripta Math. **78**, 149–163 (1993).

[90] J.-C. Sikorav, *Systèmes Hamiltoniens et topologie symplectique*, Dipartimento di Matematica dell' Università di Pisa, ETS Editrice, Pisa, 1990.

[91] J.-C. Sikorav, *Quelques propriétés des plongements lagrangiens*, Analyse globale et physique mathématique (Lyon, 1989), Mém. Soc. Math. France **46**, 151–167 (1991).

[92] T. Tokieda, *Isotropic isotopy and symplectic null sets*, Proc. Nat. Acad. Sci. U.S.A. **94**, 13407–13408 (1997).

[93] L. Traynor, *Symplectic homology via generating functions*, Geom. Funct. Anal. **4**, 718–748 (1994).

[94] L. Traynor, *Symplectic packing constructions*, J. Differential Geom. **42**, 411–429 (1995).

[95] C. Viterbo, *Capacités symplectiques et applications (d'après Ekeland–Hofer, Gromov)*, Séminaire Bourbaki, Vol. 1988/89. Astérisque **177–178** (1989), Exp. No. 714, 345–362.

[96] C. Viterbo, *Plongements lagrangiens et capacités symplectiques de tores dans* $\mathbb{R}^{2n}$, C. R. Acad. Sci. Paris Sér. I Math. **311**, 487–490 (1990).

[97] C. Viterbo, *Symplectic topology as the geometry of generating functions*, Math. Ann. **292**, 685–710 (1992).

[98] C. Viterbo, *Metric and isoperimetric problems in symplectic geometry*, J. Amer. Math. Soc. **13**, 411–431 (2000).

KAI CIELIEBAK
MATHEMATISCHES INSTITUT
LUDWIG-MAXIMILIANS-UNIVERSITÄT MÜNCHEN
THERESIENSTRASSE 39
D-80333 MÜNCHEN
GERMANY
kai@mathematik.uni-muenchen.de

HELMUT HOFER
COURANT INSTITUTE
NEW YORK UNIVERSITY
251 MERCER STREET
NEW YORK, NY 10012
UNITED STATES
hofer@cims.nyu.edu

JANKO LATSCHEV
MATHEMATISCHES INSTITUT
LUDWIG-MAXIMILIANS-UNIVERSITÄT MÜNCHEN
THERESIENSTRASSE 39
D-80333 MÜNCHEN
GERMANY
latschev@mathematik.uni-muenchen.de

FELIX SCHLENK
DÉPARTEMENT DE MATHÉMATIQUES
UNIVERSITÉ LIBRE DE BRUXELLES
BOULEVARD DU TRIOMPHE
1050 BRUXELLES
BELGIUM
fschlenk@ulb.ac.be

Recent Progress in Dynamics
MSRI Publications
Volume **54**, 2007

Local rigidity of group actions:
past, present, future

DAVID FISHER

To Anatole Katok on the occasion of his 60th birthday.

ABSTRACT. This survey aims to cover the motivation for and history of the study of local rigidity of group actions. There is a particularly detailed discussion of recent results, including outlines of some proofs. The article ends with a large number of conjectures and open questions and aims to point to interesting directions for future research.

1. Prologue

Let Γ be a finitely generated group, D a topological group, and $\pi : \Gamma \to D$ a homomorphism. We wish to study the space of deformations or perturbations of π. Certain trivial perturbations are always possible as soon as D is not discrete, namely we can take $d\pi d^{-1}$ where d is a small element of D. This motivates the following definition:

DEFINITION 1.1. Given a homomorphism $\pi : \Gamma \to D$, we say π is *locally rigid* if any other homomorphism π' which is close to π is conjugate to π by a small element of D.

We topologize $\mathrm{Hom}(\Gamma, D)$ with the compact open topology which means that two homomorphisms are close if and only if they are close on a generating set for Γ. If D is path connected, then we can define *deformation rigidity* instead, meaning that any continuous path of representations π_t starting at π is conjugate to the trivial path $\pi_t = \pi$ by a continuous path d_t in D with d_0 being

Author partially supported by NSF grant DMS-0226121 and a PSC-CUNY grant.

the identity in D. If D is an algebraic group over $\mathbb{R}$ or $\mathbb{C}$, it is possible to prove that deformation rigidity and local rigidity are equivalent since $\mathrm{Hom}(\Gamma, D)$ is an algebraic variety and the action of D by conjugation is algebraic; see [Mu], for example. For D infinite dimensional and path-connected, this equivalence is no longer clear.

The study of local rigidity of lattices in semisimple Lie groups is probably the beginning of the general study of rigidity in geometry and dynamics, a subject that is by now far too large for a single survey. See [Sp1] for the last attempt at a comprehensive survey and [Sp2] for a more narrowly focused updating of that survey. Here we abuse language slightly by saying a subgroup is locally rigid if the defining embedding is locally rigid as a homomorphism. See subsection 3.1 for a brief history of local rigidity of lattices and some discussion of subsequent developments that are of particular interest in the study of rigidity of group actions.

In this article we will focus on a survey of local rigidity when $D = \mathrm{Diff}^{\infty}(M)$ or occasionally $D = \mathrm{Diff}^{k}(M)$ for some finite k. Here we often refer to a homomorphism $\pi : \Gamma \to \mathrm{Diff}^{\infty}(M)$ as an action, since it can clearly be thought of as C^{∞} action $\pi : \Gamma \times M \to M$. We will consistently use the same letter π to denote either the action or the homomorphism to $\mathrm{Diff}^{\infty}(M)$. The title of this article refers to this interpretation of π as defining a group action. In this context, one considers rigidity of actions of connected groups as well as of discrete groups. In cases where Γ has any topology, we will always only study continuous actions, i.e. ones for which the defining homomorphism π is a continuous map.

One can, in this context, develop more refined notions of local rigidity, since the topology on $\mathrm{Diff}^{\infty}(M)$ is an inverse limit topology. This means that two C^{∞} diffeomorphisms of M are close precisely when they are C^{k} close for some large value of k. The most exhaustive definition of local rigidity is probably the following:

DEFINITION 1.2. Let Γ be a discrete group and $\pi : \Gamma \to \mathrm{Diff}^{k}(M)$ a homomorphism where k is either a positive integer or ∞. We say that π is $C^{k,l,i,j,m}$ *locally rigid* if any $\pi' : \Gamma \to \mathrm{Diff}^{l}(M)$ which is close to π in the C^{i} topology is conjugate to π by a C^{j} diffeomorphism ϕ which is C^{m} small. Here l, i, j, m are all either nonnegative integers or ∞ and the only a priori constraint are $i \leq \min(k, l)$ and $m \leq j$. When $j = 0$, we will call the action *stable* or *structurally stable*. When $j > 0$, we will call the action *locally rigid* or simply *rigid*.

We will avoid using this cumbersome notation when at all possible. There is a classical, dynamical notion of structural stability which is equivalent to $C^{1,1,1,0,0}$ local rigidity. I.e. a C^{1} action π of a group Γ is *structurally stable* if any C^{1} close C^{1} action of Γ is conjugate to π by a small homeomorphism. For actions of $\mathbb{Z}$ this notion arose in hyperbolic dynamics in the work of Anosov and Smale [An; Sm]. From a dynamical point of view structural stability is important since it allows one to control dynamical properties of an open set of actions in $\text{Diff}^1(M)$. Local rigidity can be viewed as a strengthening of this property in that it shows that an open set of actions is exhausted by smooth conjugates of a single action.

Though actions of $\mathbb{Z}$ and free groups on k generators are often structurally stable, they are never locally rigid, and it is an interesting question as to how "large" a group needs to be in order to have actions which are locally rigid. Many of the original questions and theorems concerning local rigidity were for lattices in higher rank semisimple Lie groups, where here higher rank means that all simple factors have real rank at least 2. (See subsection 2.1 for a definition of rank.) Fairly early in the theory it became clear that local rigidity often held, and was in fact easier to prove, for certain actions of higher rank abelian groups, i.e. $\mathbb{Z}^k$ for $k \geq 2$, see [KL1]. In addition, local rigidity results have been proven for actions of a wider variety of groups, including

(1) certain non-volume-preserving actions of lattices in $SO(1, n)$ in [Kan1]
(2) all isometric actions of groups with property (T) in [FM2],
(3) certain affine actions of lattices in $SP(1, n)$ in [Hi].

There is extremely interesting related work of Ghys, older than the work just mentioned, which shows that the space of deformations of certain actions of surface groups on the circle is finite dimensional [Gh1; Gh2; Gh3]. Ghys also proved some very early results on local and global rigidity of very particular actions of connected solvable groups, see [GhS; Gh1] and subsection 4.2.

The study of local rigidity of group actions has had three primary historical motivations, one from the theory of lattices in Lie groups, one from dynamical systems theory and a third from the theory of foliations. (This statement is perhaps a bit coarse, and there is heavy overlap between these motivations, particularly the second and the third.) The first is the general study of rigidity of actions of large groups, as discussed in [Z3; Z4], see [La; FK] for more up to date surveys. This area is motivated by the study of rigidity of lattices in semisimple Lie groups, particularly by Margulis' superrigidity theorem and it's nonlinear generalization by Zimmer to a cocycle superrigidity theorem, see subsection 3.1 and [Z4] for more discussion. This motivation also stems from an analogy between semisimple Lie groups and diffeomorphism groups. When M is a compact manifold, not only is $\text{Diff}^\infty(M)$ an infinite dimensional Lie group, but its connected component is simple. Simplicity of the connected component of $\text{Diff}^\infty(M)$ was proven by Thurston using results of Epstein and Herman [Th2; Ep; Hr]. Herman had used Epstein's work to see that the connected component of $\text{Diff}^\infty(\mathbb{T}^n)$ is simple and Thurston's proof of the general case uses this. See also further work on the topic by Banyaga and Mather [Ba1; Mt1; Mt2; Mt3], as well as Banyaga's book [Ba2].

The dynamical motivation for studying rigidity of group actions comes from the study of structural stability of diffeomorphisms and flows in hyperbolic dynamics, see the introduction of [KS2]. This area provides many of the basic techniques by which results in the area have been proven, particularly early in the history of the field. Philosophically, hyperbolic diffeomorphisms are structurally stable, group actions generated by structurally stable diffeomorphisms are quite often structurally stable, and the presence of a large group action frequently allows one to improve the regularity of the conjugacy. See subsection 3.2 for a brief history of relevant results on structural stability and subsections 4.1, 4.2, and 5.1 for some applications of these ideas.

The third motivation for studying rigidity of group actions comes from the theory of foliations. Many techniques and ideas in this area are also related to work on hyperbolic dynamics, and many of the foliations of interest are dynamical foliations of hyperbolic dynamical systems. A primary impetus in this area is the theory of codimension one foliations, and so many of the ideas here were first developed either for groups acting on the circle or for actions of connected groups on manifolds only one dimension larger then the acting group. See particularly [GhS; Gh1] for the early history of these developments.

Some remarks on biases and omissions. Like any survey of this kind, this work is informed by it's authors biases and experiences. The most obvious of these is that my point of view is primarily motivated by the study of rigidity properties of semisimple Lie groups and their lattices, rather than primarily motivated by hyperbolic dynamics or foliation theory. This informs the biases of this article and a very different article would result from different biases.

There are two large omissions in this article. The first omission is that it is primarily occupied with local rigidity of discrete group actions. When similar results are known for actions of Lie groups, they are mentioned, though frequently only special cases are stated. This is partially because results in this context are often complicated by the need to consider time changes, and I did not want to dwell on that issue here. The second omission is that little to no care is taken to state optimal results relating the various constants in $C^{k,l,i,j,m}$ local rigidity. Dwelling on issues of regularity seemed likely to obscure the main line of the developments, so many results are stated without any explicit mention of regularity. Usually this is done only when the action can be shown to be *locally rigid* in $\mathrm{Diff}^{\infty}(M)$ in the sense of Definition 1.1. This implicitly omits both the degree of regularity to which the perturbed and unperturbed actions are close and the degree of regularity with which the size of the conjugacy is small. In other words local rigidity is $C^{\infty,\infty,i,\infty,m}$ local rigidity for some unspecified i and m, and I always fail to specify i and m even when they are known. Occasionally a result is stated that only produces a finite regularity conjugacy, with this issue only remarked on following the statement of the result. It seems quite likely that most results of this kind can be improved to produce C^∞ conjugacies using the techniques of [FM2; FM3], see discussion at the end of Section 5.1.

Lastly we remark that the study of local rigidity of group actions is often closely intertwined with the study of *global rigidity* of group actions. The meaning of the phrase global rigidity is not entirely precise, but it is typically used to cover settings in which one can classify all group actions satisfying certain hypotheses on a certain manifold or class of manifolds. The study of global rigidity is too broad and interesting to summarize briefly, but some examples are mentioned below when they are closely related to work on local rigidity. See [FK; La] for recent surveys concerning both local and global rigidity.

2. A brief digression: some examples of groups and actions

In this section we briefly describe some of the groups that will play important roles in the results discussed here. The reader already familiar with semisimple Lie groups and their lattices may want to skip to the second subsection where we give descriptions of group actions.

2.1. Semisimple groups and their lattices. By a simple Lie group, we mean a connected Lie group all of whose normal subgroups are discrete, though we make the additional convention that $\mathbb{R}$ and S^1 are not simple. By a semisimple Lie group we mean the quotient of a product of simple Lie groups by some subgroup of the product of their centers. Note that with our conventions, the center of a simple Lie group is discrete and is in fact the maximal normal subgroup. There is an elaborate structure theory of semisimple Lie groups and the groups are completely classified, see [He] or [Kn] for details. Here we merely describe some examples, all of which are matrix groups. All connected semisimple Lie groups are discrete central extensions of matrix groups, so the reader will lose very little by always thinking of matrix groups.

(1) The groups $SL(n, \mathbb{R})$, $SL(n, \mathbb{C})$ and $SL(n, \mathbb{H})$ of n by n matrices of determinant one over the real numbers, the complex numbers or the quaternions.
(2) The group $SP(2n, \mathbb{R})$ of $2n$ by $2n$ matrices of determinant one which preserve a real symplectic form on $\mathbb{R}^{2n}$.
(3) The groups $SO(p, q)$, $SU(p, q)$ and $SP(p, q)$ of matrices which preserve inner products of signature (p, q) where the inner product is real linear on $\mathbb{R}^{p+q}$, hermitian on $\mathbb{C}^{p+q}$ or quaternionic hermitian on $\mathbb{H}^{p+q}$ respectively.

Let G be a semisimple Lie group which is a subgroup of $GL(n, \mathbb{R})$. We say that G has *real rank k* if G has a k dimensional abelian subgroup which is conjugate to a subgroup of the real diagonal matrices and no $k + 1$ dimensional abelian subgroups with the same property. The groups in (1) have rank $n - 1$, the groups in (2) have rank n and the groups in (3) have rank $\min(p, q)$.

Since this article focuses primarily on finitely generated groups, we are more interested in discrete subgroups of Lie groups than in the Lie groups themselves. A discrete subgroup Γ in a Lie group G is called a lattice if G/Γ has finite Haar measure. The lattice is called *cocompact* or *uniform* if G/Γ is compact and *nonuniform* or simply not cocompact otherwise. If $G = G_1 \times \cdots \times G_n$ is a product then we say a lattice $\Gamma < G$ is *irreducible* if it's projection to each G_i is dense. More generally we make the same definition for an *almost direct product*, by which we mean a direct product G modulo some subgroup of the center $Z(G)$. Lattices in semisimple Lie groups can always be constructed by arithmetic methods, see [Bo] and also [Mr] for more discussion. In fact, one of the most important results in the theory of semisimple Lie groups is that if G is a semisimple Lie group without compact factors, then all irreducible lattices in G are arithmetic unless G is locally isomorphic to $SO(1, n)$ or $SU(1, n)$. For G of real rank at least two, this is Margulis' arithmeticity theorem, which he deduced from his superrigidity theorems [Ma2; Ma3; Ma4]. For nonuniform lattices, Margulis had an earlier proof which does not use the superrigidity theorems, see [Ma1; Ma2]. This earlier proof depends on the study of dynamics of unipotent elements on the space G/Γ, and particularly on what is now known as the "nondivergence of unipotent flows". Special cases of the superrigidity theorems were then proven for $Sp(1, n)$ and F_4^{-20} by Corlette and Gromov–Schoen, which sufficed to imply the statement on arithmeticity given above [Co2; GS]. As we will be almost exclusively concerned with arithmetic lattices, we do not give examples of nonarithmetic lattices here, but refer the reader to [Ma4] and [Mr] for more discussion. A formal definition of arithmeticity, at least when G is algebraic is:

DEFINITION 2.1. Let G be a semisimple algebraic Lie group and $\Gamma < G$ a lattice. Then Γ is arithmetic if there exists a semisimple algebraic Lie group H defined over $\mathbb{Q}$ such that

(1) there is a homomorphism $\pi : H^0 \to G$ with compact kernel,
(2) there is a rational structure on H such that the projection of the integer points of H to G are commensurable to Γ, i.e. $\pi(H(\mathbb{Z})) \cap \Gamma$ is of finite index in both $H(\mathbb{Z})$ and Γ.

We now give some examples of arithmetic lattices. The simplest is to take the integer points in a simple (or semisimple) group G which is a matrix group, e.g. $SL(n, \mathbb{Z})$ or $Sp(n, \mathbb{Z})$. This exact construction always yields lattices, but also always yields nonuniform lattices. In fact the lattices one can construct in this way have very special properties because they will contain many unipotent matrices. If a lattice is cocompact, it will necessarily contain no unipotent matrices. The standard trick for understanding the structure of lattices in G which become integral points after passing to a compact extension is called *change of base*. For much more discussion see [Ma4; Mr; Z2]. We give one example to illustrate the process. Let $G = SO(m, n)$ which we view as the set of matrices in $SL(n + m, \mathbb{R})$ which preserve the inner product

$$\langle v, w \rangle = \left(-\sqrt{2} \sum_{i=1}^{m} v_i w_i \right) + \left(\sum_{i=m+1}^{n+m} v_i w_i \right)$$

where v_i and w_i are the ith components of v and w. This form, and therefore G, are defined over the field $\mathbb{Q}(\sqrt{2})$ which has a Galois conjugation σ defined by $\sigma(\sqrt{2}) = -\sqrt{2}$. If we looks at the points $\Gamma = G(\mathbb{Z}[\sqrt{2}])$, we can define an embedding of Γ in $SO(m, n) \times SO(m + n)$ by taking γ to $(\gamma, \sigma(\gamma))$. It is straightforward to check that this embedding is discrete. In fact, this embeds Γ in $H = SO(m, n) \times SO(m + n)$ as integral points for the rational structure on H where the rational points are exactly the points $(m, \sigma(m))$ where $m \in G(\mathbb{Q}(\sqrt{2}))$. This makes Γ a lattice in H and it is easy to see that Γ projects to a lattice in G, since G is cocompact in H. What is somewhat harder to verify is that Γ is cocompact in H, for which we refer the reader to the list of references above.

Similar constructions are possible with $SU(m, n)$ or $SP(m, n)$ in place of $SO(m, n)$ and also with more simple factors and fields with more Galois automorphisms. There are also a number of other constructions of arithmetic lattices using division algebras. See [Mr] for a comprehensive treatment.

We end this section by defining a key property of many semisimple groups and their lattices. This is property (T) of Kazhdan, and was introduced by Kazhdan in [Ka1] in order to prove that nonuniform lattices in higher rank semisimple Lie groups are finitely generated and have finite abelianization. It has played a fundamental role in many subsequent developments. We do not give Kazhdan's original definition, but one which was shown to be equivalent by work of Delorme and Guichardet [De; Gu].

DEFINITION 2.2. A group Γ has property (T) of Kazhdan if $H^1(\Gamma, \pi) = 0$ for every continuous unitary representation π of Γ on a Hilbert space. This is equivalent to saying that any continuous isometric action of Γ on a Hilbert space has a fixed point.

REMARKS. (1) Kazhdan's definition is that the trivial representation is isolated in the unitary dual of Γ.
(2) If a continuous group G has property (T) so does any lattice in G. This result was proved in [Ka1].

(3) Any semisimple Lie group has property (T) if and only if it has no simple factors locally isomorphic to $SO(1, n)$ or $SU(1, n)$. For a discussion of this fact and attributions, see [HV]. For groups with all simple factors of real rank at least three, this is proven in [Ka1].

(4) No noncompact amenable group, and in particular no noncompact abelian group, has property (T). An easy averaging argument shows that all compact groups have property (T).

Groups with property (T) play an important role in many areas of mathematics and computer science.

2.2. Some actions of groups and lattices.
Here we define and give examples of the general classes of actions for which local rigidity results have been proven. Let H be a Lie group and $L < H$ a closed subgroup. Then a diffeomorphism f of H/L is called *affine* if there is a diffeomorphism $\tilde{f}$ of H such that $f([h]) = \tilde{f}(h)$ where $\tilde{f} = A \circ \tau_h$ with A an automorphism of H with $A(L) = L$ and τ_h is left translation by some h in H. Two obvious classes of affine diffeomorphisms are left translations on any homogeneous space and either linear automorphisms of tori or more generally automorphisms of nilmanifolds. A group action is called *affine* if every element of the group acts by an affine diffeomorphism. It is easy to check that the full group of affine diffeomorphisms $\mathrm{Aff}(H/L)$ is a finite dimensional Lie group and an affine action of a group D is a homomorphism $\pi : D \rightarrow \mathrm{Aff}(H/L)$. The structure of $\mathrm{Aff}(H/L)$ is surprisingly complicated in general, it is a quotient of a subgroup of the group $\mathrm{Aut}(H) \ltimes H$ where $\mathrm{Aut}(H)$ is a the group of automorphisms of H. For a more detailed discussion of this relationship, see [FM1, Section 6]. While it is not always the case that any affine action of a group D on H/L can be described by a homomorphism $\pi : D \rightarrow \mathrm{Aut}(H) \ltimes H$, this is true for two important special cases:

(1) D is a connected semisimple Lie group and L is a cocompact lattice in H,

(2) D is a lattice in a semisimple Lie group G where G has no compact factors and no simple factors locally isomorphic to $SO(1, n)$ or $SU(1, n)$, and L is a cocompact lattice in H.

These facts are [FM1, Theorem 6.4 and 6.5] where affine actions as in (1) and (2) above are classified.

The most obvious examples of affine actions of large groups are of the following forms, which are frequently referred to as *standard actions*:

(1) Actions of groups by automorphisms of nilmanifolds. I.e. let N be a simply connected nilpotent group, $\Lambda < N$ a lattice (which is necessarily cocompact) and assume a finitely generated group Γ acts by automorphisms of N preserving Λ. The most obvious examples of this are when $N = \mathbb{R}^n$, $\Lambda = \mathbb{Z}^n$ and $\Gamma < SL(n, \mathbb{Z})$, in which case we have a linear action of Γ on $\mathbb{T}^n$.

(2) Actions by left translations. I.e. let H be a Lie group and $\Lambda < H$ a co-compact lattice and $\Gamma < H$ some subgroup. Then Γ acts on H/Λ by left translations. Note that in this case Γ need not be discrete.

(3) Actions by isometries. Here K is a compact group which acts by isometries on some compact manifold M and $\Gamma < K$ is a subgroup. Note that here Γ is either discrete or a discrete extension of a compact group.

We now briefly define a few more general classes of actions, for which local rigidity results are either known or conjectured. We first fix some notations. Let A and D be topological groups, and $B < A$ a closed subgroup. Let $\rho : D \times A/B \to A/B$ be a continuous affine action.

DEFINITION 2.3. (1) Let A, B, D and ρ be as above. Let C be a compact group of affine diffeomorphisms of A/B that commute with the D action. We call the action of D on $C \backslash A/B$ a *generalized affine action*.

(2) Let A, B, D and ρ be as in 1 above. Let M be a compact Riemannian manifold and $\iota : D \times A/B \to \text{Isom}(M)$ a C^1 cocycle. We call the resulting skew product D action on $A/B \times M$ a *quasiaffine action*. If C and D are as in 2, and $\alpha : D \times C \backslash A/B \to \text{Isom}(M)$ is a C^1 cocycle, then we call the resulting skew product D action on $C \backslash A/B \times M$ a *generalized quasiaffine action*.

For many of the actions we consider here, there will be a foliation of particular importance. If ρ is an action of a group D on a manifold N, and ρ preserves a foliation $\mathfrak{F}$ and a Riemannian metric along the leaves of $\mathfrak{F}$, we call $\mathfrak{F}$ a *central foliation* for ρ. For quasiaffine and generalized quasiaffine actions on manifolds of the form $C \backslash A/B \times M$ the foliation by leaves of the $\{[a]\} \times M$ is always a central foliation. There are also actions with more complicated central foliations. For example if H is a Lie group, $\Lambda < H$ is discrete and a subgroup $G < H$ acts on H/Λ by left translations, then the foliation of H/Λ by orbits of the centralizer $Z_H(G)$ of G in H is a central foliation. It is relatively easy to construct examples where this foliation has dense leaves. Another example of an action which has a foliation with dense leaves is to embed the $\mathbb{Z}[\sqrt{2}]$ points of $SO(m, n)$ into $SL(2(m + n), \mathbb{Z})$ as described in the preceding subsection and then let this group act on $\mathbb{T}^{2(m+n)}$ linearly. It is easy to see in this case that the maximal central foliation for the action is a foliation by densely embedded leaves none of which are compact.

We end this section by describing briefly the standard construction of an *induced or suspended action*. This notion can be seen as a generalization of the construction of a flow under a function or as an analogue of the more algebraic notion of inducing a representation. Given a group H, a (usually closed) subgroup L, and an action ρ of L on a space X, we can form the space $(H \times X)/L$

where L acts on $H \times X$ by $h \cdot (l, x) = (lh^{-1}, \rho(h)x)$. This space now has a natural H action by left multiplication on the first coordinate. Many properties of the L action on X can be studied more easily in terms of properties of the H action on $(H \times X)/L$. This construction is particularly useful when L is a lattice in H.

3. Prehistory

3.1. Local and global rigidity of homomorphisms into finite dimensional groups.
The earliest work on local rigidity in the context of Definition 1.1 was contained in series of works by Calabi–Vesentini, Selberg, Calabi and Weil, which resulted in the following:

THEOREM 3.1. *Let G be a semisimple Lie group and assume that G is not locally isomorphic to $SL(2, \mathbb{R})$. Let $\Gamma < G$ be an irreducible cocompact lattice, then the defining embedding of Γ in G is locally rigid.*

REMARKS. (1) If $G = SL(2, \mathbb{R})$ the theorem is false and there is a large, well studied space of deformation of Γ in G, known as the Teichmüller space.

(2) There is an analogue of this theorem for lattices that are not cocompact. This result was proven later and has a more complicated history which we omit here. In this case it is also necessary to exclude G locally isomorphic to $SL(2, \mathbb{C})$.

This theorem was originally proven in special cases by Calabi, Calabi–Vesentini and Selberg. In particular, Selberg gives a proof for cocompact lattices in $SL(n, \mathbb{R})$ for $n \geq 3$ in [S], Calabi–Vesentini give a proof when the associated symmetric space $X = G/K$ is Kähler in [CV] and Calabi gives a proof for $G = SO(1, n)$ where $n \geq 3$ in [C]. Shortly afterwards, Weil gave a complete proof of Theorem 3.1 in [We1; We2].

In all of the original proofs, the first step was to show that any perturbation of Γ was discrete and therefore a cocompact lattice. This is shown in special cases in [C; CV; S] and proven in a somewhat broader context than Theorem 3.1 in [W1].

The different proofs of cases of Theorem 3.1 are also interesting in that there are two fundamentally different sets of techniques employed and this dichotomy continues to play a role in the history of rigidity. Selberg's proof essentially combines algebraic facts with a study of the dynamics of iterates of matrices. He makes systematic use of the existence of singular directions, or Weyl chamber walls, in maximal diagonalizable subgroups of $SL(n, \mathbb{R})$. Exploiting these singular directions is essential to much later work on rigidity, both of lattices in higher rank groups and of actions of abelian groups. It seems possible to generalize Selberg's proof to the case of G an $\mathbb{R}$-split semisimple Lie group with rank at least 2. Selberg's proof, which depended on asymptotics at infinity of iterates of matrices, inspired Mostow's explicit use of boundaries in his proof of strong rigidity [Mo2]. Mostow's work in turn provided inspiration for the use of boundaries in later work of Margulis, Zimmer and others on rigidity properties of higher rank groups.

The proofs of Calabi, Calabi–Vesentini and Weil involve studying variations of geometric structures on the associated locally symmetric space. The techniques are analytic and use a variational argument to show that all variations of the geometric structure are trivial. This work is a precursor to much work in geometric analysis studying variations of geometric structures and also informs later work on proving rigidity/vanishing of harmonic forms and maps. The dichotomy between approaches based on algebra/dynamics and approaches that are in the spirit of geometric analysis continues through much of the history of rigidity and the history of local rigidity of group actions in particular.

Shortly after completing this work, Weil discovered a new criterion for local rigidity [We3]. In the context of Theorem 3.1, this allows one to avoid the step of showing that a perturbation of Γ remains discrete. In addition, this result opened the way for understanding local rigidity of more general representations of discrete groups.

THEOREM 3.2. *Let Γ be a finitely generated group, G a Lie group and π : $\Gamma \to G$ a homomorphism. Then π is locally rigid if $H^1(\Gamma, \mathfrak{g}) = 0$. Here $\mathfrak{g}$ is the Lie algebra of G and Γ acts on $\mathfrak{g}$ by $Ad_G \circ \pi$.*

Weil's proof of this result uses only the implicit function theorem and elementary properties of the Lie group exponential map. The same theorem is true if G is an algebraic group over any local field of characteristic zero. In [We3], Weil remarks that if $\Gamma < G$ is a cocompact lattice and G satisfies the hypothesis of Theorem 3.1, then the vanishing of $H^1(\Gamma, \mathfrak{g})$ can be deduced from the computations in [We2]. The vanishing of $H^1(\Gamma, \mathfrak{g})$ is proven explicitly by Matsushima and Murakami in [MM].

Motivated by Weil's work and other work of Matsushima, conditions for vanishing of $H^1(\Gamma, \mathfrak{g})$ were then studied by many authors. See particularly [MM] and [Rg1]. The results in these papers imply local rigidity of many linear representations of lattices.

To close this section, I will briefly discuss some subsequent developments concerning rigidity of lattices in Lie groups that motivated the study of both local and global rigidity of group actions.

The first remarkable result in this direction is Mostow's rigidity theorem, see [Mo1] and references there. Given G as in Theorem 3.1, and two irreducible cocompact lattices Γ_1 and Γ_2 in G, Mostow proves that any isomorphism from Γ_1 to Γ_2 extends to an isomorphism of G with itself. Combined with the prin-

cipal theorem of [We1] which shows that a perturbation of a lattice is again a
lattice, this gives a remarkable and different proof of Theorem 3.1, and Mostow
was motivated by the desire for a "more geometric understanding" of Theorem
3.1 [Mo1]. Mostow's theorem is in fact a good deal stronger, and controls not
only homomorphisms $\Gamma \rightarrow G$ near the defining homomorphism, but any homomorphism into any other simple Lie group G' where the image is lattice.
As mentioned above, Mostow's approach was partially inspired by Selberg's
proof of certain cases of Theorem 3.1, [Mo2]. A key step in Mostow's proof
is the construction of a continuous map between the geometric boundaries of
the symmetric spaces associated to G and G'. Boundary maps continue to play
a key role in many developments in rigidity theory. A new proof of Mostow
rigidity, at least for G_i of real rank one, was provided by Besson, Courtois and
Gallot. Their approach is quite different and has had many other applications
concerning rigidity in geometry and dynamics; see [BCG; CF], for example.

The next remarkable result in this direction is Margulis' superrigidity theorem. Margulis proved this theorem as a tool to prove arithmeticity of irreducible
uniform lattices in groups of real rank at least 2. For irreducible lattices in
semisimple Lie groups of real rank at least 2, the superrigidity theorems clas-
sifies all finite dimensional linear representations. Margulis' theorem holds for
irreducible lattices in semisimple Lie groups of real rank at least two. Given
a lattice $\Gamma < G$ where G is simply connected, one precise statement of some
of Margulis results is to say that any linear representation σ of Γ *almost extends* to a linear representation of G. By this we mean that there is a linear
representation $\tilde{\sigma}$ of G and a bounded image representation $\bar{\sigma}$ of Γ such that
$\sigma(\gamma) = \tilde{\sigma}(\gamma)\bar{\sigma}(\gamma)$ for all γ in G. Margulis' theorems also give an essentially
complete description of the representations $\bar{\sigma}$, up to some issues concerning
finite image representations. The proof here is partially inspired by Mostow's
work: a key step is the construction of a measurable "boundary map". However
the methods for producing the boundary map in this case are very dynamical.
Margulis' original proof used Oseledec Multiplicative Ergodic Theorem. Later
proofs were given by both Furstenberg and Margulis using the theory of group
boundaries as developed by Furstenberg from his study of random walks on
groups [Fu1; Fu2]. Furstenberg's probabilistic version of boundary theory has
had a profound influence on many subsequent developments in rigidity theory.
For more discussion of Margulis' superrigidity theorem, see [Ma2; Ma3; Ma4].

A main impetus for studying rigidity of group actions on manifolds came
from Zimmer's theorem on superrigidity for cocycles. This theorem and it's
proof were strongly motivated by Margulis' work. In fact, Margulis' theorem
is Zimmer's theorem for a certain cocycle $\alpha : G \times G/\Gamma \rightarrow \Gamma$ and the proof of
Zimmer's theorem is quite similar to the proof of Margulis'. In order to avoid

technicalities, we describe only a special case of this result. Let M be a compact manifold, H a matrix group and $P = M \times H$. Now let a group Γ act on M and P continuously, so that the projection from P to M is equivariant. Further assume that the action on M is measure preserving and ergodic. If Γ is a lattice in a simply connected, semisimple Lie group G all of whose simple factors have real rank at least two then there is a measurable map $s : M \to H$, a representation $\pi : G \to H$, a compact subgroup $K < H$ which commutes with $\pi(G)$ and a measurable map $\Gamma \times M \to K$ such that

$$\gamma \cdot s(m) = k(m, \gamma)\pi(\gamma)s(\gamma \cdot m). \tag{3--1}$$

It is easy to check from this equation that the map K satisfies a certain equation that makes it into a cocycle over the action of Γ. One should view s as providing coordinates on P in which the Γ action is *almost a product*. For more discussion of this theorem the reader should see any of [Fe1; Fe2; FM1; Fu3; Z2]. (The version stated here is only proven in [FM1], previous proofs all yielded somewhat more complicated statements.) As a sample application, let $M = \mathbb{T}^n$ and let P be the frame bundle of M, i.e. the space of frames in the tangent bundle of M. Since $\mathbb{T}^n$ is parallelizable, we have $P = \mathbb{T}^n \times GL(n, \mathbb{R}^n)$. The cocycle superrigidity theorem then says that "up to compact noise", the derivative of any measure preserving Γ action on $\mathbb{T}^n$ looks measurably like a constant linear map. In fact, the cocycle superrigidity theorems apply more generally to continuous actions on any principal bundle P over M with fiber H, an algebraic group, and in this context produces a measurable section $s : M \to P$ satisfying equation (3--1). So in fact, cocycle superrigidity implies that for any action preserving a finite measure on any manifold the derivative cocycle looks measurably like a constant cocycle, up to compact noise. That cocycle superrigidity provides information about actions of groups on manifolds through the derivative cocycle was first observed in [Fu3]. Zimmer originally proved cocycle superrigidity in order to study orbit equivalence of group actions. For a recent survey of subsequent developments concerning orbit equivalence rigidity and other forms of superrigidity for cocycles, see [Sl2].

3.2. Stability in hyperbolic dynamics. A diffeomorphism f of a manifold X is said to be *Anosov* if there exists a continuous f invariant splitting of the tangent bundle $TX = E_f^u \oplus E_f^s$ and constants $a > 1$ and $C, C' > 0$ such that for every $x \in X$,

(1) $\| Df^n(v^u) \| \geq Ca^n \| v^u \|$ for all $v^u \in E_f^u(x)$ and,
(2) $\| Df^n(v^s) \| \leq C'a^{-n} \| v^s \|$ for all $v^s \in E_f^s(x)$.

We note that the constants C and C' depend on the choice of metric, and that a metric can always be chosen so that $C = C' = 1$. There is an analogous notion for a flow f_t, where $TX = T\mathbb{O} \oplus E^u_{f_t} \oplus E^s_{f_t}$ where $T\mathbb{O}$ is the tangent space to the flow direction and vectors in $E^u_{f_t}$ (resp. $E^s_{f_t}$) are uniformly expanded (resp. uniformly contracted) by the flow. This notion was introduced by Anosov and named after Anosov by Smale, who popularized the notion in the United States [An; Sm]. One of the earliest results in the subject is Anosov's proof that Anosov diffeomorphisms are *structurally stable*, or, in our language $C^{1,1,1,0,0}$ locally rigid. There is an analogous result for flows, though this requires that one introduce a notion of time change that we will not consider here. Since Anosov also showed that C^2 Anosov flows and diffeomorphisms are ergodic, structural stability implies that the existence of an open set of "chaotic" dynamical systems.

The notion of an Anosov diffeomorphism has had many interesting generalizations, for example: Axiom A diffeomorphisms, nonuniformly hyperbolic diffeomorphisms, and diffeomorphisms admitting a dominated splitting. The notion that has been most useful in the study of local rigidity is the notion of a partially hyperbolic diffeomorphism as introduced by Hirsch, Pugh and Shub. Under strong enough hypotheses, these diffeomorphisms have a weaker stability property similar to structural stability. More or less, the diffeomorphisms are hyperbolic relative to some foliation, and any nearby action is hyperbolic to some nearby foliation. To describe more precisely the class of diffeomorphisms we consider and the stability property they enjoy, we require some definitions.

The use of the word *foliation* varies with context. Here a *foliation by C^k leaves* will be a continuous foliation whose leaves are C^k injectively immersed submanifolds that vary continuously in the C^k topology in the transverse direction. To specify transverse regularity we will say that a foliation is transversely C^r. A foliation by C^k leaves which is transversely C^k is called simply a C^k foliation. (Note our language does not agree with that in the reference [HPS].)

Given an automorphism f of a vector bundle $E \rightarrow X$ and constants $a > b \geq 1$, we say f is (a, b)-*partially hyperbolic* or simply *partially hyperbolic* if there is a metric on E, a constant and $C \geq 1$ and a continuous f invariant, nontrivial splitting $E = E^u_f \oplus E^c_f \oplus E^s_f$ such that for every x in X:

(1) $\|f^n(v^u)\| \geq C a^n \|v^u\|$ for all $v^u \in E^u_f(x)$,

(2) $\|f^n(v^s)\| \leq C^{-1} a^{-n} \|v^s\|$ for all $v^s \in E^s_f(x)$ and

(3) $C^{-1} b^{-n} \|v^0\| < \|f^n(v^0)\| \leq C b^n \|v^0\|$ for all $v^0 \in E^c_f(x)$ and all integers n.

A C^1 diffeomorphism f of a manifold X is (a, b)-*partially hyperbolic* if the derivative action Df is (a, b)-partially hyperbolic on TX. We remark that for any partially hyperbolic diffeomorphism, there always exists an *adapted metric* for which $C = 1$. Note that E^c_f is called the *central distribution* of f, E^u_f is called the *unstable distribution* of f and E^s_f the *stable distribution* of f.

Integrability of various distributions for partially hyperbolic dynamical systems is the subject of much research. The stable and unstable distributions are always tangent to invariant foliations which we call the stable and unstable foliations and denote by $\mathcal{W}^s_f$ and $\mathcal{W}^u_f$. If the central distribution is tangent to an f invariant foliation, we call that foliation a *central foliation* and denote it by $\mathcal{W}^c_f$. If there is a unique foliation tangent to the central distribution we call the central distribution *uniquely integrable*. For smooth distributions unique integrability is a consequence of integrability, but the central distribution is usually not smooth. If the central distribution of an (a, b)-partially hyperbolic diffeomorphism f is tangent to an invariant foliation $\mathcal{W}^c_f$, then we say f is r-*normally hyperbolic to* $\mathcal{W}^c_f$ for any r such that $a > b^r$. This is a special case of the definition of r-normally hyperbolic given in [HPS].

Before stating a version of one of the main results of [HPS], we need one more definition. Given a group G, a manifold X, two foliations $\mathfrak{F}$ and $\mathfrak{F}'$ of X, and two actions ρ and ρ' of G on X, such that ρ preserves $\mathfrak{F}$ and ρ' preserves $\mathfrak{F}'$, following [HPS] we call ρ and ρ' *leaf conjugate* if there is a homeomorphism h of X such that:

(1) $h(\mathfrak{F}) = \mathfrak{F}'$ and
(2) for every leaf $\mathfrak{L}$ of $\mathfrak{F}$ and every $g \in G$, we have $h(\rho(g)\mathfrak{L}) = \rho'(g)h(\mathfrak{L})$.

The map h is then referred to as a *leaf conjugacy* between $(X, \mathfrak{F}, \rho)$ and $(X, \mathfrak{F}', \rho')$. This essentially means that the actions are conjugate modulo the central foliations.

We state a special case of some the results of Hirsch–Pugh–Shub on perturbations of partially hyperbolic actions of $\mathbb{Z}$, see [HPS]. There are also analogous definitions and results for flows. As these are less important in the study of local rigidity, we do not discuss them here.

THEOREM 3.3. *Let f be an (a, b)-partially hyperbolic C^k diffeomorphism of a compact manifold M which is k-normally hyperbolic to a C^k central foliation $\mathcal{W}^c_f$. Then for any $\delta > 0$, if f' is a C^k diffeomorphism of M which is sufficiently C^1 close to f we have the following:*

(1) *f' is (a', b')-partially hyperbolic, where $|a - a'| < \delta$ and $|b - b'| < \delta$, and the splitting $TM = E^u_{f'} \oplus E^c_{f'} \oplus E^s_{f'}$, for f' is C^0 close to the splitting for f;*
(2) *there exist f' invariant foliations by C^k leaves $\mathcal{W}^c_{f'}$, tangent to $E^c_{f'}$, which is close in the natural topology on foliations by C^k leaves to $\mathcal{W}^c_f$,*
(3) *there exists a (nonunique) homeomorphism h of M with $h(\mathcal{W}^c_f) = \mathcal{W}^c_{f'}$, and h is C^k along leaves of $\mathcal{W}^c_f$, furthermore h can be chosen to be C^0 small and C^k small along leaves of $\mathcal{W}^c_f$*

(4) *the homeomorphism h is a leaf conjugacy between the actions $(M, \mathcal{W}^c_f, f)$ and $(M, \mathcal{W}^c_{f'}, f')$.*

Conclusion (1) is easy and probably older than [HPS]. One motivation for Theorem 3.3 is to study stability of dynamical properties of partially hyperbolic diffeomorphisms. See the survey, [BPSW], for more discussion of that and related issues.

4. History

In this section, we describe the history of the subject roughly to the year 2000. More recent developments will be discussed below. Here we do not treat the subject entirely chronologically, but break the discussion into four subjects: first, the study of local rigidity of volume preserving actions, second the study of local rigidity of certain non-volume-preserving actions called boundary actions, third the existence of (many) deformations of (many) actions of groups that are typically quite rigid, and lastly a brief discussion of infinitesimal rigidity. This is somewhat ahistorical as the first results on smooth conjugacy of perturbations of group actions appear in [Gh1], which we describe in subsection 4.2. While those results are not precisely local rigidity results, they are clearly related and the techniques involved inform some later approaches to local rigidity.

4.1. Volume preserving actions. In this subsection, we discuss local rigidity of volume preserving actions. The acting groups will usually be lattices in higher rank semisimple Lie groups or higher rank free abelian groups. Many of the results discussed here were motivated by conjectures of Zimmer in [Z4; Z5]. The first result we mention, due to Zimmer, does not prove local rigidity, but did motivate much later work on the subject.

THEOREM 4.1. *Let Γ be a group with property (T) of Kazhdan and let ρ be a Riemannian isometric action of Γ on a compact manifold M. Further assume the action is ergodic. Then any C^k action ρ' which is C^k close to ρ, volume preserving and ergodic, preserves a C^{k-3} Riemannian metric.*

REMARKS. (1) Zimmer first proved this theorem in [Z1], but only for Γ a lattice in a semisimple group, all of whose simple factors have real rank at least two, and only producing a C^0 invariant metric for ρ'. In [Z2], he gave the proof of the regularity stated here and in [Z4] he extended the theorem to all Kazhdan groups.
(2) In this theorem if ρ' is C^∞, the invariant metric for ρ' can also be chosen C^∞.

The first major result that actually produced a conjugacy between the perturbed and unperturbed actions was due to Hurder, [H1]. Again, this result is not quite a local rigidity theorem, but only a *deformation rigidity* theorem. Hurder's work is the first place where hyperbolic dynamics is used in the theory, and is the beginning of a long development in which hyperbolic dynamics play a key role.

THEOREM 4.2. *The standard action of any finite index subgroup of $SL(n, \mathbb{Z})$ on the n dimensional torus is deformation rigid when $n \geq 3$.*

Hurder actually proves a much more general result. His theorem proves deformation rigidity of any group Γ acting on the n dimensional torus by linear transformations provided:

(1) the set of periodic points for the Γ action is dense, and
(2) the first cohomology of any finite index subgroup of Γ in any n dimensional representation vanishes, and
(3) the action contains "enough" Anosov elements.

Here we intentionally leave the meaning of (3) vague, as the precise notion needed by Hurder is quite involved. To produce a continuous path of continuous conjugacies, Hurder only need conditions (1) and (2) and the existence of an Anosov diffeomorphism in the stabilizer of every periodic point for the action. The additional Anosov elements needed in (3) are used to improve regularity of the conjugacy, and better techniques for this are now available. Hurder's work has some applications to actions of irreducible lattices in products of rank 1 Lie groups, which we discuss below in subsection 6.2. These applications do not appear to be accessible by later techniques.

A key element in Hurder's argument is to use results of Stowe on persistence of fixed points under perturbations of actions [St1; St2]. To use Stowe's result one requires that the cohomology in the derivative representation at the fixed point vanishes. This is where (2) above is used. Hurder constructs his conjugacy by using the theorem of Anosov, mentioned above in subsection 3.2, that any Anosov diffeomorphism is structurally stable. This produces a conjugacy h for an Anosov element $\rho(\gamma_0)$ which one then needs to see is a conjugacy for the entire group action. Hurder uses Stowe's results to show that h is conjugacy for the Γ actions at all of the periodic points for the Γ action, and since periodic points are dense it is then a conjugacy for the full actions. The precise argument using Stowe's theorem is quite delicate and we do not attempt to summarize it here. This argument applies much more generally, see [H1, Theorem 2.9]. That the conjugacy depends continuously, and in fact even smoothly, on the original action is deduced from results on hyperbolic dynamics in [dlLMM].

The first major development after Hurder's theorem was a theorem of Katok and Lewis [KL1] of which we state a special case:

THEOREM 4.3. *Let $\Gamma < SL(n, \mathbb{Z})$ be a finite index subgroup, $n > 3$. Then the linear action of Γ on $\mathbb{T}^n$ is locally rigid.*

It is worth noting that this theorem does not cover the case of $n = 3$. A major ingredient in the proof is studying conjugacies produced by hyperbolic dynamics for certain $\mathbb{Z}$ actions generated by hyperbolic and partially hyperbolic diffeomorphisms in both the original action and the perturbation. The strategy is to find a hyperbolic generating set and to show that the conjugacies produced by the stability of those diffeomorphisms agree. A key ingredient idea is to show that they agree on the set of periodic orbits. Periodic orbits are then studied via Theorem 3.3 for elements of Γ with large centralizers and large central foliations. Periodic orbits are detected as intersections of central foliations for different elements, and this allows the authors to show that the structure of the periodic set persists under deformations. Elements with large centralizers had previously been exploited by Lewis for studying infinitesimal rigidity of similar actions. The authors also exploit their methods to prove the following remarkable result:

THEOREM 4.4. *Let $\mathbb{Z}^n$ be a maximal diagonalizable (over $\mathbb{R}$) subgroup of $SL(n + 1, \mathbb{Z})$ where $n \geq 2$. The linear action of $\mathbb{Z}^n$ on $\mathbb{T}^{n+1}$ is locally rigid.*

This result is the first in a long series of results showing that many actions of higher rank *abelian* groups are locally rigid. See Theorems 4.6 and 5.8 below for more instances of this remarkable behavior.

The next major development occurs in a paper of Katok, Lewis and Zimmer where Theorem 4.3 is extended to cover the case of $n = 3$ as well as some more general groups acting on tori. Though this does not seem, on the face of it, to be a very dramatic development, an important idea is introduced in this paper. The authors proceed by comparing the measurable data coming from cocycle superrigidity to the continuous data provided by hyperbolic dynamics. In this context, this essentially allows the authors to show that the map s in equation (3–1), described in the statement of cocycle superrigidity given in subsection 3.1, is continuous. This idea of comparing the output of cocycle superrigidity to information provided by hyperbolic dynamics has played a major role in the development of both local and global rigidity of group actions.

The results in the papers [H1; KL1; KLZ] are all proven for particular actions of particular groups, and in particular are all proven for actions on tori. The next sequence of developments was a generalization of the ideas and methods contained in these papers to fairly general Anosov actions of higher rank lattices on nilmanifolds. Part of this development takes place in the works [Q1; Q2; QY]. A key difficulty in generalizing the early approaches to rigidity of groups of toral automorphims is in adapting the methods from hyperbolic dynamics which are used to improve the regularity of the conjugacy. In [KS1; KS2], Katok and Spatzier developed a broadly applicable method for smoothing conjugacies which depends on the theory of nonstationary normal forms as developed by Guysinsky and Katok in [GK; G]. Two main consequences of this method are:

THEOREM 4.5. *Let G be a semisimple Lie group with all simple factors of real rank at least two, $\Gamma < G$ a lattice, N a nilpotent Lie group and $\Lambda < N$ a lattice. Then any affine Anosov action of Γ on N/Λ is locally rigid.*

Here by *Anosov action*, we mean that some element of Γ acts on N/Λ as an Anosov diffeomorphism.

THEOREM 4.6. *Let N a nilpotent Lie group and $\Lambda < N$ a lattice. Let $\mathbb{Z}^d$ be a group of affine transformations of N/Λ such that the derivative action (on a subgroup of finite index) is simultaneously diagonalizable over $\mathbb{R}$ with no eigenvalues on the unit circle (i.e. on subgroup of finite index, each element of $\mathbb{Z}^d$ is an Anosov diffeomorphism which has semisimple derivative). Then the $\mathbb{Z}^d$ action on N/Λ is locally rigid.*

Katok and Spatzier also apply their method to show that for certain standard Anosov $\mathbb{R}^d$ actions the orbit foliation is locally rigid. I.e. any nearby action has conjugate orbit foliation. This yields interesting applications to rigidity of boundary actions, see Theorem 4.16 below. Also, combined with other results of the same authors on rigidity of cocycles over actions of Abelian groups, this yields local rigidity of certain algebraic actions of $\mathbb{R}^d$, [KS3; KS4]. We state a special case of these results here:

THEOREM 4.7. *Let G be an $\mathbb{R}$-split semisimple Lie group of real rank at least two. Let $\Lambda < G$ be a cocompact lattice and let $\mathbb{R}^d < G$ be a maximal $\mathbb{R}$-split subgroup. Then the $\mathbb{R}^d$ action on G/Λ is locally rigid.*

REMARKS. (1) Here "local rigidity" has a slightly different meaning than above. Since the automorphism group of $\mathbb{R}^d$ has nontrivial connected component, it is possible to perturb the action by taking a small automorphism of $\mathbb{R}^d$. What is proven in this theorem is that any small enough perturbation is conjugate to one obtained in this way.

(2) Another approach to related cocycle rigidity results is developed in the paper [KNT].

(3) The actual theorem in [KS2] is much more general.

A key ingredient in the Katok–Spatzier method is to find foliations which are orbits of transitive, isometric, smooth group actions for both the perturbed and unperturbed action. To show smoothness of the conjugacy, one constructs such group actions that

(1) are intertwined by a continuous conjugacy and
(2) exist on enough foliations to span all directions in the space.

This proves that the conjugacy is "smooth along many directions" and one then uses a variety of analytic methods to prove that the conjugacy is actually globally smooth. The fact that the transitive group exists and acts smoothly on the leaves of some foliation for the unperturbed action is typically obvious. One then uses the continuous conjugacy to define the group action along leaves for the perturbed action and the fact that the resulting action is smooth along leaves is verified using the normal form theory. The foliations along which one builds transitive group actions are typically central foliations for certain special elements of the suspension of the action. If the original action is a $\mathbb{Z}^k$ action by automorphisms on some nilmanifold N/Λ, the suspension of the action is the left action of $\mathbb{R}^k$ on the solv-manifold $M = (\mathbb{R}^k \ltimes N)/(\mathbb{Z}^k \ltimes \Lambda)$. A typical one parameter subgroup of $\mathbb{R}^k$ acts hyperbolically M, but certain special directions in $\mathbb{R}^k$, those in so-called *Weyl chamber walls* give rise to one parameter subgroups with nontrivial central direction. A key fact used in the argument is that one can find another subgroup of $\mathbb{R}^k$ for which the central foliation for some one parameter subgroup $\rho(t)$ is also a dynamically defined, contracting foliation for the some other element $a \in \mathbb{R}^k$.

All results quoted so far have strong assumptions on hyperbolicity of the action. For actions of semisimple groups and their lattices, the ultimate result on local rigidity in hyperbolic context was proven by Margulis and Qian in [MQ]. This result is for so-called weakly hyperbolic actions, which we define below. This work proceeds by first using a comparison between hyperbolic data and data from cocycle superrigidity to produce a continuous conjugacy between C^1 close actions and then uses an adaptation of the Katok–Spatzier smoothing method mentioned above. A key technical innovation in this work is the choice of cocycle to which cocycle superrigidity is applied. In all work to this point, it was applied to the derivative cocycle. Here it is applied to a cocycle that measures the difference between the action and the perturbation. To illustrate the idea, we give the definition of this cocycle, which we refer to as the *Margulis–Qian cocycle*, in the special case of actions by left translations. As this construction is quite general, we will let D be the acting group. Let the D action ρ on H/Λ be defined via a homomorphism $\pi_0 : D \rightarrow H$. Let ρ' be a perturbation of ρ. If D is connected it is clear that the action lifts to $\tilde{H}$ and therefore to H. If D is discrete, this lifting still occurs, since the obstacle to lifting is a cohomology class in $H^2(D, \pi_1(H/\Lambda))$ which does not change under a small perturbation of the action. (A direct justification without reference to group cohomology can be found in [MQ] section 2.3.) Write the lifted actions of D on H by $\tilde{\rho}$ and $\tilde{\rho}'$ respectively. We can now define a cocycle $\alpha : D \times H \rightarrow H$ by

$$\tilde{\rho}'(g)x = \alpha(g, x)x$$

for any g in D and any x in H. It is easy to check that this is a cocycle and that it is right Λ invariant, and so defines a cocycle $\alpha : D \times H/\Lambda \rightarrow H$. See [MQ] section

2 or [FM1] section 6 for more discussion as well as for more general variants on this definition. We remark that the use of this cocycle allowed Margulis and Qian to prove the first local rigidity results for volume preserving actions of lattices that have no global fixed point. The construction of this cocycle is inspired by a cocycle used by Margulis in his first proof of superrigidity. This is the cocycle $G \times G / \Gamma \to \Gamma$ defined by the choice of a fundamental domain for Γ in G. See Example 4 in subsection 6.2 for a more explicit description.

The work of Margulis and Qian applies to actions which satisfy the following condition. This condition essentially says that the action is hyperbolic in all possible directions, at least for some element of the acting group. It is easy to construct weakly hyperbolic actions of lattices, in particular Γ acting on G / Λ where G is a simple Lie group and $\Lambda < G$ is a cocompact lattice and Γ is any other lattice in G. It is important to note that for this example, no element acts as an Anosov diffeomorphism and, with an appropriate choice of Γ and Λ, there are no finite Γ orbits.

DEFINITION 4.8. An action ρ of a group D on a manifold M is called *weakly hyperbolic* if there exist elements $d_1, \ldots, d_k$ and constants $a_i > b_i \geq 1$ for $i = 1, \ldots, k$ such that each $\rho(d_i)$ is (a_i, b_i)-partially hyperbolic in the sense of subsection 3.2 and we have $TM = \sum E^s_{\rho(d_i)}$. I.e. there are partially hyperbolic elements whose stable (or unstable) directions span the tangent space at any point.

THEOREM 4.9. *Let H be a real algebraic Lie group and $\Lambda < H$ a cocompact lattice. Assume G is a semisimple Lie group with all simple factors of real rank at least two and $\Gamma < G$ is a lattice. Then any weakly hyperbolic, affine algebraic action of Γ or G on H / Λ is locally rigid.*

REMARK. This is somewhat more general than the result claimed in [MQ], as they only work with certain special classes of affine actions, which they call standard. This result can proven by the methods of [MQ], and a proof in precisely this generality can be read out of [FM1; FM3], simply by assuming that the common central direction for the acting group is trivial.

The next major result was a remarkable theorem of Benveniste concerning isometric actions. This is a stronger result than Theorem 4.1 because it actually produces a conjugacy, but is weaker in that it requires much stronger assumptions on the acting group.

THEOREM 4.10. *Let Γ be a cocompact lattice in a semisimple Lie group with all simple factors of real rank at least two. Let ρ be an isometric Γ action on a compact manifold M. Then ρ is locally rigid.*

The proof of this theorem is inspired by the work of Calabi, Vesentini and Weil in the original proofs of Theorem 3.1 and is based on showing that certain deformations of foliated geometric structures are trivial. The argument is much more difficult than the classical case and uses Hamilton's implicit function theorem. This is the first occasion on which analytic methods like KAM theory or hard implicit function theorems appear in work on local rigidity of group actions. More recently these kinds of methods have been applied more systematically, see subsection 5.2.

The theorems described so far concern actions that are either isometric or weakly hyperbolic. There are many affine actions which satisfy neither of these dynamical hypotheses, but are genuinely partially hyperbolic. Local rigidity results for actions of this kind first arise in work of Nitica and Torok. We state special cases of two of their theorems:

THEOREM 4.11. *Let $\Gamma < SL(n, \mathbb{Z})$ be a finite index subgroup with $n \geq 3$. Let ρ_1 be the standard Γ action on $\mathbb{T}^n$ and let ρ be the diagonal Γ action on $\mathbb{T}^n \times \mathbb{T}^m$ defined by ρ_1 on the first factor and the trivial action on the second factor. The action ρ is deformation rigid.*

THEOREM 4.12. *Let Γ, ρ_1, ρ be as above and further assume that $m = 1$. The action ρ is locally rigid.*

REMARKS. (1) Nitica and Torok prove more general theorems in which both Γ and ρ_1 can be more general. The exact hypotheses required are different in the two theorems.
(2) We are being somewhat ahistorical here, Theorem 4.11 predates the work of Margulis and Qian.
(3) The conjugacy produced in the papers [NT1; NT2; T] is never C^∞, but only C^k for some choice of k. The choice of k is essentially free and determines the size of perturbations or deformations that can be considered. It should be possible to produce a C^∞ conjugacy by combining the arguments in these papers with arguments in [FM2; FM3], see the end of subsection 5.1 for some discussion.

The work of Nitica and Torok is quite complex, using several different ideas. The most novel is to study rigidity of cocycles over hyperbolic dynamical systems taking values in diffeomorphism groups. The dynamical system is either the action ρ_1 in Theorem 4.11 or 4.12 or it's restriction to *any* sufficiently generic subgroup containing an Anosov diffeomorphism of $\mathbb{T}^n$, and the target group is the group of diffeomorphisms of $\mathbb{T}^m$. This part of the work is inspired by a classical theorem of Livsic and the proof his modelled on his proof. To reduce the rigidity question to the cocycle question is quite difficult and depends on an adaptation of the work of [HPS] discussed in subsection 3.2 as well as use of results of Stowe [St1; St2]. Regrettably, the technology seems to limit the applicability of the ideas to diagonal actions $\rho_1 \times \text{Id}$ on products $M \times N$ where the action on M has many periodic points and the action on N is trivial. Theorem 4.12 also depends on the acting group having property (T) of Kazhdan. The method of proof of Theorem 4.11 has additional applications, see particularly subsection 6.2 below.

To close this section, we remark that local rigidity is often considerably easier in the analytic setting. Not much work has been done in this direction, but there is an interesting note of Zeghib [Zg]. A sample result is the following:

THEOREM 4.13. *Let $\Gamma < SL(n, \mathbb{Z})$ be a subgroup of finite index and let ρ be the standard action of Γ on $\mathbb{T}^n$. Then any analytic action close enough to ρ is analytically conjugate to ρ. Furthermore, if M is a compact analytic manifold on which Γ acts trivially and we let $\tilde{\rho}$ be the diagonal action of Γ on $\mathbb{T}^n \times M$, then $\tilde{\rho}$ is also locally rigid in the analytic category.*

Zeghib also proves a number of other interesting results for both volume preserving and non-volume-preserving actions and it is clear that his method has applications not stated in his note. The key point for all of his arguments is a theorem of Ghys and Cairns that says that one can linearize an analytic action of a higher rank lattice in a neighborhood of a fixed point. Zeghib proves his results by using results of Stowe [St1; St2] to find fixed points for the perturbed action and then studying the largest possible set to which the linearization around this point can be extended. We end this section by stating the theorem of Cairns and Ghys from [CGh] which Zeghib uses.

THEOREM 4.14. *Let G be a semisimple Lie group of real rank at least two with no compact factors and finite center and let $\Gamma < G$ be a lattice. Then every analytic action of Γ with a fixed point p is analytically linearizable in a neighborhood of p.*

REMARKS. (1) By analytically linearizable in a neighborhood of p, we mean that there exists a neighborhood U of p and an analytic diffeomorphism ϕ of U into the ambient manifold M such that the action of Γ, conjugated by ϕ is the restriction of a linear action to $\phi(U)$.

(2) In the same paper, Ghys and Cairns give an example of a C^∞ action of $SL(3, \mathbb{Z})$ on $\mathbb{R}^8$ fixing the origin, which is not C^0 linearizable in any neighborhood of the origin. So the assumption of analyticity in the theorem is necessary.

4.2. Actions on boundaries. In this subsection we discuss rigidity results for groups acting on homogeneous spaces known as "boundaries". In contrast to the last section, the actions we describe here never preserve a volume form, or even a Borel measure. We will not discuss here all the geometric, function theoretic or probabalistic reasons why these spaces are termed boundaries, but merely describe examples. For us, if G is a semisimple Lie group, then a *boundary* of G is a space of the form G/P where P is a connected Lie subgroup of G such that the quotient G/P is compact. The groups P having this property are often called *parabolic* subgroups. The space G/P is also considered a boundary for any lattice $\Gamma < G$. For more precise motivation for this terminology, see [Fu1; Fu2; Mo1; Ma4].

The simplest example of a boundary is for the group $SL(2,\mathbb{R})$ in which case the only choice of P resulting in a nontrivial boundary is the group of upper (or lower) triangular matrices. The resulting quotient is naturally diffeomorphic to the circle and $SL(2,\mathbb{R})$ acts on this circle by the action on rays through the origin in $\mathbb{R}^2$. We can restrict this $SL(2,\mathbb{R})$ action to any lattice Γ in $SL(2,\mathbb{R})$. The following remarkable theorem was first proved by Ghys in [Gh1]:

THEOREM 4.15. *Let $\Gamma < SL(2,\mathbb{R})$ be a cocompact lattice and let ρ be the action of Γ on S^1 described above. If ρ' is any perturbation of ρ, then ρ' is smoothly conjugate to an action defined by another embedding π' of Γ in $SL(2,\mathbb{R})$ close to the original embedding. In particular $\pi'(\Gamma)$ is a cocompact lattice in $SL(2,\mathbb{R})$.*

Ghys gives two proofs of this fact, one in [Gh1] and another different one in [Gh2]. A third and also different proof is in later work of Kononenko and Yue [KY]. Ghys' first proof derives from a remarkable global rigidity result for actions on certain three dimensional manifolds by the affine group of the line, while his second derives from rigidity results concerning certain Anosov flows on three dimensional manifolds. We remark that the fact that ρ' is continuously conjugate to an action defined by a nearby embedding into $SL(2,\mathbb{R})$ was known and so Theorem 4.15 can be viewed as a regularity theorem though this is not how the proof proceeds.

Both of Ghys' proofs pass through a statement concerning local rigidity of foliations. This uses the following variant on the construction of the induced action. Let Γ be a cocompact lattice in $SL(2,\mathbb{R})$, and let ρ be the Γ action on S^1 defined by the action of $SL(2,\mathbb{R})$ there. There is also a Γ action on the hyperbolic plane $\mathbb{H}^2$. We form the manifold $(\mathbb{H}^2 \times S^1)/\Gamma_\rho$, where Γ acts diagonally. This manifold is diffeomorphic to the unit tangent bundle of $\mathbb{H}^2/\Gamma$ which is also diffeomorphic to $SL(2,\mathbb{R})/\Gamma$, and the foliation by planes of the form $\mathbb{H}^2 \times \{point\}$ is the weak stable foliation for the geodesic flow and also the orbit foliation for the action of the affine group. Given a C^r perturbation ρ' of the Γ action on S^1, we can form the corresponding bundle $(\mathbb{H}^2 \times S^1)_{\rho'}/\Gamma_{\rho'}$, and the foliation by planes of the form $\mathbb{H}^2 \times \{point\}$ is C^r close to analogous foliation in $(\mathbb{H}^2 \times S^1)\Gamma_\rho$. To show that ρ and ρ' are conjugate, it suffices to find a diffeomorphism of $(\mathbb{H}^2 \times S^1)/\Gamma_\rho$ conjugating the two foliations. Both of Ghys' proofs proceed by constructing such a conjugacy of foliations. This reduction to studying local rigidity of foliations has further applications in slightly different settings, see Theorems 4.16 and 4.17 below.

In later work, Ghys proved a remarkable result which characterized an entire connected component of the space of actions of Γ on S^1. Let X be the component of $\mathrm{Hom}(\Gamma, \mathrm{Diff}(S^1))$ containing the actions described in Theorem 4.15. In [Gh3], Ghys showed that this component consisted entirely of actions conjugate to actions defined by embeddings π' of Γ into $SL(2, \mathbb{R})$ where $\pi'(\Gamma)$ is a cocompact lattice. This result builds on earlier work of Ghys where a similar result was proven concerning $\mathrm{Hom}(\Gamma, \mathrm{Homeo}(S^1))$. A key ingredient is the use of the Euler class of the action, viewed as a bounded cocycle.

As mentioned above, Ghys' method of reducing local rigidity of an action to local rigidity of a foliation has had two more applications. The first of these is due to Katok and Spatzier [KS2].

THEOREM 4.16. *Let G be a semisimple Lie group with no compact factors and real rank at least two. Let $\Gamma < G$ be a cocompact lattice and $B = G/P$ a boundary for Γ. Then the Γ action on G/P is locally rigid.*

The proof of this result uses an argument similar to Ghys' to reduce to a need to study regularity of foliations for perturbations of the action of certain connected abelian subgroups of G on G/Γ. The result used in the proof here is the same as the one used in the proof of Theorem 4.7. For Γ a lattice but not cocompact, some partial results are obtained by Yaskolko in his Ph.D. thesis [Yk].

Following a similar outline, Kanai proved the following:

THEOREM 4.17. *Let $G = SO(n, 1)$ and $\Gamma < G$ be a cocompact lattice. Then the action of Γ on the boundary G/P is locally rigid.*

Partial results in this direction were proven earlier by Chengbo Yue. Yue also proves partial results in the case where $SO(1, n)$ is replaced by any rank 1 noncompact simple Lie group.

In somewhat earlier work, Kanai had also proven a special case of Theorem 4.16. More precisely:

THEOREM 4.18. *Let $\Gamma < SL(n, \mathbb{R})$ be a cocompact lattice where $n \geq 21$ and let ρ be the Γ action on S^{n-1} by acting on the space of rays in $\mathbb{R}^n$. Then ρ is locally rigid.*

Kanai's proof proceeds in two steps. In the first step, he uses Thomas' notion of a projective connection to reduce the question to one concerning vanishing of certain cohomology groups. In the second step, he uses stochastic calculus to prove a vanishing theorem for the relevant cohomology groups. The first step is rather special, and is what restricts Kanai's attention to spheres, rather than other boundaries, which are Grassmanians. The method in the second step seems a good deal more general and should have further applications, perhaps in the context of Theorem 5.12 below. While the approach here is similar in spirit to the work of Benveniste in [Be1], it should be noted that Kanai does not use a hard implicit function theorem.

We end this subsection by recalling a construction due to Stuck, which shows that much less rigidity should be expected from actions which do not preserve volume [Sk]. Let G be a semisimple Lie group and $P < G$ a minimal parabolic. Then there always exists a homomorphism $\pi : P \rightarrow \mathbb{R}$. Given any manifold M and any action s of $\mathbb{R}$ on M, we can then form the induced G action ρ_s on $(G \times M)/P$ where P acts on G on the left and on M by $\pi \circ s$. Varying the action s varies the action ρ_s. It is easy to see that if ρ_s and $\rho_{s'}$ are two such actions, then they are conjugate as G actions if and only if s and s' are conjugate. If one picks an irreducible lattice Γ in $G \times G$, project Γ to G and restricts the actions to Γ, then it is also easy to see that the restriction of ρ_s and $\rho_{s'}$ to Γ are conjugate if and only if s and s' are conjugate. The author does not know a proof that this is also true if one simply takes a lattice Γ in G, but believes that this is also true and may even be known.

4.3. "Flexible" actions of rigid groups. In this subsection, I discuss a sequence of results concerning flexible actions of large groups. More or less, the sequence of examples provides counterexamples to most naive conjectures of the form "all of actions of some lattice Γ are locally rigid." There are some groups, for example compact groups and finite groups, all of whose smooth actions are locally rigid. It seems likely that there should be infinite discrete groups with this property as well, but the constructions in this subsection show that one must look beyond lattices in Lie groups for examples.

Essentially all of the examples given here derive from the simple construction of "blowing up" a point or a closed orbit, which was introduced to this subject in [KL2]. The further developments after that result are all further elaborations on one basic construction. The idea is to use the "blow up" construction to introduce distinguished closed invariant sets which can be varied in some manner to produce deformations of the action. The "blow up" construction is a classical tool from algebraic geometry which takes a manifold N and a point p and constructs from it a new manifold N' by replacing p by the space of directions at p. Let $\mathbb{R}P^l$ be the l dimensional projective space. To blow up a point, we take the product of $N \times \mathbb{R}P^{\dim(N)}$ and then find a submanifold where the projection to N is a diffeomorphism off of p and the fiber of the projection over p is $\mathbb{R}P^{\dim(N)}$. For detailed discussion of this construction we refer the reader to any reasonable book on algebraic geometry.

The easiest example to consider is to take the action of $SL(n, \mathbb{Z})$, or any subgroup $\Gamma < SL(n, \mathbb{Z})$ on the torus $\mathbb{T}^n$ and blow up the fixed point, in this case the equivalence class of the origin in $\mathbb{R}^n$. Call the resulting manifold M. Provided Γ is large enough, e.g. Zariski dense in $SL(n, \mathbb{R})$, this action of Γ does not preserve the measure defined by any volume form on M. A clever construction introduced in [KL2] shows that one can alter the standard blowing up procedure in order to produce a one parameter family of $SL(n, \mathbb{Z})$ actions on M, only one of which preserves a volume form. This immediately shows that this action on M admits perturbations, since it cannot be conjugate to the nearby, non-volume-preserving actions. Essentially, one constructs different differentiable structures on M which are diffeomorphic but not equivariantly diffeomorphic.

After noticing this construction, one can proceed to build more complicated examples by passing to a subgroup of finite index, and then blowing up several fixed points. One can also glue together the "blown up" fixed points to obtain an action on a manifold with more complicated topology. See [KL2; FW] for discussion of the topological complications one can introduce.

In [Be2] it is observed that a similar construction can be used for the action of a simple group G by left translations on a homogeneous space H/Λ where H is a Lie group containing G and $\Lambda < H$ is a cocompact lattice. Here we use a slightly more involved construction from algebraic geometry, and "blow up" the directions normal to a closed submanifold. I.e. we replace some closed submanifold N in H/Λ by the projective normal bundle to N. In all cases we consider here, this normal bundle is trivial and so is just $N \times \mathbb{R}P^l$ where $l = \dim(H) - \dim(N)$.

Benveniste used his construction to produce more interesting perturbations of actions of higher rank simple Lie group G or a lattice Γ in G. In particular, he produced volume preserving actions which admit volume preserving perturbations. He does this by choosing $G < H$ such that not only are there closed G orbits but so that the centralizer $Z = Z_H(G)$ of G in H has no-trivial connected component. If we take a closed G orbit N, then any translate zN for z in Z is also closed and so we have a continuum of closed G orbits. Benveniste shows that if we choose two closed orbits N and zN to blow up and glue, and then vary z in a small open set, the resulting actions can only be conjugate for a countable set of choices of z.

This construction is further elaborated in [F1]. Benveniste's construction is not optimal in several senses, nor is his proof of rigidity. In [F1], I give a modification of the construction that produces nonconjugate actions for every choice of z in a small enough neighborhood. By blowing up and gluing more pairs of closed orbits, this allows me to produce actions where the space of deformations contains a submanifold of arbitrarily high, finite dimension. Further, Benveniste's proof that the deformation are nontrivial is quite involved and applies only to higher rank groups. In [F1], I give a different proof of nontriviality of the deformations, using consequences of Ratner's theorem to due Witte and Shah [R; Sh; W2]. This shows that the construction produces nontrivial perturbations for any semisimple G and any lattice Γ in G.

In [BF] we show that none of these actions preserve any rigid geometric structure in the sense of Gromov. It is possible that any action of a higher rank lattice which preserves a rigid geometric structure is locally rigid. It is also possible that any such action is generalized quasiaffine.

4.4. Infinitesimal rigidity. In [Z4], Zimmer introduced a notion of infinites-imal rigidity motivated by Weil's Theorem 3.2 and the analogy between finite dimensional Lie algebras and vector fields. Let ρ be a smooth action of a group Γ on a manifold M, then ρ is *infinitesimally rigid* if $H^1(\Gamma, \text{Vect}^\infty(M)) = 0$. Here the Γ action on $\text{Vect}^\infty(M)$ is given by the derivative of ρ. The notion of infinitesimal rigidity was introduced with the hope that one could prove an analogue of Weil's Theorem 3.2 and then results concerning infinitesimal rigid-ity would imply results concerning local rigidity. Many infinitesimal rigidity results were then proven, see [H2; Ko; L; LZ; Q3; Z6]. For some more results on infinitesimal rigidity, see Theorems 5.10 and 5.11. Also see subsection 5.2 for a discussion of known results on the relation between infinitesimal and local rigidity.

5. Recent developments

In this section we discuss the most recent dramatic developments in the field. The first subsection discusses work of the author and Margulis on rigidity of actions of higher rank groups and lattices. Our main result is that if H is the real points of an algebraic group defined over $\mathbb{R}$ and $\Lambda < H$ is a cocompact lattice, then any affine action of G or Γ on H/Λ is locally rigid. This work is quite involved and spans a sequence of three long papers [FM1; FM2; FM3]. One of the main goals of subsection 5.1 is to provide something of a "reader's guide" to those papers.

The second subsection discusses some recent developments involving more geometric and analytic approaches to local rigidity. Till this point, the study of local rigidity of group actions has been dominated by algebraic ideas and hyper-bolic dynamics with the exception of [Ka1] and [Be2]. The results described in subsection 5.2 represent (the beginning of) a dramatic development in analytic and geometric techniques. The first of these is the work of Damjanovich and Katok on local rigidity of certain partially hyperbolic affine actions of abelian groups on tori using a KAM approach [DK1; DK2]. The second is the author's

proof of a criterion for local rigidity of groups actions modelled on Weil's Theorem 3.2 and proven using Hamilton's implicit function theorem [F2]. This result
currently has an unfortunate "side condition" on second cohomology that makes
it difficult to apply.

The final subsection concerns a few other very recent results and developments that the author feels point towards the future development of the field.

5.1. The work of Margulis and the author. Let G be a (connected) semisimple
Lie group with all simple factors of real rank at least two, and $\Gamma < G$ is a lattice.
The main result of the papers [FM1; FM2; FM3] is:

THEOREM 5.1. *Let ρ be a volume preserving quasiaffine action of G or Γ on a
compact manifold X. Then the action locally rigid.*

REMARKS. (1) This result subsumes essentially all of the theorems in subsection 4.1, excepting those concerning actions of abelian groups.
(2) In [FM3] we also achieve some remarkable results for perturbations of very
low regularity. In particular, we prove that any perturbation which is a C^3
close C^3 action is conjugate back to the original action by a C^2 diffeomorphism.
(3) The statement here is slightly different than that in [FM3]. Here $X =
H/L \times M$ with L cocompact, while there $X = H/\Lambda \times M$ with Λ discrete and
cocompact. An essentially algebraic argument using results in [W1], shows
that possibly after changing H and M, these hypotheses are equivalent.

Another main result of the research resulting in Theorem 5.1 is the following:

THEOREM 5.2. *Let Γ be a discrete group with property (T). Let X be a compact smooth manifold, and let ρ be a smooth action of Γ on X by Riemannian
isometries. Then ρ is locally rigid.*

REMARKS. (1) A key step in the proof of Theorem 5.1 is a foliated version of
Theorem 5.2.
(2) As in Theorem 5.1, there is a finite regularity version of Theorem 5.2 and
it's foliated generalization, we refer the reader to [FM2] for details.

The remainder of this subsection will consist of a sketch of the proof of Theorem
5.1. The intention is essentially to provide a reader's guide to the three papers
[FM1; FM2; FM3]. Throughout the remainder of this subsection, to simplify
notation, we will discuss only the case of affine Γ actions on H/Λ with Λ a
cocompact lattice. The proof for connected groups and quasiaffine action on
$X = H/\Lambda \times M$ is similar. To further simplify the discussion, we assume that ρ'
is a C^∞ perturbation of ρ.

Step 1: An invariant "central" foliation for the perturbed action and leaf conjugacy. To begin the discussion and the proof, we need some knowledge of the structure of the affine actions considered. By [FM1, Theorem 6.4], there is a finite index subgroup $\Gamma' < \Gamma$ such that the action of Γ' on H/Λ is given by a homomorphism $\sigma : \Gamma' \to \mathrm{Aut}(H) \ltimes H$. We simplify the discussion by assuming $\Gamma = \Gamma'$ throughout. Using Margulis superrigidity theorems, which are also used in the proof of [FM1, Theorem 6.4], it is relatively easy to understand the maximal central foliation $\mathfrak{F}$ for ρ: there is a subgroup $Z < H$ whose orbit foliation is exactly the central foliation. For example, if $G < H$ acts on H/Λ by left translations and ρ is restriction of that action to Γ, then $Z = Z_H(G)$. For details on what Z is more generally, see [FM1].

Given a perturbation ρ' of ρ, we begin by finding a ρ' invariant foliation $\mathfrak{F}'$ and a leaf conjugacy ϕ from $(H/\Lambda, \rho, \mathfrak{F})$ to $(H/\Lambda, \rho', \mathfrak{F}')$. To do this, we apply a result concerning local rigidity of cocycles over actions of higher rank groups and lattices to the Margulis–Qian cocycle defined by the perturbation. As the statements of the local rigidity results for cocycles are somewhat technical, we refer the reader to [FM1, Theorems 1.1 and 5.1]. Those Theorems are proven in Section 5 of that paper using results in Section 4 concerning orbits in representation varieties as well as the cocycle superrigidity theorems. The construction of the leaf conjugacy is completed in [FM3, Section 2.2] using [FM1, Theorem 1.8]. We remark that we actually construct ϕ^{-1} rather than ϕ. The paper [FM1] also contains a proof of superrigidity for cocycles that results in many technical improvements to that result.

Step 2: Smoothness of the central foliation, reduction to a foliated perturbation. The next step in the proof is to show that $\mathfrak{F}'$ is a foliation by smooth leaves. In fact, it is only possible to show at this point that it is a foliation by C^k leaves for some k depending on ρ and ρ' and particularly on the C^1 size of the perturbation. This is done using the work of Hirsch, Pugh and Shub described in subsection 3.2. If the central foliation for ρ is the central foliation for $\rho(\gamma)$ for some single element γ in Γ, this amounts to showing that the foliation $\mathfrak{F}'$ constructed in step one is the same foliation as the central foliation for $\rho'(\gamma)$ constructed in the proof of Theorem 3.3. To prove this, one needs to analyze the proof of Theorem 3.3. More generally, we show, in [FM3, Section 3.2] that there is a finite collection of elements $\gamma_1, \ldots, \gamma_k$ in Γ such that each leaf of the foliation $\mathfrak{F}$ is a transverse intersection of central leaves of $\rho(\gamma_1), \ldots, \rho(\gamma_k)$. One then needs to combine an analysis of the proof of Theorem 3.3 with some arguments concerning persistence of transversality under certain kinds of perturbations. This argument is carried out in [FM3, Section 3.3].

Once we know that $\mathfrak{F}'$ is C^k, it is easy to see that the leaf conjugacy ϕ is C^k and C^k small along leaves of $\mathfrak{F}$ though all derivatives are only continuous in the transverse direction. Conjugating the action ρ' by ϕ, we obtain a new action ρ'' on H/Λ which preserves $\mathfrak{F}$. This action is only continuous, but it is C^0 close to ρ and C^k and C^k close to ρ along leaves of $\mathfrak{F}$. In [FM2; FM3] we refer to perturbations of this type as *foliated perturbations*.

Step 3: Conjugacy along the central foliation. The next step is to apply a foliated generalization of Theorem 5.2 to the actions ρ and ρ''. The exact result we apply is [FM2, Theorem 2.11] which produces a semiconjugacy ψ between ρ and ρ'. This result is somewhat involved to state and the regularity of ψ is hard to describe. The map ψ is $C^{k-1-\varepsilon}$ along the leaves of $\mathfrak{F}$ at almost all points in H/Λ, for ε depending on the size of the perturbation, but only transversely measurable. In addition, the map ψ satisfies a certain Sobolev estimate, that implies that it is $C^{k-1-\varepsilon}$ small in a small ball in $\mathfrak{F}$ at most points, and that the $C^{k-1-\varepsilon}$ norm is only large on very small sets. Rather than try to make this precise here, we include a sketch of the proof of Theorem 5.2. Before doing so we remark that the map $\varphi = \phi \circ \psi$ is a semiconjugacy from between the Γ action ρ and the Γ action ρ'. The last step in the argument is to improve the regularity of φ which we will discuss following the sketch of the proof of Theorem 5.2

We recall two definitions and another theorem from [FM2].

DEFINITION 5.3. Let $\varepsilon \geq 0$ and Z and Y be metric spaces. Then a map $h : Z \to Y$ is an *ε-almost isometry* if

$$(1 - \varepsilon)d_Z(x, y) \leq d_Y(h(x), h(y)) \leq (1 + \varepsilon)d_Z(x, y)$$

for all $x, y \in Z$.

The reader should note that an ε-almost isometry is a bilipschitz map. We prefer this notation and vocabulary since it emphasizes the relationship to isometries.

DEFINITION 5.4. Given a group Γ acting on a metric space X, a compact subset S of Γ and a point $x \in X$. The number $\sup_{k \in S} d(x, k \cdot x)$ is called the S-displacement of x and is denoted $\mathrm{disp}_S(x)$.

THEOREM 5.5. *Let Γ be a locally compact, σ-compact group with property (T) and S a compact generating set. There exist positive constants ε and D, depending only on Γ and S, such that for any continuous action of Γ on a Hilbert space $\mathcal{H}$ where S acts by ε-almost isometries there is a fixed point x; furthermore for any y in X, the distance from y to the fixed set is not more than $D \mathrm{disp}_S(y)$.*

We now sketch the proof of Theorem 5.2 for Theorem 5.5. Given a compact Riemannian manifold X, there is a canonical construction of a Sobolev inner product on $C^k(X)$ such that the Sobolev inner product is invariant under isometries of the Riemannian metric, see [FM2, Section 4]. We call the completion of $C^k(X)$ with respect to the metric induced by the Sobolev structure $L^{2,k}(X)$. Given an isometric Γ action ρ on a manifold M there may be no nonconstant Γ invariant functions in $L^{2,k}(X)$. However, if we pass to the diagonal Γ action on $X \times X$, then any function of the distance to the diagonal is Γ invariant and, if C^k, is in $L^{2,k}(X \times X)$.

We choose a smooth function f of the distance to the diagonal in $X \times X$ which has a unique global minimum at x on $\{x\} \times X$ for each x, and such that any function C^2 close to f also has a unique minimum on each $\{x\} \times X$. This is guaranteed by a condition on the Hessian and the function is obtained from $d(x, y)^2$ by renormalizing and smoothing the function away from the diagonal. This implies f is invariant under the diagonal Γ action defined by ρ. Let ρ' be another action C^k close to ρ. We define a Γ action on $X \times X$ by acting on the first factor by ρ and on the second factor by ρ'. For the resulting action $\bar{\rho}'$ of Γ on $L^{2,k}(X \times X)$ and every $\gamma \in S$, we show that $\bar{\rho}'(k)$ is an ε-almost isometry and that the S-displacement of f is a small number δ, where both ε and δ can be made arbitrarily small by choosing ρ' close enough to ρ. Theorem 5.5 produces a $\bar{\rho}'$ invariant function f' close to f in the $L^{2,k}$ topology. Then f' is $C^{k-\dim(X)}$ close to f by the Sobolev embedding theorems and if $k - \dim(X) \geq 2$, then f has a unique minimum on each fiber $\{x\} \times X$ which is close to (x, x). We verify that the set of minima is a $C^{k-\dim(X)-1}$ submanifold and, in fact, the graph of a conjugacy between the Γ actions on X defined by ρ and ρ'.

Note that this argument yields worse regularity than we discussed in the foliated context or than is stated in Theorem 5.2. There are considerable difficulties involved in achieving lower loss of regularity or in producing a C^∞ conjugacy and we do not dwell on these here, but refer the reader to [FM2] and also to some discussion in the next step.

Step 4: Regularity of the conjugacy. We improve the regularity of φ in three stages. First we show it is a homeomorphism in [FM3, Section 5.2]. The key to this argument is proving that there is a finite collection $\gamma_1, \ldots, \gamma_k$ of elements of Γ such that φ takes stable foliations for $\rho(\gamma_i)$ to stable foliations for $\rho'(\gamma_i)$. If φ were continuous, this is both easy and classical. In our context, we require a density point argument to prove this, which is given in [FM3, Section 5.1]. Once we have that stable foliations go to stable foliations, we use this to show that φ is actually uniformly continuous along central foliations and then patch together continuity along various foliations. (In [FM3, Section 5.3], we show how to remove the assumption, made above, that the Γ action lifts from H/Λ to H. This cannot be done until we have produced a continuous conjugacy.)

The next stage is to show that φ is a finite regularity diffeomorphism. To show this, we show that φ is smooth (with estimates) along certain foliations which span the tangent space. This step is essentially an implementation of the method of Katok and Spatzier described in subsection 4.1. A few technical difficulties occur as we need to keep careful track of estimates in the method for use at later steps and because we need to identify ergodic components of the measure for certain elements in the unperturbed action. After applying the Katok–Spatzier method, we have that φ is smooth along many contracting directions and smooth along the central foliation, and then use a fairly standard argument involving elliptic operators to show that it is actually smooth, and even $C^{k'}$ small, for $k' = k - 1 - \varepsilon - \frac{1}{2}\dim(H)$. It would be interesting to see if one could lose less regularity at this step, for example by a method like Journé's. A key difficulty in adapting the method of [Jn] is that we only have a Sobolev estimate along central leaves and not a uniform one.

The last stage of the argument is to show that φ is smooth. There are two parts to this argument. The first is to show that if ρ and ρ' are C^k close, we can actually show that φ is C^l for some $l \geq k$. The main difficulty here is obtaining better regularity in the foliated version of Theorem 5.2. This requires the use of estimates on convexity of derivatives and estimates on compositions of diffeomorphisms, as well as an iterative method of constructing the semiconjugacy ψ, for this we refer the reader to [FM2, Sections 6 and 7.3]. Once we know we can produce a conjugacy of greater regularity, we can then approximate φ in the C^l topology by a C^∞ map and smoothly conjugate ρ' to a very small C^l perturbation of ρ. The point is to iterate this procedure while obtaining estimates on the size of the conjugacy produced at each step. We then show that the iteration converges to produce a smooth conjugacy. We give here a general theorem whose proof follows from arguments in [FM2; FM3].

It is convenient to fix right invariant metrics d_l on the connected components of $\mathrm{Diff}^l(X)$ with the additional property that if φ is in the connected component of $\mathrm{Diff}^\infty(X)$, then $d_l(\varphi, \mathrm{Id}) \leq d_{l+1}(\varphi, \mathrm{Id})$. To fix d_l, it suffices to define inner products $\langle\ ,\ \rangle_l$ on $\mathrm{Vect}^l(X)$ which satisfy $\langle V, V \rangle_l \leq \langle V, V \rangle_{l+1}$ for $V \in \mathrm{Vect}^\infty(X)$. As remarked in [FM2, Section 6], after fixing a Riemannian metric g on X, it is straightforward to introduce such metrics using the methods of [FM2, Section 4].

DEFINITION 5.6. Let Γ be a group, M a compact manifold and assume $\rho : \Gamma \to \mathrm{Diff}^\infty(M)$. We say ρ is *strongly $C^{k,l,n}$ locally rigid* if for every $\varepsilon > 0$ there exists $\delta > 0$ such that if ρ' is an action of Γ on M with $d_k(\rho'(g)\rho(g)^{-1}, \mathrm{Id}) < \delta$ for all $g \in K$ then there exists a C^l conjugacy φ between ρ and ρ' such that $d_{k-n}(\varphi, \mathrm{Id}) < \varepsilon$.

We are mainly interested in the case where $l > k$.

THEOREM 5.7. *Let Γ be a group, M a compact manifold and assume $\rho : \Gamma \to \mathrm{Diff}^\infty(M)$. Assume that there are constants $n > 0$ and $k_0 \geq 0$ and that for* *every $k > k_0$ there exists an $l > k$ such that ρ is strongly $C^{k,l,n}$ locally rigid. Then ρ is C^∞ locally rigid.*

The proof of Theorem 5.7 follows [FM3, Corollary 7.2], though the result is not stated in this generality there.

5.2. KAM, implicit function theorems: work of Damjanovich–Katok and the author.

In this subsection, we discuss some new results that use more geometric/analytic methods to approach questions of local rigidity. These methods are entirely independent of methods using hyperbolic dynamics and appear likely to be robustly applicable.

We begin with a theorem of Katok and Damjanovich concerning abelian groups of toral automorphisms. Here we consider actions $\pi : \mathbb{Z}^n \to \mathrm{Diff}^\infty(\mathbb{T}^m)$ where $\pi(\mathbb{Z}^n)$ lies in $GL(m, Z)$ acting on $\mathbb{T}^m$ by linear automorphisms or more generally where $\pi(\mathbb{Z}^n)$ acts affinely on $\mathbb{T}^m$. An *affine factor* π' of π is another affine action $\pi' : \mathbb{Z}^n \to \mathbb{T}^l$ and there is an affine map $\phi : \mathbb{T}^n \to \mathbb{T}^l$ such that $\phi \circ \pi(v) = \pi'(v) \circ \phi$ for every $v \in \mathbb{Z}^n$. We say a factor π' has *rank one* if $\pi'(\mathbb{Z}^n)$ has a finite index cyclic subgroup.

THEOREM 5.8. *Let $\pi : \mathbb{Z}^n \to \mathrm{Aff}(\mathbb{T}^m)$ have no rank one factors. Then π is locally rigid.*

This theorem is proven by a KAM method. One should note that all the theorems on actions of abelian groups by toral automorphisms stated in subsection 4.1 are special cases of this theorem. (This is not quite literally true. Those theorems apply to perturbations that are only C^1 close, while the result currently under discussion only applies to actions that are close to very high order.) It is also worthwhile to note that Theorem 5.8 can be proven using no techniques of hyperbolic dynamics.

We now briefly describe the KAM method. Let Γ be a finitely generated group and $\pi : \Gamma \to \mathrm{Diff}^\infty(M)$ a homomorphism. To apply a KAM-type argument, define $L : \mathrm{Diff}^\infty(M)^k \times \mathrm{Diff}^\infty(M) \to \mathrm{Diff}^\infty(M)^k$ by taking

$$L(\phi_1, \ldots, \phi_k, f) = (\pi(\gamma_1) \circ f \circ \phi_1 \circ f^{-1}, \ldots, \pi(\gamma_1) \circ f \circ \phi_1 \circ f^{-1})$$

where $\pi : \Gamma \to \mathrm{Diff}^\infty(M)$ is a homomorphism. If π' is another Γ action on M then $L(\pi'(\gamma_1), \ldots, \pi'(\gamma_k), f) = (\mathrm{Id}, \ldots, \mathrm{Id})$ implies f is a conjugacy between π and π', so the problem of finding a conjugacy is the same as finding a diffeomorphism f which solves $L(\pi'(\gamma_1), \ldots, \pi'(\gamma_k), f) = (\mathrm{Id}, \ldots, \mathrm{Id})$ subject to the constraint that π' is a Γ action.

The KAM method proceeds by taking the derivative DL of L at (π, f) and solving the resulting linear equation instead subject to a linear constraint that is the derivative of the condition that π' is a Γ action. This produces an "approximate solution" to the nonlinear problem and one proceeds by an iteration. If the original perturbation is of size ε then the perturbation obtained after one step in the iteration is of size ε^2 at least to whatever order one can control the size of the solution of DL. This allows one to show that the iteration converges even under conditions where there is some dramatic "loss of regularity", i.e. when solutions of DL are only small to much lower order than the initial data. A standard technique used to combat this loss of regularity is to alter the equation given by DL by inserting smoothing operators. That one can solve the linearized equation modified by smoothing operators in place of the original linearized equation and still expect to prove convergence of the iterative procedure depends heavily on the quadratic convergence just described. The main difficulty in applying this outline is obtaining so-called tame estimates on inverses of linearized operators. For a definition of a tame estimate, see following Theorem 5.12.

The KAM method is often presented as a method for proving hard implicit function theorems. The paradigmatic theorem of this kind is due to Hamilton [Ha1; Ha2], and is used by the author in the proof of the following theorem. For a brief discussion of the relation between this work and that of Katok and Damjanovich, see the end of this subsection.

THEOREM 5.9. *Let Γ be a finitely presented group, (M, g) a compact Riemannian manifold and $\pi : \Gamma \to \mathrm{Isom}(M, g) \subset \mathrm{Diff}^\infty(M)$ a homomorphism. If $H^1(\Gamma, \mathrm{Vect}^\infty(M)) = 0$ and $H^2(\Gamma, \mathrm{Vect}^\infty(M))$ is Hausdorff in the tame topology, the homomorphism π is locally rigid as a homomorphism into $\mathrm{Diff}^\infty(M)$.*

I believe the condition on $H^2(\Gamma, \mathrm{Vect}^\infty(M))$ should hold automatically under the other hypotheses of the theorem. If this is true, then one has a new proof of Theorem 5.2 using a result in [LZ]. There are some other infinitesimal rigidity results that would then yield more novel applications. For example:

THEOREM 5.10. *Let Γ be an irreducible lattice in a semisimple Lie group G with real rank at least two. Then for any Riemannian isometric action of Γ on a compact manifold $H^1(\Gamma, \mathrm{Vect}^\infty(M)) = 0$.*

Theorem 5.10 naturally applies in greater generality, in particular to irreducible S-arithmetic lattices and to irreducible lattices in products of more general locally compact groups.

To give another result on infinitesimal rigidity, we require a definition. For certain cocompact arithmetic lattices Γ in a simple group G, the arithmetic structure of Γ comes from a realization of Γ as the integer points in $G \times K$ where K is a compact Lie group. In this case it always true that the projection to G is a lattice and the projection to K is dense. We say a Γ action is *arithmetic isometric* if it is defined by projecting Γ to K, letting K act by C^∞ diffeomorphisms on a compact manifold M and restricting the K action to Γ.

THEOREM 5.11. *For certain congruence lattices $\Gamma < SU(1, n)$, any arithmetic isometric action of Γ has $H^1(\Gamma, \text{Vect}^\infty(M)) = 0$.*

For a description of which lattices this applies to, we refer the reader to [F2]. The theorem depends on deep results of Clozel [Cl].

Theorem 5.9 actually follows from the following, more general result.

THEOREM 5.12. *Let Γ be a finitely presented group, M a compact manifold, and $\pi : \Gamma \to \text{Diff}^\infty(M)$ a homomorphism. If $H^1(\Gamma, \text{Vect}^\infty(M)) = 0$ and the sequence*

$$C^0(\Gamma, \text{Vect}^\infty(M)) \xrightarrow{d_1} C^1(\Gamma, \text{Vect}^\infty(M)) \xrightarrow{d_2} C^2(\Gamma, \text{Vect}^\infty(M))$$

admits a tame splitting then the homomorphism π is locally rigid. I.e. π is locally rigid provided there exist tame linear maps

$$V_1 : C^1(\Gamma, \text{Vect}^\infty(M)) \to C^0(\Gamma, \text{Vect}^\infty(M))$$

and

$$V_2 : C^2(\Gamma, \text{Vect}^\infty(M)) \to C^1(\Gamma, \text{Vect}^\infty(M))$$

such that $d_1 \circ V_1 + V_2 \circ d_2$ is the identity on $C^1(\Gamma, \text{Vect}^\infty(M))$.

Here $C^i(\Gamma, Vect^\infty)$ is the group of i-cochains with values in $\text{Vect}^\infty(M)$ and d_i are the standard coboundary maps where we have identified the cohomology of Γ with the cohomology of a $K(\Gamma, 1)$ space with one vertex, one edge for each generator in our presentation of Γ and one 2 cell for each relator in our presentation of Γ. A map L is called tame if there is an estimate of the type $\|Lv\|_k \leq C_k \|v\|_{k+r}$ for a fixed choice of r. Here the $\|\cdot\|_l$ can be taken to be the C^l norm on cochains with values in $\text{Vect}^\infty(M)$. This notion clearly formalizes the notion of being able to solve an equation with some loss of regularity.

The proof of Theorem 5.12 proceeds by reducing the question to Hamilton's implicit function theorem for short exact sequences and is similar in outline to Weil's proof of Theorem 3.2.

Fix a finitely presented group Γ and a presentation of Γ. This is a finite collection S of generators $\gamma_1, \ldots, \gamma_k$ and finite collection R of relators $w_1, \ldots, w_r$ where each w_i is a finite word in the γ_j and their inverses. More formally, we can view each w_i as a word in an alphabet on k letters. Let $\pi : \Gamma \to \text{Diff}^\infty(M)$ be a homomorphism, which we can identify with a point in $\text{Diff}^\infty(M)^k$ by taking the images of the generators. We have a complex:

$$\text{Diff}^\infty(M) \xrightarrow{P} \text{Diff}^\infty(M)^k \xrightarrow{Q} \text{Diff}^\infty(M)^r \qquad (5\text{--}1)$$

Where P is defined by taking ψ to $(\psi \pi(\gamma_1) \psi^{-1}, \ldots, \psi \pi(\gamma_k) \psi^{-1})$ and Q is defined by viewing each w_i as a word in k letters and taking $(\psi_1, \ldots, \psi_k)$ to

$(w_1(\psi_1, \ldots, \psi_k), \ldots, w_r(\psi_1, \ldots, \psi_k))$. To this point this is simply Weil's proof where $\mathrm{Diff}^\infty(M)$ is replacing a finite dimensional Lie group H. Letting Id be the identity map on M, it follows that $P(\mathrm{Id}) = \pi$ and $Q(\pi) = (\mathrm{Id}, \ldots, \mathrm{Id})$. Also note that $Q^{-1}(\mathrm{Id}_M, \ldots, \mathrm{Id}_M)$ is exactly the space of Γ actions. Note that P and Q are $\mathrm{Diff}^\infty(M)$ equivariant where $\mathrm{Diff}^\infty(M)$ acts on itself by left translations and on $\mathrm{Diff}^\infty(M)^k$ and $\mathrm{Diff}^\infty(M)^r$ by conjugation. Combining this equivariance with Hamilton's implicit function theorem, I show that local rigidity is equivalent to producing a tame splitting of the sequence

$$\mathrm{Vect}^\infty(M) \xrightarrow{DP_{\mathrm{Id}}} \mathrm{Vect}^\infty(M)^k \xrightarrow{DQ_\pi} \mathrm{Vect}^\infty(M)^r \qquad (5\text{–}2)$$

To complete the proof of Theorem 5.12 requires that one compute DP_{Id} and DQ_π in order to relate the sequence in equation (5–2) to the cohomology sequence in Theorem 5.12.

We remark here that the information needed to split the sequence in Theorem 5.12 is quite similar to the information one would need to apply a KAM method. This is not particularly surprising as Hamilton's implicit function theorem is a formalization of the KAM method. In particular, to prove Theorem 5.8, one can apply Theorem 5.12 using estimates and constructions from [DK2, Section 3] to produce the required tame splitting. This avoids the use of the explicit KAM argument in [DK2, Section 4].

Finally, we remark that there is a theorem of Fleming that is an analogue of Theorems 5.9 and 5.12 in the setting of finite, or finite Sobolev, regularity [Fl]. This is proven using an infinite dimensional variant of Stowe's fixed point theorems, though it has recently been reproven by An and Neeb using a new implicit function theorem [AN]. With either proof, this result also has a similar condition on second cohomology. We remark here that due to the nature of the respective topologies on spaces of vector fields, the condition on second cohomology in the work of Fleming or An–Neeb is considerably stronger than what is needed in Theorems 5.9 and 5.12. As an illustration, no version of Theorem 5.8 can be proven using these results. This is because a cohomological equation can have solutions with tame estimates, i.e. with some loss of regularity, without having solutions with an estimate at any fixed regularity.

5.3. Further results. In this subsection, we describe a few more recent developments related to the results discussed so far. These results are either very recent or somewhat removed from the main stream of research.

The first result we discuss concerns actions of lattices in $Sp(1, n)$ or F_4^{-20} and is due to T. J. Hitchman. We state here only a special case of his results. For this result, we assume that Γ is an arithmetic subgroup of $Sp(1, n)$ or F_4^{-20} in the standard $\mathbb{Q}$ structures on those groups. This means that Γ is a finite index subgroup of the integer points in the standard matrix representation of these groups. This means that in the defining representation of $Sp(1, n)$ or F_4^{-20} on $\mathbb{R}^m$ the action of Γ preserves the integer lattice $\mathbb{Z}^m$ and therefore defines an action ρ of Γ on $\mathbb{T}^m$.

THEOREM 5.13. *The action ρ defined in the last paragraph is deformation rigid.*

The proof proceeds in two steps. Building a path of C^0 conjugacies follows more or less as in [H1], see subsection 4.1 above. The main novelty in [Hi] is the proof that these conjugacies are in fact smooth. Theorem 5.13 is a special case of the results obtained in [Hi].

Another recent development should lead to a common generalization of Theorems 5.8 and 4.7. For example, one can consider actions of an abelian subgroup $\mathbb{R}^k$ of the full diagonal group $\mathbb{R}^{n-1}$ in $SL(n, \mathbb{R})$ on $SL(n, \mathbb{R})/\Lambda$ where Λ is a cocompact lattice. In this context, Damjanovich and Katok are developing a more geometric approach in contrast to the analytic method of [DK1; DK2]. In [DK3], under a natural nondegeneracy condition on the subgroup $\mathbb{R}^k < \mathbb{R}^{n-1}$, the authors prove the cocycle rigidity result required to generalize Theorem 4.7 to the natural result for the $\mathbb{R}^k$ action. Here there are "additional trivial perturbations" of the action arising from $\operatorname{Hom}(\mathbb{R}^k, \mathbb{R}^{n-1})$. A rigidity theorem in this context is work in progress, see [DK3] for some discussion.

To close this section, we mention two other recent works. The first is a paper by Burslem and Wilkinson which investigates local and global rigidity questions for actions of certain solvable groups on the circle. Particularly striking is their construction of group actions which admit C^r perturbations but no C^{r+1} perturbations for every integer r. The second is a paper by M. Einsiedler and T. Fisher in which the method of proof of Theorem 4.6 is extended to affine actions of $\mathbb{Z}^d$ where the matrices generating the group action have nontrivial Jordan blocks. For perturbations of the group action which are close to very high order, this result follows from Theorem 5.8, but in [EF] the result only requires that the perturbation be C^1 close to the original action.

6. Directions for future research and conjectures

In this section, I mention a few conjectures and point a few directions for future research. These are particularly informed by my taste.

6.1. Actions of groups with property (T). Lattices in $SP(1, n)$ and F_4^{-20} share many of the rigidity properties of higher rank lattices. In light of Theorems 5.1, 5.2 and 5.13, it seems natural to conjecture:

CONJECTURE 6.1. *Let G be a semisimple Lie group with no compact factors and no simple factors isomorphic to $SO(1,n)$ or $SU(1,n)$, and let $\Gamma < G$ be a lattice. Then any volume preserving generalized quasiaffine action of G or Γ on a compact manifold is locally rigid.*

Note that for G with no rank one factors and quasiaffine actions, this is just Theorem 5.1. To apply the methods of [FM1; FM2; FM3] in the setting of Conjecture 6.1 there are essentially three difficulties:

(1) If the action is generalized quasiaffine and not quasiaffine, then one cannot use the construction of the Margulis–Qian cocycle described above. This is easiest to see for a generalized affine action ρ on some $K\backslash H/\Lambda$. The action ρ lifts to H/Λ, but a perturbation ρ' need not. If K is finite, this difficulty can be overcome by passing to a subgroup of finite index $\Gamma' < \Gamma$ and arguments in [FM3] can be used prove rigidity of Γ from rigidity of Γ'. If K is compact and connected, this is a genuine and surprisingly intractable difficulty.

(2) Proving a version of Zimmer's cocycle superrigidity for groups as in the assumptions of the conjecture. Partial results in this direction were obtained by Corlette–Zimmer and Korevaar–Schoen, but their results all require hypotheses that are obviously restrictive or simply difficult to verify. Very recently, the author and Hitchman have proven a complete version of cocycle superrigidity, at least for so-called L^2-cocycles. In light of our work, this difficulty is already overcome.

(3) Replace the method of Katok–Spatzier in the proof that the conjugacy is actually smooth. The proof of Theorem 5.13 gives some progress in this direction, but there are significant technical difficulties to overcome in applying Hitchman's methods at this level of generality. There is some progress on this question by Gorodnik, Hitchman and Spatzier.

The author and Hitchman have another approach to Conjecture 6.1 based on Theorem 5.12, some estimates proven in [FH1], and using heat flow and those estimates to produce a tame splitting of the short exact sequence in Theorem 5.12. It is not yet clear how generally applicable this method will be.

The following question is also interesting in this context:

QUESTION 6.2. *Let G and Γ be as in Conjecture 6.1. Let ρ be a non-volume-preserving affine action of Γ on a compact manifold M. When is ρ locally rigid?*

By the work of Stuck discussed at the end of Section 4.2, it is clear that local rigidity will not hold in full generality here. In particular for the product of the action of Γ on the boundary G/P with the trivial action of Γ on any manifold, there is already strong evidence against local rigidity. We give two particularly interesting special cases where we expect local rigidity to occur. The first is to take a lattice Γ in $SP(1, n)$ and ask if the Γ action on the boundary $SP(1, n)/P$ is locally rigid. This is analogous to Theorems 4.16 and 4.18. Another direction worth pursuing is to see if the action of say $Sl(n, \mathbb{Z})$ on $\mathbb{R}P^{n+k}$ is locally rigid for $n \geq 3$ and any $k \geq 0$. For cocompact lattices instead of $SL(n, \mathbb{Z})$ and $k = -1$ it is Theorem 4.16. One can ask a wide variety of similar questions for both compact and noncocompact lattices acting on homogeneous spaces that are "larger than" any natural boundary for the group as long as one avoids settings in which Stuck's examples can occur. We remark that in some instances, partial results for analytic perturbations can be obtained by Zeghib's method [Zg].

We remark here that the action of $SL(n, \mathbb{Z})$ on $\mathbb{T}^n$ is not locally rigid in Homeo($\mathbb{T}^n$) by a construction of Weinberger. It would be interesting to understand local rigidity in low regularity for other actions.

6.2. Actions of irreducible lattices in products. In this subsection, we formulate a general conjecture concerning local rigidity of actions of irreducible lattices in products. We begin by making a few remarks on other rigidity properties of irreducible lattices and by describing a few examples where rigidity might hold, as well as some examples where it does not.

Rigidity properties of irreducible lattices have traditionally been studied together with rigidity of lattices in simple groups, and irreducible lattices enjoy many of the same rigidity properties. We list a few here to motivate our conjectures on rigidity of actions of these lattices. The properties we list also rule out certain trivial constructions of perturbations and deformations of actions. In the following, Γ is an irreducible lattice in $G = (\prod_{i=1}^{k} G_i)/Z$ where each G_i is a noncompact semisimple Lie group and Z is a subgroup of the center of $\prod_{i=1}^{k} G_i$. Many of these results hold more generally, see below.

Properties of irreducible lattices:

(1) There are no nontrivial homomorphisms $\Gamma \to \mathbb{Z}$ (and therefore no nontrivial homomorphisms to any abelian or nonabelian free group).
(2) All linear representations of Γ are classified. In particular, given any representation ρ of Γ into $GL(V)$, where V is a finite dimensional vector space, then $H^1(\Gamma, V) = 0$.
(3) All normal subgroups of Γ are either finite or finite index.

The first property, for Γ cocompact, was originally proven by Bernstein and Kazhdan [BK]. Some special cases of this result were proven earlier by Matsushima and Shimura [MS]. All other properties are originally due to Margulis, see [Ma4] and historical references there. The original proofs all go through with only minor adaptations if some of the G_i are replaced with the k-points of a k-algebraic group over some other local field k. These properties have been shown to hold for appropriate classes of lattices in even more general products of locally compact groups by Bader, Monod and Shalom, [Sm; Md1; Md2; BSh].

Example 1: Let $\Gamma = SL(2, \mathbb{Z}[\sqrt{2}])$. We embed Γ in $SL(2, \mathbb{R}) \times SL(2, \mathbb{R})$ by taking γ to $(\gamma, \sigma(\gamma))$ where σ is the Galois automorphism of $\mathbb{Q}(\sqrt{2})$ taking $\sqrt{2}$ to $-\sqrt{2}$. This embedding defines an action of Γ on $\mathbb{T}^4$ where $\mathbb{T}^4 = \mathbb{R}^4/\mathbb{Z}^4$ where we identity $\mathbb{Z}^4$ with the image in $\mathbb{R}^4$ of $\mathbb{Z}[\sqrt{2}]^2$ via the embedding $v \rightarrow (v, \sigma(v))$. We first note that the list of properties given above imply one can prove deformation rigidity of this action using Hurder's argument from [H1] to produce a continuous conjugacy and using the method of Katok–Spatzier [KS1; KS2] to show that the conjugacy is smooth. Anatole Katok has suggested one might be able to show local rigidity of this action by using the methods in [KL1]. This example is just the first in a large class of Anosov actions of irreducible lattices, all of which should be locally rigid. We leave the general construction to the interested reader.

Example 2: We take the action of $\Gamma = SL(2, \mathbb{Z}[\sqrt{2}]$ and let Γ act on $\mathbb{T}^5 = \mathbb{T}^4 \times \mathbb{T}^1$ by a diagonal action where the action on $\mathbb{T}^4$ is as defined in Example 1 and the action on $\mathbb{T}^1$ is trivial. Once again, this is merely the first example of a large class of partially hyperbolic actions of irreducible lattices. In this instance, the central foliation for the Γ action consists of compact tori. For this type of example, many of the argument of [NT1; NT2; T] carry over, but the fact that Γ does not have property (T) prevents one from using those outlines to prove local rigidity. On the other hand, the methods of [NT1] can be adapted to prove deformation rigidity again replacing their argument for smoothness of the conjugacy by the Katok–Spatzier method. We remark that it is also possible to give many examples of actions of irreducible lattices in products where the central foliation is by dense leaves, and the methods of Nitica and Torok cannot be applied. See discussion below for the obstructions to applying the methods of [FM1; FM2; FM3].

Example 3: We now give an example of a family of actions which extend to an action of $G = G_1 \times G_2$. Let H be a simple Lie group with $G < H$, for example, $H = SL(4, \mathbb{R})$ or $H = Sp(4, \mathbb{R})$ and $\Lambda < H$ a cocompact lattice. Then both G and Γ act by left translations on H/Λ.

Example 4: We end with a family of examples for which there exists a large family of deformations. Let Γ be as above and let $\Lambda < SL(2, \mathbb{R}) = G_1$ be an irreducible lattice. Then Γ acts by left translations on $M = SL(2, \mathbb{R})/\Lambda$. Call this action $\bar{\rho}$. I do not currently know whether it is possible to deform this action, but one can use this action to build actions with perturbations on a slightly larger manifold. Let Γ act trivially on any manifold N and take the diagonal action ρ on $M \times N$. It is well-known that there exists a nontrivial homomorphism $\sigma : \Lambda \rightarrow \mathbb{Z}$. There is also a standard construction of a cocycle $\alpha : G_1 \times G_1/\Lambda \rightarrow \Lambda$

over the G_1 action on G_1/Λ. The cocycle is defined by taking a fundamental domain X for Λ in G_1, identifying G_1/Λ with X and letting $\alpha(g, x)$ be the element of Λ such that $gx\alpha(g, x)^{-1}$, as an element of G_1, is in X. Taking any vector field V on N and let s_ε be the $\mathbb{Z}$ action on N defined by having 1 act by flowing to time ε along V. We then define a Γ action on $M \times N$ by taking

$$\rho_\varepsilon(\gamma)(m, n) = (\bar{\rho}(\gamma)m, s_\varepsilon(\sigma(\alpha(g, m))(n))).$$

We leave it to the interested reader to show that the actions ρ_ε are not conjugate to ρ for essentially any choice of V.

The key point in example 4, which is not present in the first three examples, is that the action factors through a projection of Γ into a simple factor of G. Motivated by the examples so far, by the results in Theorem 5.10, and by analogy with results on actions of higher rank abelian groups, we make the following definition and conjecture:

DEFINITION 6.3. Let $G = G_1 \times \cdots \times G_k$ be a semisimple Lie group where all the G_i are noncompact. Let $\Gamma < G$ be an irreducible lattice. Let ρ be an affine action of Γ on some H/Λ where H is a Lie group and $\Lambda < H$ is a cocompact lattice. Then we say ρ has *rank one factors* if there exists

(1) an action $\bar{\rho}$ of Γ on a some space X which is a factor of ρ
(2) and a rank one factor G_i of G

such that $\bar{\rho}$ is the restriction of a G_i action. I.e. Γ acts on X by projecting Γ to G_i and restricting a G_i action.

CONJECTURE 6.4. *Let G, Γ, H, Λ and ρ be as in Definition 6.3. Then if ρ has no rank one factors ρ is locally rigid.*

REMARK. It is a consequence of Ratner's measure rigidity theorem, see [R; Sh; W2], that any rank one factor of an affine action for these groups is in fact affine. This implies that any rank one factor is a left translation action on some H'/Λ'. So a special case of the conjecture is that any affine Γ action on a torus or nilmanifold is locally rigid.

There is a variant of Conjecture 6.4 for G actions. We recall that a measure preserving action of $G = G_1 \times \cdots \times G_k$ action is *irreducible* if each G_i acts ergodically. We extend this notion to nonergodic G actions by saying that the action is *weakly irreducible* if every ergodic component of the volume measure for the action of any G_i is an ergodic component of the volume measure for the action of G.

CONJECTURE 6.5. *Let G, H, Λ be as in Definition 6.3 and let $G < H$, so that we have a left translation action ρ of G on H/Λ. Then if ρ is weakly irreducible, ρ is locally rigid.*

The relation of the two conjectures follows from the following easy lemma.

LEMMA 6.6. *Let G, Γ, H, Λ and ρ be as in Definition 6.3, then the Γ action on H/Λ has no rank one factors if and only if the induced G action on $(G \times H/\Lambda)/\Gamma$ is weakly irreducible.*

To prove the lemma requires both some algebraic untangling of induced actions and a use of Margulis' superrigidity theorem to describe affine actions of G and Γ along the lines of [FM1, Theorems 6.4 and 6.5]. We leave this as an exercise for the interested reader. It is fairly easy to check the lemma for any particular affine Γ action.

We end this subsection by pointing out the difficulty in approaching this conjecture by means of the methods of [FM1; FM2; FM3]. A central difficulty is that the lattices in question do not have property (T) and so the foliated generalization of Theorem 5.2 does not apply. However, even in the case of weakly hyperbolic actions, there are significant difficulties. To begin the argument, one would like to apply cocycle superrigidity to the Margulis–Qian cocycle. To do this requires the existence of an invariant measure which is usually established using property (T) by an argument of Seydoux [Sy]. In this setting, where property (T) does not hold, one might try instead to use the work of Nevo and Zimmer, [NZ1; NZ2; NZ3], but there are nontrivial difficulties here as well. One cannot apply their theorems without first showing that the perturbed action satisfies some irreducibility assumption. Even if one were to obtain an invariant measure, the precise form of cocycle superrigidity required is not known for products of rank one groups or their irreducible lattices. And the strongest possible forms of cocycle superrigidity in this context again require a kind of irreducibility of the perturbed action. So to proceed by this method one would need to show that perturbations of the actions in Conjecture 6.4 and 6.5 still satisfied some irreducibility conditions. This seems quite difficult. It may also be possible to approach these questions by using Theorem 5.12, but even proving that the relevant cohomology groups vanish seems subtle.

6.3. Other questions and conjectures. We end this article by discussing some other questions and conjectures.

In the context of Theorem 5.11 it is interesting to ask if isometric actions of lattices in $SU(1, n)$ are locally rigid. For some choices of lattice, the answer is trivially no. Namely some cocompact lattices in $SU(1, n)$ have homomorphisms ρ to $\mathbb{Z}$ [Ka2; BW], and so have arithmetic actions with deformations provided the centralizer Z of K in $\text{Diff}^\infty(M)$ is nontrivial. Having centralizer allows one to deform the action along the image of the homomorphism $\rho \circ \sigma_t : F \to Z$ where $\sigma_t : \mathbb{Z} \to Z$ is any one parameter family of homomorphisms. It seems reasonable to conjecture:

CONJECTURE 6.7. *Let ρ be an arithmetic isometric action of a lattice in $SU(1, n)$. Then any sufficiently small perturbation of ρ is of the form described in the previous paragraph.*

This conjecture is in a certain sense an infinite dimensional analogue of work of Goldman–Millson and Corlette [Co1; GM]. Another conjecture concerning complex hyperbolic lattices, for which work of Yue provides significant evidence [Yu], is:

CONJECTURE 6.8. *Is the action of any lattice in $SU(1, n)$ on the boundary of complex hyperbolic space locally rigid?*

There are also many interesting questions concerning the failure of local rigidity for lattices in $SO(1, n)$. The only rigidity theorem we know of in this context is Kanai's, Theorem 4.17, and it would be interesting to extend Kanai's theorem to nonuniform lattices. In [F1; F3] various deformations of lattices in $SO(1, n)$ are constructing for affine and isometric actions. These constructions both adapt the bending construction of Johnson and Millson, [JM]. It seems likely that in some cases one should be able to prove results concerning the structure of the representation space and, in particular, to show that it is "singular" in an appropriate sense. See [F3] for more discussion.

Two other paradigmatic examples of large groups are the outer automorphism group of the free group, Out(F_n), and the mapping class group of a surface S, $MCG(S)$. These groups do not admit many natural actions on compact manifolds, but there are some natural interesting actions quite analogous to those we have already discussed. For $MCG(S)$, the question we raise here is already raised in [La]. The actions we consider are "nonlinear" analogues of the standard actions of $SL(n, \mathbb{Z})$ on $\mathbb{T}^n$ and $SP(2n, \mathbb{Z})$ on $\mathbb{T}^{2n}$. The spaces acted upon are moduli spaces of representations of either the free group or the fundamental group of a surface S, where the representations take values in compact groups. More precisely, we have an action of Out(F_n) on Hom(F_n, K)/K and an action of $MCG(S)$ on Hom($\pi(S), K$)/K where K is a compact group. It is natural to ask whether these actions are locally rigid, though the meaning of the question is somewhat obscured by the fact that the representation varieties are not smooth. For $K = S^1$, one obtains actions on manifolds, and in fact tori, and one might begin by considering that case.

We end with a question motivated by the recent work of Damjanovic and Katok. We only give a special case here. Let G be a real split, simple Lie group of real rank at least two. Let $\Gamma < G \times G$ be an irreducible lattice. Let $K < G$ be a maximal compact subgroup and view K as a subgroup of $G \times G$ by viewing it as a subgroup of the second factor. The quotient $K \backslash (G \times G) / \Gamma$ has a natural G

action on the left on the first factor. We can restrict this action to the action of a maximal split torus A in G. Note that A is isomorphic to $\mathbb{R}^d$ for some $d \geq 2$.

QUESTION 6.9. *Is the action of $\mathbb{R}^d$ described in the paragraph above locally rigid?*

References

[An] Anosov, D. V. *Geodesic flows on closed Riemann manifolds with negative curvature.* Proceedings of the Steklov Institute of Mathematics, **No. 90** (1967). Translated from the Russian by S. Feder. American Mathematical Society, Providence, RI, 1969.

[AN] An, J.; Neeb, K. An implicit function theorem for Banach spaces and some applications. Preprint.

[BSh] Bader, U.; Shalom, Y. Factor and normal subgroup theorems for lattices in products of groups. Preprint.

[Ba1] Banyaga, A. Sur la structure du groupe des difféomorphismes qui préservent une forme symplectique. *Comment. Math. Helv.* **53** (1978), No. 2, 174–227.

[Ba2] Banyaga, A. *The structure of classical diffeomorphism groups.* Mathematics and its Applications, **No. 400**. Kluwer, Dordrecht, 1997.

[Be1] Benveniste, E. J. Rigidity of isometric lattice actions on compact Riemannian manifolds. *GAFA* **10** (2000), 516–542.

[Be2] Benveniste, E. J. Exotic actions of semisimple groups and their deformations. Unpublished preprint and Chapter 2 of University of Chicago Ph.D. dissertation, 1996.

[BF] Benveniste, E. J.; Fisher, D. Non-existence of invariant rigid structures and invariant almost rigid structures. To appear in *Comm. Anal. Geom.*, 21 pages. http://comet.lehman.cuny.edu/fisher/

[BK] Bernstein, I. N.; Kazhdan, D. A. The one-dimensional cohomology of discrete subgroups. (Russian). *Funkcional. Anal. i Priložen.* **4** (1970), No. 1, 1–5.

[BCG] Besson, G.; Courtois, G.; Gallot, S. Minimal entropy and Mostow's rigidity theorems. *Ergodic Theory Dynam. Systems* **16** (1996), No. 4, 623–649.

[Bo] Borel, A. Compact Clifford–Klein forms of symmetric spaces. *Topology* **2** (1963), 111–122.

[BW] Borel, A.; Wallach, N. *Continuous cohomology, discrete subgroups, and representations of reductive groups.* Second edition. Math. Surveys Monogr., **No. 67**. American Mathematical Society, Providence, RI, 2000.

[BuW] Burslem, E.; Wilkinson, A. global rigidity of solvable group actions on S^1. *Geom. Topol.* **8** (2004), 877–924.

[BPSW] Burns, Keith; Pugh, Charles; Shub, Michael; Wilkinson, Amie. Recent results about stable ergodicity. *Smooth ergodic theory and its applications* (Seattle, WA, 1999), 327–366. Proc. Sympos. Pure Math., **No. 69**. American Mathematical Society, Providence, RI, 2001.

[CGh] Cairns, Grant; Ghys, Étienne. The local linearization problem for smooth SL(*n*)-actions. *Enseign. Math. (2)* **43** (1997), No. 1–2, 133–171.

[C] Calabi, E. On compact, Riemannian manifolds with constant curvature, I. Proc. Sympos. Pure Math., **No. 3**, 155–180. American Mathematical Society, Providence, RI, 1961.

[CV] Calabi, E.; Vesentini, E. On compact, locally symmetric Kähler manifolds. *Ann. of Math. (2)* **71** (1960), 472–507.

[Cl] Clozel, Laurent. On the cohomology of Kottwitz's arithmetic varieties. *Duke Math. J.* **72** (1993), No. 3, 757–795.

[CF] Connell, Christopher; Farb, Benson. Some recent applications of the barycenter method in geometry. *Topology and geometry of manifolds* (Athens, GA, 2001), 19–50. Proc. Sympos. Pure Math., **No. 71**. American Mathematical Society, Providence, RI, 2003.

[Co1] Corlette, Kevin. Flat G-bundles with canonical metrics. *J. Differential Geom.* **28** (1988), No. 3, 361–382.

[Co2] Corlette, Kevin. Archimedean superrigidity and hyperbolic geometry. *Ann. of Math. (2)* **135** (1992), No. 1, 165–182.

[DK1] Damjanovich, D.; Katok, A. Local rigidity of partially hyperbolic actions of $\mathbb{Z}^k$ and $\mathbb{R}^k$, $k \geq 2$, I. KAM method and actions on the torus. Preprint. http://www.math.psu.edu/katok_a/papers.html

[DK2] Damjanović, Danijela; Katok, Anatole. Local rigidity of actions of higher rank abelian groups and KAM method. *Electron. Res. Announc. Amer. Math. Soc.* **10** (2004), 142–154.

[DK3] Damjanović, Danijela; Katok, Anatole. Periodic cycle functions and cocycle rigidity for certain partially hyperbolic $\mathbb{R}^k$ actions. *Discrete Contin. Dynam. Systems* **13** (2005), No. 4, 985–1005.

[dlLMM] de la Llave, R.; Marco, J. M.; Moriyón, R. Canonical perturbation theory of Anosov systems and regularity results for the Livšic cohomology equation. *Ann. of Math. (2)* **123** (1986), No. 3, 537–611.

[De] Delorme, Patrick. 1-cohomologie des représentations unitaires des groupes de Lie semi-simples et résolubles. Produits tensoriels continus de représentations. *Bull. Soc. Math. France* **105** (1977), No. 3, 281–336.

[EF] Einsiedler, M.; Fisher, T. Differentiable rigidity for hyperbolic toral actions. Preprint.

[Ep] Epstein, D. B. A. The simplicity of certain groups of homeomorphisms. *Compositio Math.* **22** (1970), 165–173.

[Fe1] Feres, Renato. *Dynamical systems and semisimple groups: an introduction.* Cambridge Tracts in Math., **No. 126**. Cambridge University Press, Cambridge, 1998.

[Fe2] Feres, Renato. An introduction to cocycle super-rigidity. *Rigidity in dynamics and geometry* (Cambridge, 2000), 99–134. Springer, Berlin, 2002.

[FK] R. Feres; Katok, A. Ergodic theory and dynamics of G-spaces (with special emphasis on rigidity phenomena). *Handbook of dynamical systems, 1A*, edited by B. Hasselblatt and A. Katok, 665–763. Elsevier, Amsterdam, 2002. http://www.math.wustl.edu/ -feres/publications.html.

[F1] Fisher, D. Deformations of group actions. Preprint, 14 pages, to appear in *Transactions AMS*. http://comet.lehman.cuny.edu/fisher/

[F2] Fisher, D. First cohomology and local rigidity of group actions. In preparation.

[F3] Fisher, D. Bending group actions and cohomology of arithmetic lattices in $SO(1, n)$. In preparation.

[FH1] Fisher, D.; Hitchman, T. J. Strengthening property (T) by Bochner methods. Preprint.

[FH2] Fisher, D.; Hitchman, T. J. Superrigidity for cocycles via Bochner methods. In preparation.

[FM1] Fisher, D.; Margulis, G. A. Local rigidity for cocycles. *Papers in honor of Calabi, Lawson, Siu and Uhlenbeck*, 191–234. Surv. Diff. Geom., **No. VIII**. S.-T. Yau (ed.). International Press, Somerville, MA, 2003.

[FM2] Fisher, D.; Margulis, G. A. Almost isometries, property (T) and local rigidity. To appear in *Invent. Math.*

[FM3] Fisher, D.; Margulis, G. A. Local rigidity for affine actions of higher rank Lie groups and their lattices. Preprint, 59 pages. http://comet.lehman.cuny.edu/fisher/

[FW] Fisher, D.; Whyte, K. Continuous quotients for lattice actions on compact spaces. *Geom. Dedicata* **87** (2001), 181–189.

[Fl] Fleming, Philip J. Structural stability and group cohomology. *Trans. Amer. Math. Soc.* **275** (1983), No. 2, 791–809.

[Fu1] Furstenberg, Harry. A Poisson formula for semi-simple Lie groups. *Ann. of Math. (2)* **77** (1963), 335–386.

[Fu2] Furstenberg, Harry. Boundaries of Lie groups and discrete subgroups. *Actes du Congrès International des Mathématiciens* (Nice, 1970), **Tome 2**, 301–306. Gauthier-Villars, Paris, 1971.

[Fu3] Furstenberg, Harry. Rigidity and cocycles for ergodic actions of semisimple Lie groups (after G. A. Margulis and R. Zimmer). *Bourbaki Seminar*, 1979/80, 273–292. Lecture Notes in Math., **No. 842**. Springer, Berlin, 1981.

[GhS] Ghys, E.; Sergiescu, V. Stabilité et conjugaison différentiable pour certains feuilletages. *Topology* **19** (1980), No. 2, 179–197.

[Gh1] Ghys, Étienne. Actions localement libres du groupe affine. [Locally free actions of the affine group]. *Invent. Math.* **82** (1985), No. 3, 479–526.

[Gh2] Ghys, Étienne. Déformations de flots d'Anosov et de groupes fuchsiens. [Deformations of Anosov flows and Fuchsian groups]. *Ann. Inst. Fourier (Grenoble)* **42** (1992), No. 1–2, 209–247.

[Gh3] Ghys, Étienne. Rigidité différentiable des groupes fuchsiens. [Differentiable rigidity of Fuchsian groups]. *Publ. Math. IHES* **78** (1993), 163–185.

[GM] Goldman, W. M.; Millson, J. J. Local rigidity of discrete groups acting on complex hyperbolic space. *Invent. Math.* **88** (1987), No. 3, 495–520.

[GS] Gromov, Mikhail; Schoen, Richard. Harmonic maps into singular spaces and *p*-adic superrigidity for lattices in groups of rank one. *Publ. Math. IHES* **76** (1992), 165–246.

[Gu] Guichardet, Alain. Sur la cohomologie des groupes topologiques, II. *Bull. Sci. Math. (2)* **96** (1972), 305–332.

[G] Guysinsky, M. The theory of nonstationary normal forms. *Ergodic Theory Dynam. Systems* **22** (2002), No. 3, 845–862.

[GK] Guysinsky, M.; Katok, A. Normal forms and invariant geometric structures on transverse contracting foliations. *Math. Res. Lett.* **5** (1998), 149–163.

[HV] de la Harpe, P.; Valette, A. *La propriete (T) de Kazhdan pour les groupes localement compacts*. Asterisque, **No. 175**. Soc. Math. de France, Paris, 1989.

[Ha1] Hamilton, Richard S. The inverse function theorem of Nash and Moser. *Bull. Amer. Math. Soc.* **7** (1982), No. 1, 65–222.

[Ha2] Hamilton, Richard S. Deformation of complex structures on manifolds with boundary, I: The stable case. *J. Differential Geom.* **12** (1977), No. 1, 1–45.

[He] Helgason, Sigurdur. *Differential geometry, Lie groups, and symmetric spaces*. Corrected reprint of the 1978 original. Grad. Stud. Math., **No. 34**. American Mathematical Society, Providence, RI, 2001.

[Hr] Herman, Michael-Robert. Simplicité du groupe des difféomorphismes de classe C^∞, isotopes à l'identité, du tore de dimension *n*. *C. R. Acad. Sci. Paris Sér. A-B* **273** (1971), A232–A234.

[HPS] Hirsch, M. W.; Pugh, C. C.; Shub, M. *Invariant Manifolds*. Lecture Notes in Math., **No. 583**. Springer, New York, 1977.

[Hi] Hitchman, T. J. Rigidity theorems for large dynamical systems with hyperbolic behavior. Ph. D. Thesis, University of Michigan, 2003.

[H1] Hurder, Steven. Rigidity for Anosov actions of higher rank lattices. *Ann. of Math. (2)* **135** (1992), No. 2, 361–410.

[H2] Hurder, Steven. Infinitesimal rigidity for hyperbolic actions. *J. Differential Geom.* **41** (1995), No. 3, 515–527.

[JM] Johnson, Dennis; Millson, John J. Deformation spaces associated to compact hyperbolic manifolds. *Discrete groups in geometry and analysis* (New Haven, CT, 1984), 48–106. Progr. Math., **No. 67**. Birkhäuser, Boston, 1987.

[Jn] Journé, J.-L. A regularity lemma for functions of several variables. *Rev. Mat. Iberoamericana* **4** (1988), No. 2, 187–193.

[Kan1] Kanai, M. A new approach to the rigidity of discrete group actions. *Geom. Funct. Anal.* **6** (1996), No. 6, 943–1056.

[Kan2] Kanai, Masahiko. A remark on local rigidity of conformal actions on the sphere. *Math. Res. Lett.* **6** (1999), No. 5–6, 675–680.

[KL1] Katok, A.; Lewis, J. Local rigidity for certain groups of toral automorphisms. *Israel J. Math.* **75** (1991), No. 2–3, 203–241.

[KL2] Katok, A.; Lewis, J. Global rigidity results for lattice actions on tori and new examples of volume-preserving actions. *Israel J. Math.* **93** (1996), 253–280.

[KLZ] Katok, A.; Lewis, J.; Zimmer, R. Cocycle superrigidity and rigidity for lattice actions on tori. *Topology* **35** (1996), No. 1, 27–38.

[KNT] Katok, Anatole; Niţică, Viorel; Török, Andrei. Non-abelian cohomology of abelian Anosov actions. *Ergodic Theory Dynam. Systems* **20** (2000), No. 1, 259–288.

[KS1] Katok, A.; Spatzier, R. J. Nonstationary normal forms and rigidity of group actions. *Electron. Res. Announc. Amer. Math. Soc.* **2** (1996), No. 3, 124–133.

[KS2] Katok, A.; Spatzier, R. J. Differential rigidity of Anosov actions of higher rank abelian groups and algebraic lattice actions. *Din. Sist. i Smezhnye Vopr.*, 292–319. Tr. Mat. Inst. Steklova, **No. 216** (1997). Translation in *Proc. Steklov Inst. Math.* **216** (1997), No. 1, 287–314.

[KS3] Katok, Anatole; Spatzier, Ralf J. First cohomology of Anosov actions of higher rank abelian groups and applications to rigidity. *Publ. Math. IHES* **79** (1994), 131–156.

[KS4] Katok, A.; Spatzier, R. J. Subelliptic estimates of polynomial differential operators and applications to rigidity of abelian actions. *Math. Res. Lett.* **1** (1994), No. 2, 193–202.

[Ka1] Kazhdan, D. A. On the connection of the dual space of a group with the structure of its closed subgroups. (Russian). *Funkcional. Anal. i Priložen.* **1** (1967), 71–74.

[Ka2] Kazhdan, David. Some applications of the Weil representation. *J. Analyse Mat.* **32** (1977), 235–248.

[Kn] Knapp, Anthony W. *Lie groups beyond an introduction.* Second edition. Prog. Math., **No. 140**. Birkhäuser, Boston, 2002.

[Ko] Kononenko, A. Infinitesimal rigidity of boundary lattice actions. *Ergodic Theory Dynam. Systems* **19** (1999), No. 1, 35–60.

[KY] Kononenko, A.; Yue, C. B. Cohomology and rigidity of Fuchsian groups. *Israel J. Math.* **97** (1997), 51–59.

[La] Labourie, François. Large groups actions on manifolds. Proceedings of the International Congress of Mathematicians, II (Berlin, 1998). *Doc. Math.* Extra Vol. **II** (1998), 371–380 (electronic).

[L] Lewis, J. Infinitesimal rigidity for the action of $SL_n(\mathbb{Z})$ on $\mathbb{T}^n$. *Trans. Amer. Math. Soc.* **324** (1991), No. 1, 421–445.

[LZ] Lubotzky, Alexander; Zimmer, Robert J. Variants of Kazhdan's property for subgroups of semisimple groups. *Israel J. Math.* **66** (1989), No. 1–3, 289–299.

[Ma1] Margulis, G. A. On the action of unipotent groups in the space of lattices. *Lie groups and their representations* (Budapest, 1971), 365–370. Halsted, New York, 1975.

[Ma2] Margulis, G. A. Arithmetic properties of discrete subgroups. (Russian). *Uspehi Mat. Nauk* **29** (1974), No. 1 (175), 49–98.

[Ma2] Margulis, G. A. Discrete groups of motions of manifolds of non-positive curvature. Proc. Int. Cong. Math. (Vancouver, 1974). *Amer. Math. Soc. Transl.* **109** (1977), 33–45.

[Ma3] Margulis, G. A. Arithmeticity of irreducible lattices in the semisimple groups of rank greater than 1. Appendix to the Russian translation of M. Ragunathan, *Discrete subgroups of Lie groups*, Mir, Moscow, 1977 (in Russian). English translation in *Invent. Math.* **76** (1984), 93–120.

[Ma4] Margulis, G. A. *Discrete subgroups of semisimple Lie groups*. Springer, New York, 1991.

[MQ] Margulis, G.; Qian, N. Local rigidity of weakly hyperbolic actions of higher real rank semisimple Lie groups and their lattices. *Ergodic Theory Dynam. Systems* **21** (2001), 121–164.

[Mt1] Mather, John N. Integrability in codimension 1. *Comment. Math. Helv.* **48** (1973), 195–233.

[Mt2] Mather, John N. Commutators of diffeomorphisms. *Comment. Math. Helv.* **49** (1974), 512–528.

[Mt3] Mather, John N. Foliations and local homology of groups of diffeomorphisms. Proceedings of the International Congress of Mathematicians (Vancouver, BC, 1974), Vol. 2, 35–37. Canadian Mathematical Congress, Montreal, 1975.

[MM] Matsushima, Yozō; Murakami, Shingo. On vector bundle valued harmonic forms and automorphic forms on symmetric riemannian manifolds. *Ann. of Math. (2)* **78** (1963), 365–416.

[MS] Matsushima, Yozō; Shimura, Goro. On the cohomology groups attached to certain vector valued differential forms on the product of the upper half planes. *Ann. of Math. (2)* **78** (1963), 417–449.

[Md1] Monod, Nicolas. Superrigidity for irreducible lattices and geometric splitting. Preprint.

[Md2] Monod, Nicolas. Arithmeticity vs. non-linearity for irreducible lattices. To appear in *Geom. Dedicata*.

[Mr] Morris, Dave Witte. *An introduction to arithmetic groups*. Book in preprint form.

[Mo1] Mostow, G. D. *Strong rigidity of locally symmetric spaces*. Ann. of Math. Stud., **No. 78**. Princeton University Press, Princeton, NJ, 1973.

[Mo2] Mostow, G. D. Personal communication.

[Mu] Mumford, D.; Fogarty, J.; Kirwan, F. *Geometric invariant theory*. Third edition. Ergeb. Math. Grenzgeb. (3), **No. 34**. Springer, Berlin, 1994.

[NZ1] Nevo, Amos; Zimmer, Robert J. Homogenous projective factors for actions of semi-simple Lie groups. *Invent. Math.* **138** (1999), No. 2, 229–252.

[NZ2] Nevo, Amos; Zimmer, Robert J. Rigidity of Furstenberg entropy for semisimple Lie group actions. *Ann. Sci. École Norm. Sup. (4)* **33** (2000), No. 3, 321–343.

[NZ3] Nevo, Amos; Zimmer, Robert J. A structure theorem for actions of semisimple Lie groups. *Ann. of Math. (2)* **156** (2002), No. 2, 565–594.

[NT1] Nitica, Viorel; Török, Andrei. Cohomology of dynamical systems and rigidity of partially hyperbolic actions of higher-rank lattices. *Duke Math. J.* **79** (1995), No. 3, 751–810.

[NT2] Nitica, Viorel; Török, Andrei. Local rigidity of certain partially hyperbolic actions of product type. *Ergodic Theory Dynam. Systems* **21** (2001), No. 4, 1213–1237.

[Q1] Qian, Nan Tian. Topological deformation rigidity of higher rank lattice actions. *Math. Res. Lett.* **1** (1994), No. 4, 485–499.

[Q2] Qian, Nantian. Anosov automorphisms for nilmanifolds and rigidity of group actions. *Ergodic Theory Dynam. Systems* **15** (1995), No. 2, 341–359.

[Q3] Qian, Nantian. Infinitesimal rigidity of higher rank lattice actions. *Comm. Anal. Geom.* **4** (1996), No. 3, 495–524.

[QY] Qian, Nantian; Yue, Chengbo. Local rigidity of Anosov higher-rank lattice actions. *Ergodic Theory Dynam. Systems* **18** (1998), No. 3, 687–702.

[Rg1] Raghunathan, M. S. On the first cohomology of discrete subgroups of semisimple Lie groups. *Amer. J. Math.* **87** (1965), 103–139.

[R] Ratner, M. On Raghunathan's measure conjectures. *Ann. of Math. (2)* **134** (1991), No. 3, 545–607.

[S] Selberg, A. On discontinuous groups in higher-dimensional symmetric spaces. *Contributions to function theory* (Bombay, 1960), 147–164. Internat. Colloq. Function Theory. Tata Institute of Fundamental Research, Bombay, 1960.

[Sy] Seydoux, G. Rigidity of ergodic volume-preserving actions of semisimple groups of higher rank on compact manifolds. *Trans. Amer. Math. Soc.* **345** (1994), 753–776.

[Sh] Shah, N. Invariant measures and orbit closures on homogeneous spaces for actions of subgroups generated by unipotent elements. *Lie groups and ergodic theory* (Mumbai, 1996), 229–271. Tata Inst. Fund. Res. Stud. Math., **No. 14**. S. G. Dani (ed.). Tata Institute of Fundamental Research, Bombay, 1998.

[Sl1] Shalom, Yehuda. Rigidity of commensurators and irreducible lattices. *Invent. Math.* **141** (2000), No. 1, 1–54.

[Sl2] Shalom, Yehuda. Measurable group theory. Preprint.

[Sm] Smale, S. Differentiable dynamical systems. *Bull. Amer. Math. Soc.* **73** (1967), 747–817.

[Sp1] Spatzier, R. J. Harmonic analysis in rigidity theory. *Ergodic theory and its connections with harmonic analysis* (Alexandria, 1993), 153–205. London Math. Soc. Lecture Note Ser., **No. 205**. Cambridge University Press, Cambridge, 1995.

[Sp2] Spatzier, R. J. An invitation to rigidity theory. *Modern dynamical systems and applications*, 211–231. Cambridge University Press, Cambridge, 2004.

[St1] Stowe, Dennis. The stationary set of a group action. *Proc. Amer. Math. Soc.* **79** (1980), No. 1, 139–146.

[St2] Stowe, Dennis. Stable orbits of differentiable group actions. *Trans. Amer. Math. Soc.* **277** (1983), No. 2, 665–684.

[Sk] Stuck, Garrett. Minimal actions of semisimple groups. *Ergodic Theory Dynam. Systems* **16** (1996), No. 4, 821–831.

[Th1] Thurston, W. A generalization of the Reeb stability theorem. *Topology* **13** (1974), 347–352.

[Th2] Thurston, William. Foliations and groups of diffeomorphisms. *Bull. Amer. Math. Soc.* **80** (1974), 304–307.

[T] Török, Andrei. Rigidity of partially hyperbolic actions of property (T) groups. *Discrete Contin. Dynam. Systems* **9** (2003), No. 1, 193–208.

[We1] Weil, André. On discrete subgroups of Lie groups. *Ann. of Math. (2)* **72** (1960), 369–384.

[We2] Weil, André. On discrete subgroups of Lie groups, II. *Ann. of Math. (2)* **75** (1962), 578–602.

[We3] Weil, André. Remarks on the cohomology of groups. *Ann. of Math. (2)* **80** (1964), 149–157.

[W1] Witte, Dave. Cocompact subgroups of semisimple Lie groups. *Lie algebra and related topics* (Madison, WI, 1988), 309–313. Contemp. Math., **No. 110**. American Mathematical Society, Providence, RI, 1990.

[W2] Witte, D. Measurable quotients of unipotent translations on homogeneous spaces. *Trans. Amer. Math. Soc.* **354** (1994), No. 2, 577–594.

[Yk] Yaskolko, S. Penn State Ph. D. dissertation, 1999.

[Yu] Yue, Cheng Bo. Smooth rigidity of rank-1 lattice actions on the sphere at infinity. *Math. Res. Lett.* **2** (1995), No. 3, 327–338.

[Zg] Zeghib, Abdelghani. Quelques remarques sur les actions analytiques des réseaux des groupes de Lie de rang supérieur. [Some remarks on analytic actions of lattices of Lie groups of higher rank]. *C. R. Acad. Sci. Paris Sér. I Math.* **328** (1999), No. 9, 799–804.

[Z1] Zimmer, R. J. Volume preserving actions of lattices in semisimple groups on compact manifolds. *Publ. Math. IHES* **59** (1984), 5–33.

[Z2] Zimmer, R. J. *Ergodic theory and semisimple groups*. Birkhäuser, Boston, 1984.

[Z3] Zimmer, R. J. Lattices in semisimple groups and distal geometric structures. *Invent. Math.* **80** (1985), 123–137.

[Z4] Zimmer, R. J. Actions of semisimple groups and discrete subgroups. Proc. Internat. Cong. Math. (Berkeley, 1986), 1247–1258.

[Z5] Zimmer, R. J. Lattices in semisimple groups and invariant geometric structures on compact manifolds. *Discrete groups in geometry and analysis: Papers in honor of G. D. Mostow* (New Haven, CT, 1984), 152–210. Roger Howe (ed.). Progr. Math., **No. 67**. Birkhäuser, Boston, 1987.

[Z6] Zimmer, Robert J. Infinitesimal rigidity for smooth actions of discrete subgroups of Lie groups. *J. Differential Geom.* **31** (1990), No. 2, 301–322.

DAVID FISHER
DEPARTMENT OF MATHEMATICS
RAWLES HALL
INDIANA UNIVERSITY
BLOOMINGTON, IN 47405
fisherdm@indiana.edu

Recent Progress in Dynamics
MSRI Publications
Volume **54**, 2007

Le lemme d'Ornstein–Weiss
d'après Gromov

FABRICE KRIEGER

Dédié à Anatole Katok pour son 60ème anniversaire

RÉSUMÉ. Dans cette note on démontre un théorème de convergence pour les fonctions sous-additives invariantes définies sur les parties finies d'un groupe dénombrable moyennable. Ce théorème peut être déduit d'un résultat général dû à D. S. Ornstein et B. Weiss. La démonstration que l'on présente ici suit une preuve esquissée par M. Gromov.

1. Introduction

Soit G un groupe dénombrable. On note $\mathscr{P}(G)$ l'ensemble des parties de G.

DÉFINITION 1.1. On dit que G est *moyennable* s'il existe une fonction

$$\mu : \mathscr{P}(G) \to [0, 1]$$

vérifiant les conditions suivantes :

(i) $\mu(G) = 1$,

(ii) $\mu(A \cup B) = \mu(A) + \mu(B)$ quelles que soient $A, B \in \mathscr{P}(G)$ telles que $A \cap B = \varnothing$,

(iii) $\mu(gA) = \mu(A)$ quels que soient $g \in G$ et $A \in \mathscr{P}(G)$.

Mathematics Subject Classification: 43A07, 20F65, 20F19, 28C10.

Mots clés: Groupe moyennable, fonction sous-additive, lemme d'Ornstein–Weiss, entropie, dimension topologique moyenne.

Notons $\mathscr{F}(G)$ l'ensemble des parties finies non vides de G. D'après un théorème de Følner [Føl], le groupe G est moyennable si et seulement si il existe une suite $(F_i)_{i \in \mathbb{N}}$ d'éléments de $\mathscr{F}(G)$ telle que

$$\lim_{i \to \infty} \frac{|(gF_i) \bigtriangleup F_i|}{|F_i|} = 0 \text{ quel que soit } g \in G,$$

où $A \bigtriangleup B = (A \cup B) \setminus (A \cap B)$ désigne la différence symétrique entre les ensembles A et B, et $|A|$ est le cardinal de A. Une telle suite est appelée une *suite de Følner* de G. Voici quelques résultats concernant la classe des groupes moyennables (voir par exemple [Gre]) :

(i) tout sous-groupe d'un groupe moyennable est moyennable ;

(ii) tout quotient d'un groupe moyennable est moyennable ;

(iii) toute extension d'un groupe moyennable par un groupe moyennable est moyennable ;

(iv) si $G = \bigcup_{i \in \mathbb{N}} G_i$ avec $G_0 \subset G_1 \subset \ldots$ une suite croissante de sous-groupes moyennables de G, alors G est moyennable ;

(v) les groupes finis et $\mathbb{Z}$ sont moyennables.

Le but de cette note est de démontrer le théorème suivant :

THÉORÈME 1.1 (ORNSTEIN–WEISS). *Soit G un groupe dénombrable moyennable et $h : \mathscr{F}(G) \to \mathbb{R}$ une fonction vérifiant les propriétés suivantes :*

(a) *h est sous-additive, c'est-à-dire*

$$h(A \cup B) \le h(A) + h(B) \text{ quelles que soient } A, B \in \mathscr{F}(G) \, ;$$

(b) *h est invariante à droite, c'est-à-dire*

$$h(Ag) = h(A) \text{ quels que soient } g \in G \text{ et } A \in \mathscr{F}(G).$$

Alors il existe un réel $\lambda = \lambda(G, h) \ge 0$ tel que

$$\lim_{i \to \infty} \frac{h(F_i)}{|F_i|} = \lambda$$

pour toute suite de Følner $(F_i)_{i \in \mathbb{N}}$ de G.

Une démonstration de ce résultat (énoncé avec des hypothèses plus fortes sur la fonction h) à partir d'un théorème dû à Ornstein et Weiss sur les quasi-pavages [OrW, Section I.2, Th. 6] se trouve dans [LiW, Th. 6.1]. Dans [Gro, Section 1.3], Gromov esquisse une preuve directe du théorème 1.1 en laissant au lecteur la vérification de certains passages. Il utilise des notions introduites par Ornstein et Weiss dans [OrW]. La démonstration qui est présentée ici suit l'approche de Gromov.

Le théorème 1.1 est à la base de la construction d'invariants d'actions de groupes moyennables comme l'entropie métrique, l'entropie topologique et la dimension topologique moyenne (voir [Gro], [LiW], [OrW], [CoK]). On trouve dans [Mou] un théorème de convergence de ce type pour des fonctions invariantes vérifiant une condition plus forte que la sous-additivité. Le résultat énoncé dans [Mou] est suffisant pour définir l'entropie métrique d'actions de groupes moyennables.

Plan de l'article. Dans la section 2, on introduit la notion de K-frontière qui permet de donner une autre caractérisation des suites de Følner. La section 3 est consacrée à la démonstration d'un lemme de remplissage (lemme 3.5), résultat qui est à la base de la démonstration du théorème 1.1. Grâce à ce lemme, on construit dans la section 4 un procédé de recouvrement partiel des parties finies $D \subset G$. Les propriétés de ce recouvrement permettent d'obtenir une bonne majoration du rapport $h(D)/|D|$.

Remerciement. Je tiens à remercier Michel Coornaert qui m'a encouragé à écrire cet article.

2. Moyennabilité relative

On définit dans cette section les notions de K-intérieur, K-extérieur et K-frontière d'une partie d'un groupe (voir [OrW]).

Soient K et A des parties d'un groupe G. On appelle K-*intérieur* (resp. K-*extérieur*) de A la partie de G notée $\mathrm{Int}_K(A)$ (resp. $\mathrm{Ext}_K(A)$) formée des éléments $g \in G$ tels que Kg soit contenu dans A (resp. dans $G \setminus A$). On définit la K-*frontière* de A par :

$$\partial_K(A) = G \setminus \big(\mathrm{Int}_K(A) \cup \mathrm{Ext}_K(A) \big).$$

La K-frontière de A est donc l'ensemble des éléments $g \in G$ tels que Kg rencontre à la fois A et $G \setminus A$. La proposition suivante est une conséquence immédiate de la définition de la K-frontière :

PROPOSITION 2.1. *Soient K, A, B des parties d'un groupe G et $g \in G$. Alors on a*

(i) $\partial_K(A) = \partial_K(G \setminus A)$;

(ii) $\partial_K(A \cup B) \subset \partial_K(A) \cup \partial_K(B)$;

(iii) $\partial_K(A \setminus B) \subset \partial_K(A) \cup \partial_K(B)$;

(iv) $\partial_K(A) \subset \partial_{K'}(A)$ si $K \subset K' \subset G$;

(v) $\partial_{Kg}(A) = g^{-1}\partial_K(A)$;

(vi) $\partial_K(Ag) = \partial_K(A)g$. $\square$

Soient K et A des parties finies d'un groupe G. Alors $\partial_K(A)$ est finie. Supposons $A \neq \varnothing$. On appelle *constante de moyennabilité relative de A par rapport à K* le rationnel $\alpha(A, K)$ défini par :

$$\alpha(A, K) = \frac{|\partial_K(A)|}{|A|}.$$

Remarquons que les égalités (v) et (vi) de la proposition 2.1 impliquent

$$\alpha(A, Kg) = \alpha(Ag, K) = \alpha(A, K) \text{ quel que soit } g \in G. \qquad (2\text{--}1)$$

LEMME 2.2. *Soient K et A des parties d'un groupe G. Supposons $K = K^{-1}$ et $1_G \in K$. Alors on a les inclusions suivantes :*

(i) *$(kA) \triangle A \subset \partial_K(A)$ quel que soit $k \in K$;*

(ii) *$\partial_K(A) \subset K((KA) \triangle A)$.*

DÉMONSTRATION. Montrons l'inclusion (i). Soit $g \in (kA) \triangle A$ avec $k \in K$. Alors soit $k^{-1}g \in A$ et $g \in G \setminus A$, soit $g \in A$ et $k^{-1}g \in G \setminus A$. Or k^{-1} et 1_G appartiennent à K. Dans les deux cas, on a donc $Kg \cap A \neq \varnothing$ et $Kg \cap (G \setminus A) \neq \varnothing$. On en déduit $g \in \partial_K(A)$.

Montrons l'inclusion (ii). Soit $g \in \partial_K(A)$. Alors on a $g \in K^{-1}A = KA$ et $Kg \cap (G \setminus A) \neq \varnothing$. Supposons tout d'abord $g \notin A$. Alors $g \in KA \setminus A = (KA) \triangle A \subset K\big((KA) \triangle A\big)$ puisque $1_G \in K$. Supposons maintenant $g \in A$. Comme $Kg \cap (G \setminus A) \neq \varnothing$, il existe $k \in K$ tel que $kg \in KA \setminus A = (KA) \triangle A$. On en déduit $g \in K((KA) \triangle A)$. □

PROPOSITION 2.3. *Soit $(F_i)_{i \in \mathbb{N}}$ une suite de parties finies non vides d'un groupe dénombrable G. Les propriétés suivantes sont équivalentes :*

(a) *La suite (F_i) est une suite de Følner de G ;*

(b) *Pour toute partie finie $K \subset G$, on a*

$$\lim_{i \to \infty} \alpha(F_i, K) = 0.$$

DÉMONSTRATION. Montrons l'implication (a) $\Rightarrow$ (b). Supposons que (F_i) est une suite de Følner de G. Soit K une partie finie de G et définissons $L \subset G$ par

$$L = K \cup K^{-1} \cup \{1_G\}.$$

Alors L est une partie finie de G contenant 1_G et vérifiant $L = L^{-1}$. Soit $i \in \mathbb{N}$. La proposition 2.1.(iv) et le lemme 2.2.(ii) impliquent

$$\partial_K(F_i) \subset \partial_L(F_i) \subset L((LF_i) \triangle F_i).$$

Puisque l'on a

$$(LF_i) \triangle F_i \subset \bigcup_{l \in L}((lF_i) \triangle F_i),$$

on en déduit

$$\alpha(F_i, K) = \frac{|\partial_K(F_i)|}{|F_i|} \leq |L| \sum_{l \in L} \frac{|(l F_i) \triangle F_i|}{|F_i|}.$$

Comme (F_i) est une suite de Følner de G, le membre de droite de la dernière inégalité tend vers 0 lorsque i tend vers l'infini. On a donc $\lim_{i \to \infty} \alpha(F_i, K) = 0$, ce qui démontre (b).

Montrons l'implication (b) $\Rightarrow$ (a). Supposons que la suite (F_i) vérifie la propriété (b). Soient $g \in G$ et $i \in \mathbb{N}$. Posons $K = \{1_G, g, g^{-1}\}$. D'après le lemme 2.2.(i), on a

$$(g F_i) \triangle F_i \subset \partial_K(F_i).$$

Il en résulte

$$\frac{|(g F_i) \triangle F_i|}{|F_i|} \leq \frac{|\partial_K(F_i)|}{|F_i|} = \alpha(F_i, K).$$

Comme $\lim_{i \to \infty} \alpha(F_i, K) = 0$, on en déduit

$$\lim_{i \to \infty} \frac{|(g F_i) \triangle F_i|}{|F_i|} = 0$$

ce qui montre que (F_i) est une suite de Følner de G. $\qquad\square$

3. Le lemme de remplissage

DÉFINITION 3.1. Soient X un ensemble et $\varepsilon > 0$. On dit qu'une famille $(A_i)_{i \in I}$ de parties finies de X est ε-*disjointe* s'il existe une famille $(B_i)_{i \in I}$ de parties disjointes de X telle que $B_i \subset A_i$ et

$$|B_i| \geq (1 - \varepsilon)|A_i|$$

pour tout $i \in I$.

LEMME 3.1. *Soient* X *un ensemble et* $(A_1, A_2, \ldots, A_n)$ *une famille* ε-*disjointe de parties finies de* X. *Alors on a*

$$(1 - \varepsilon) \sum_{i=1}^{n} |A_i| \leq \left| \bigcup_{i=1}^{n} A_i \right|.$$

DÉMONSTRATION. Comme la famille $(A_1, A_2, \ldots, A_n)$ est ε-disjointe, il existe une famille $(B_1, B_2, \ldots, B_n)$ de parties disjointes de X vérifiant $B_i \subset A_i$ et $|B_i| \geq (1 - \varepsilon)|A_i|$ pour tout $1 \leq i \leq n$. On a donc

$$(1 - \varepsilon) \sum_{i=1}^{n} |A_i| \leq \sum_{i=1}^{n} |B_i| = \left| \bigcup_{i=1}^{n} B_i \right| \leq \left| \bigcup_{i=1}^{n} A_i \right|. \qquad\square$$

LEMME 3.2. *Soient K une partie finie d'un groupe G et $0 < \varepsilon < 1$. Soit $(A_i)_{1 \le i \le n}$ une famille ε-disjointe de parties finies non vides de G vérifiant $\alpha(A_i, K) \le \eta$ pour tout $1 \le i \le n$. Alors on a*

$$\alpha\left(\bigcup_{i=1}^n A_i, K \right) \le \frac{\eta}{1-\varepsilon}.$$

DÉMONSTRATION. D'après la proposition 2.1.(ii), on a

$$\partial_K \left(\bigcup_{i=1}^n A_i \right) \subset \bigcup_{i=1}^n \partial_K(A_i).$$

Il en résulte

$$\left| \partial_K \left(\bigcup_{i=1}^n A_i \right) \right| \le \sum_{i=1}^n |\partial_K(A_i)| = \sum_{i=1}^n \alpha(A_i, K)|A_i| \le \eta \sum_{i=1}^n |A_i|.$$

Comme la famille $(A_i)_{1 \le i \le n}$ est ε-disjointe, on a d'après le lemme 3.1

$$(1-\varepsilon) \sum_{i=1}^n |A_i| \le \left| \bigcup_{i=1}^n A_i \right|.$$

On en déduit

$$\alpha\left(\bigcup_{i=1}^n A_i, K \right) = \frac{\left| \partial_K \bigcup_{i=1}^n A_i \right|}{\left| \bigcup_{i=1}^n A_i \right|} \le \frac{\eta}{1-\varepsilon}. \qquad \square$$

LEMME 3.3. *Soient K, A et Ω des parties finies d'un groupe G telles que $A \ne \varnothing$ et $A \subset \Omega$. Supposons qu'il existe $\varepsilon > 0$ tel que $|\Omega \setminus A| \ge \varepsilon|\Omega|$. Alors on a*

$$\alpha(\Omega \setminus A, K) \le \frac{\alpha(\Omega, K) + \alpha(A, K)}{\varepsilon}.$$

DÉMONSTRATION. D'après la proposition 2.1.(iii), on a

$$\partial_K(\Omega \setminus A) \subset \partial_K(\Omega) \cup \partial_K(A).$$

Donc

$$|\partial_K(\Omega \setminus A)| \le |\partial_K(\Omega)| + |\partial_K(A)| = \alpha(\Omega, K)|\Omega| + \alpha(A, K)|A|.$$

Puisque $|\Omega \setminus A| \ge \varepsilon|\Omega| \ge \varepsilon|A|$, on en déduit

$$\alpha(\Omega \setminus A, K) = \frac{|\partial_K(\Omega \setminus A)|}{|\Omega \setminus A|} \le \frac{\alpha(\Omega, K) + \alpha(A, K)}{\varepsilon}. \qquad \square$$

LEMME 3.4. *Soient A et B des parties finies d'un groupe G. On a*

$$\sum_{g \in G} |Ag \cap B| = |A||B|.$$

DÉMONSTRATION. Pour $E \subset G$, notons $\chi_E : G \to \{0, 1\}$ la fonction caractéristique de E. On a

$$\sum_{g \in G} |Ag \cap B| = \sum_{g \in G} \sum_{g' \in G} \chi_{Ag \cap B}(g') = \sum_{g \in G} \sum_{g' \in G} \chi_A(g'g^{-1})\chi_B(g').$$

En échangeant l'ordre de sommation puis par un changement de variable, on obtient

$$\sum_{g \in G} |Ag \cap B| = \sum_{g' \in G} \chi_B(g') \sum_{g \in G} \chi_A(g'g^{-1}) = |B||A|. \qquad \square$$

DÉFINITION 3.2 ((ε, K)-REMPLISSAGE). Soient K et Ω des parties d'un groupe G et $\varepsilon > 0$. Une partie $R \subset G$ est appelée un (ε, K)-*remplissage de* Ω si les conditions suivantes sont vérifiées :

(C1) $R \subset Int_K(\Omega)$;

(C2) la famille $(Kg)_{g \in R}$ est ε-disjointe.

Remarquons qu'un (ε, K)-remplissage peut être vide.

La démonstration du théorème 1.1 repose sur le lemme suivant :

LEMME 3.5 (LEMME DE REMPLISSAGE). *Soient Ω et K des parties finies non vides d'un groupe G. Soit $0 < \varepsilon \leq 1$. Alors il existe une partie finie $R \subset G$ vérifiant les conditions suivantes :*

(a) *R est un (ε, K)-remplissage de Ω ;*

(b) *$\left| \bigcup_{g \in R} Kg \right| \geq \varepsilon(1 - \alpha_0)|\Omega|$, où $\alpha_0 = \alpha(\Omega, K)$ désigne la constante de moyennabilité relative de Ω par rapport à K.*

DÉMONSTRATION. Puisque $K \neq \varnothing$, on peut supposer $1_G \in K$, quitte à remplacer K par Kk_0^{-1} où $k_0 \in K$ et en remarquant que $\alpha(\Omega, K) = \alpha(\Omega, Kk_0^{-1})$ d'après les égalités (2–1).

Comme $1_G \in K$, on a $Int_K(\Omega) \subset \Omega$ et $Ext_K(\Omega) \subset G \setminus \Omega$. On en déduit

$$\Omega \setminus \partial_K(\Omega) = Int_K(\Omega) \tag{3–1}$$

et donc

$$(1 - \alpha_0)|\Omega| \leq |\Omega \setminus \partial_K(\Omega)| = |Int_K(\Omega)|. \tag{3–2}$$

Puisque $Int_K(\Omega) \subset \Omega$, tout ($\varepsilon, K$)-remplissage de Ω est contenu dans Ω et a donc un cardinal majoré par $|\Omega|$. Choisissons $R \subset G$ un (ε, K)-remplissage de Ω de cardinal maximal et posons $A = \bigcup_{g \in R} Kg$. Dans la suite, nous allons démontrer que $|A| \geq \varepsilon(1 - \alpha_0)|\Omega|$, ce qui montrera (b). D'après le lemme 3.4, on a

$$\sum_{g \in Int_K(\Omega)} |Kg \cap A| \leq |K||A|. \tag{3–3}$$

Montrons que

$$\varepsilon|K| \le |Kg \cap A| \quad \text{quel que soit } g \in \text{Int}_K(\Omega). \tag{3-4}$$

Si $g \in R$, alors on a $Kg \cap A = Kg$ et l'inégalité (3–4) est trivialement vérifiée puisque $\varepsilon \le 1$. Supposons maintenant que $g \in \text{Int}_K(\Omega) \setminus R$ et que $|Kg \cap A| < \varepsilon|K|$. Mais alors

$$|Kg \setminus A| > (1 - \varepsilon)|Kg|,$$

ce qui implique que $R \cup \{g\}$ est un (ε, K)-remplissage de Ω et contredit la maximalité du cardinal de R. L'inégalité (3–4) est donc satisfaite. On en déduit

$$\varepsilon|K|\,|\text{Int}_K(\Omega)| \le \sum_{g \in \text{Int}_K(\Omega)} |Kg \cap A|. \tag{3-5}$$

Les inégalités (3–2), (3–3) et (3–5) impliquent alors

$$|A| \ge \varepsilon(1 - \alpha_0)|\Omega|. \qquad \square$$

4. Démonstration du théorème

Avant de démontrer le théorème 1.1, faisons les remarques suivantes :

(i) En choisissant $A = B$ dans l'hypothèse (a) du théorème 1.1, on obtient $h(A) \ge 0$ pour toute partie $A \in \mathcal{F}(G)$.

(ii) Si on montre la convergence de la suite $(h(F_i)/|F_i|)$ pour toute suite de Følner (F_i), alors on aura prouvé le théorème. En effet, si $(A_i)_{i \in \mathbb{N}}$ et $(B_i)_{i \in \mathbb{N}}$ sont des suites de Følner, il en est de même pour la suite

$$(F_i)_{i \in \mathbb{N}} = (A_0, B_0, A_1, B_1, \ldots).$$

L'existence de $\lim_{i \to \infty} h(F_i)/|F_i|$ implique alors que

$$\lim_{i \to \infty} \frac{h(A_i)}{|A_i|} = \lim_{i \to \infty} \frac{h(B_i)}{|B_i|}.$$

DÉMONSTRATION DU THÉORÈME 1.1. Soient $(F_i)_{i \in \mathbb{N}}$ une suite de Følner de G et $\varepsilon \in]0, \frac{1}{2}]$. Posons

$$\lambda = \liminf_{i \to \infty} \frac{h(F_i)}{|F_i|}$$

et remarquons que $\lambda < \infty$ puisque les propriétés de h impliquent $h(A) \le h(1_G)|A|$ pour tout $A \in \mathcal{F}(G)$. Fixons $n \in \mathbb{N}^*$. Alors il existe une suite finie $K_1, K_2, \ldots, K_n$ extraite de la suite (F_i) vérifiant les conditions suivantes :

(C1) $h(K_j)/|K_j| \le \lambda + \varepsilon$ quel que soit $1 \le j \le n$,

(C2) $\alpha(K_j, K_i) \le \varepsilon^{2n}$ quels que soient $1 \le i < j \le n$.

En effet, d'après la définition de λ, il existe une sous-suite $(F_{\varphi(i)})$ de (F_i) vérifiant

$$\frac{h(F_{\varphi(i)})}{|F_{\varphi(i)}|} \leq \lambda + \varepsilon,$$

pour tout $i \in \mathbb{N}$. Comme la suite $(F_{\varphi(i)})$ est aussi une suite de Følner de G, la proposition 2.3 permet alors de construire une sous-suite finie $K_1, K_2, \ldots, K_n$ de $(F_{\varphi(i)})$ vérifiant la condition (C2).

Soit D une partie finie non vide de G telle que

$$\alpha(D, K_j) \leq \varepsilon^{2n} \quad \text{pour tout } 1 \leq j \leq n. \tag{4–1}$$

Nous allons démontrer que pour n assez grand, il existe une famille ε-disjointe dans D formée d'un certain nombre des $K_j g$ (où $1 \leq j \leq n$ et $g \in G$) qui remplissent partiellement D, c'est-à-dire telle que la proportion de D recouverte par ces parties soit supérieure ou égale à $1 - \varepsilon$. On utilisera ensuite ce recouvrement partiel et les propriétés de la fonction h pour montrer que

$$\limsup_{i \to \infty} h(F_i)/|F_i| \leq \lambda,$$

ce qui prouvera le théorème.

Définissons par récurrence finie un procédé de recouvrement partiel de D en au plus n étapes :

Étape 1. Rappelons que l'on $\alpha(D, K_j) \leq \varepsilon^{2n}$ pour tout $1 \leq j \leq n$. D'après le lemme 3.5 appliqué à $\Omega = D$ et à $K = K_n$, il existe $R_n \subset G$ un (ε, K_n)-remplissage fini de D tel que

$$\frac{|\bigcup_{g \in R_n} K_n g|}{|D|} \geq \varepsilon\big(1 - \alpha(D, K_n)\big) \geq \varepsilon(1 - \varepsilon^{2n}).$$

Posons $D_1 = D \setminus \bigcup_{g \in R_n} K_n g$. D'après l'inégalité précédente, on a

$$|D_1| \leq |D|\big(1 - \varepsilon(1 - \varepsilon^{2n})\big). \tag{4–2}$$

On continue le procédé de recouvrement par une récurrence finie de la manière suivante. Posons $D_0 = D$. Supposons que le processus de recouvrement partiel se poursuive jusqu'à l'étape k, où $1 \leq k \leq n - 1$.

Les hypothèses de récurrence au rang k sont :

(H1) $\alpha(D_{k-1}, K_j) \leq (2(k-1) + 1)\varepsilon^{2n-k+1}$ pour tout $1 \leq j \leq n - k + 1$;

(H2) $R_{n-k+1} \subset G$ est un (ε, K_{n-k+1})-remplissage fini de D_{k-1} ;

(H3) en posant

$$D_k = D_{k-1} \setminus \bigcup_{g \in R_{n-k+1}} K_{n-k+1} g,$$

on a

$$|D_k| \leq |D| \prod_{i=0}^{k-1} \left(1 - \varepsilon\left(1 - (2i + 1)\varepsilon^{2n-i}\right)\right).$$

Remarquons que ces hypothèses sont vérifiées pour $k = 1$. Construisons l'étape $k + 1$:

Étape $k + 1$. Si $|D_k| \leq \varepsilon|D_{k-1}|$ alors $|D_k| \leq \varepsilon|D|$ et on arrête le processus de recouvrement. Supposons maintenant que $|D_k| > \varepsilon|D_{k-1}|$. Soit $1 \leq j \leq n - k$. Le lemme 3.3 implique

$$\alpha(D_k, K_j) \leq \frac{\alpha(\bigcup_{g \in R_{n-k+1}} K_{n-k+1}g, K_j)}{\varepsilon} + \frac{\alpha(D_{k-1}, K_j)}{\varepsilon}. \qquad (4\text{–}3)$$

D'après les égalités (2–1) et la condition (C2), on a

$$\alpha(K_{n-k+1}g, K_j) = \alpha(K_{n-k+1}, K_j) \leq \varepsilon^{2n}.$$

Puisque la famille $(K_{n-k+1}g)_{g \in R_{n-k+1}}$ est ε-disjointe, il résulte du lemme 3.2 que l'on a

$$\alpha\left(\bigcup_{g \in R_{n-k+1}} K_{n-k+1}g, K_j\right) \leq \frac{\varepsilon^{2n}}{1 - \varepsilon}.$$

On en déduit en utilisant l'inégalité (4–3) et l'hypothèse de récurrence (H1)

$$\alpha(D_k, K_j) \leq \frac{\varepsilon^{2n}}{(1-\varepsilon)\,\varepsilon} + \frac{\left(2(k-1) + 1\right)\varepsilon^{2n-k+1}}{\varepsilon} \leq (2k + 1)\varepsilon^{2n-k}$$

pour $1 \leq j \leq n-k$. Cette dernière inégalité est (H1) au rang $k+1$. En appliquant le lemme 3.5 à $\Omega = D_k$ et à $K = K_{n-k}$, il existe $R_{n-k} \subset G$ un (ε, K_{n-k})-remplissage fini de D_k tel que

$$\frac{|\bigcup_{g \in R_{n-k}} K_{n-k}g|}{|D_k|} \geq \varepsilon\left(1 - \alpha(D_k, K_{n-k})\right) \geq \varepsilon\left(1 - (2k+1)\varepsilon^{2n-k}\right).$$

En particulier (H2) est vérifiée au rang $k+1$. Posons

$$D_{k+1} = D_k \setminus \bigcup_{g \in R_{n-k}} K_{n-k}g.$$

Alors on a

$$|D_{k+1}| \leq |D_k|\left(1 - \varepsilon\left(1 - (2k+1)\varepsilon^{2n-k}\right)\right).$$

En utilisant l'hypothèse de récurrence (H3) et l'inégalité précédente, on obtient

$$|D_{k+1}| \leq |D| \prod_{i=0}^{k} \left(1 - \varepsilon\left(1 - (2i + 1)\varepsilon^{2n-i}\right)\right).$$

Ceci montre l'inégalité (H3) au rang $k + 1$ et achève la construction de l'étape $k + 1$.

Supposons que le processus de recouvrement partiel se poursuive jusqu'à l'étape n et que $|D_{n-1}| > \varepsilon |D_{n-2}|$. D'après (H3) au rang n, on a

$$|D_n| \leq |D| \prod_{i=0}^{n-1} \left(1 - \varepsilon\left(1 - (2i + 1)\varepsilon^{2n-i}\right)\right). \tag{4–4}$$

On va montrer que pour n assez grand (ne dépendant que de ε) on a $|D_n| \leq \varepsilon |D|$. De l'inégalité (4–4), on en déduit la majoration suivante :

$$|D_n| \leq |D|\left(1 - \varepsilon(1 - (2n - 1)\varepsilon^{n+1})\right)^n. \tag{4–5}$$

Comme $\lim_{i \to \infty}(2i - 1)\varepsilon^{i+1} = 0$ et $\lim_{i \to \infty}(1 - \frac{\varepsilon}{2})^i = 0$, il existe un entier n_0 tel que pour $i \geq n_0$, on a $(2i - 1)\varepsilon^{i+1} \leq \frac{1}{2}$ et $(1 - \frac{\varepsilon}{2})^i \leq \varepsilon$. Si $n \geq n_0$, il résulte de l'inégalité (4–5)

$$|D_n| \leq |D|(1 - \frac{\varepsilon}{2})^n \leq \varepsilon |D|.$$

Supposons à partir de maintenant que l'entier n fixé au début de la démonstration est plus grand que n_0. On vient donc de démontrer que pour toute partie finie D telle que $\alpha(D, K_j) \leq \varepsilon^{2n}$ pour tout $1 \leq j \leq n$, il existe un entier k_0 (où $1 \leq k_0 \leq n$) tel que $|D_{k_0}| \leq \varepsilon |D|$. Plus précisément, la proportion de D recouverte par les parties des familles ε-disjointes

$$(K_n g)_{g \in R_n}, (K_{n-1} g)_{g \in R_{n-1}}, \ldots, (K_{n-k_0+1} g)_{g \in R_{n-k_0+1}},$$

est supérieure ou égale à $1 - \varepsilon$.

Passons à la majoration de $h(D)/|D|$. Pour simplifier posons $J = \{n - k_0 + 1, \ldots, n\}$ et notons $K_j R_j = \bigcup_{g \in R_j} K_j g$ pour tout $j \in J$. Dans la suite on utilisera sans le préciser la sous-additivité et l'invariance à droite de h. Puisque l'on a

$$D = \bigcup_{j \in J} K_j R_j \cup D_{k_0}$$

et

$$|D_{k_0}| \leq \varepsilon |D|,$$

on en déduit

$$\frac{h(D)}{|D|} \leq \frac{h(\bigcup_{j \in J} K_j R_j)}{|D|} + \frac{h(D_{k_0})}{|D|} \leq \frac{h(\bigcup_{j \in J} K_j R_j)}{|D|} + \varepsilon h(1_G). \tag{4–6}$$

Par ailleurs, on a

$$\frac{h(\bigcup_{j \in J} K_j R_j)}{|D|} \leq \sum_{j \in J} \sum_{g \in R_j} \frac{h(K_j g)}{|D|} = \sum_{j \in J} \sum_{g \in R_j} \frac{h(K_j)}{|K_j|} \frac{|K_j g|}{|D|}.$$

En utilisant la condition (C1), on en déduit

$$\frac{h(\bigcup_{j\in J} K_j R_j)}{|D|} \le (\lambda + \varepsilon) \sum_{j\in J} \sum_{g\in R_j} \frac{|K_j g|}{|D|}. \tag{4-7}$$

Remarquons que la famille formée des $K_j g$, où $j \in J$ et $g \in R_j$, est ε-disjointe dans D. Il résulte du lemme 3.1

$$\sum_{j\in J} \sum_{g\in R_j} |K_j g| \le \frac{|D|}{1-\varepsilon}. \tag{4-8}$$

Les inégalités (4–7) et (4–8) impliquent alors

$$\frac{h(\bigcup_{j\in J} K_j R_j)}{|D|} \le \frac{\lambda + \varepsilon}{1-\varepsilon}. \tag{4-9}$$

On a donc d'après les inégalités (4–6) et (4–9)

$$\frac{h(D)}{|D|} \le \frac{\lambda + \varepsilon}{1-\varepsilon} + \varepsilon h(1_G). \tag{4-10}$$

Comme (F_i) est une suite de Følner, il existe $N \in \mathbb{N}$ tel que

$$(i \ge N) \Rightarrow \alpha(F_i, K_j) \le \varepsilon^{2n} \quad \text{pour tout } 1 \le j \le n.$$

En appliquant l'inégalité (4–10) à $D = F_i$ pour $i \ge N$, on en déduit

$$\limsup_{i\to\infty} \frac{h(F_i)}{|F_i|} \le \frac{\lambda + \varepsilon}{1-\varepsilon} + \varepsilon h(1_G).$$

Puisque cette dernière inégalité est vraie pour tout $\varepsilon \in]0, \frac{1}{2}]$, on obtient en faisant tendre ε vers 0

$$\limsup_{i\to\infty} \frac{h(F_i)}{|F_i|} \le \lambda = \liminf_{i\to\infty} \frac{h(F_i)}{|F_i|},$$

ce qui démontre le théorème. $\qquad\square$

References

[CoK] M. Coornaert and F. Krieger. *Mean topological dimension for actions of discrete amenable groups.* Discrete and Continuous Dynamical Systems, **13** :3 (2005), 779–793.

[Føl] E. Følner. *On groups with full Banach mean value.* Math. Scand. **3** (1955), 245–254.

[Gre] F. P. Greenleaf. *Invariant means on topological groups and their applications.* Van Nostrand, New York, 1969.

[Gro] M. Gromov. *Topological invariants of dynamical systems and spaces of holomorphic maps, I.* Math. Phys. Anal. Geom. **2** (1999), 323–415.

[LiW] E. Lindenstrauss and B. Weiss. *Mean topological dimension.* Israel J. Math. **115** (2000), 1–24.

[Mou] J. Moulin Ollagnier. *Ergodic theory and statistical mechanics.* Lecture Notes in Mathematics, **1115**. Springer, Berlin, 1985.

[OrW] D. S. Ornstein and B. Weiss. *Entropy and isomorphism theorems for actions of amenable groups.* J. Analyse Math. **48** (1987), 1–141.

FABRICE KRIEGER
INSTITUT DE RECHERCHE MATHÉMATIQUE AVANCÉE
UNIVERSITÉ LOUIS PASTEUR ET CNRS
7 RUE RENÉ DESCARTES
67084 STRASBOURG CEDEX
FRANCE

Current address:
UNIVERSITÉ DE GENÈVE
SECTION DE MATHÉMATIQUES
2-4, RUE DU LIÈVRE
CASE POSTALE 64
CH-1211 GENÈVE 4
SUISSE
Fabrice.Krieger@math.unige.ch
http://www.unige.ch/math/folks/krieger/

Recent Progress in Dynamics
MSRI Publications
Volume **54**, 2007

Entropy of holomorphic and rational maps: a survey

SHMUEL FRIEDLAND

Dedicated to Anatole Katok

ABSTRACT. We give a brief survey on the entropy of holomorphic self maps f of compact Kähler manifolds and rational dominating self maps f of smooth projective varieties. We emphasize the connection between the entropy and the spectral radii of the induced action of f on the homology of the compact manifold. The main conjecture for the rational maps states that modulo birational isomorphism all various notions of entropy and the spectral radii are equal.

1. Introduction

The subject of the dynamics of a map $f : X \to X$ has been studied by hundreds, or perhaps thousands, of mathematicians, physicists and other scientists in the last 150 years. One way to classify the *complexity* of the map f is to assign to it a number $h(f) \in [0, \infty]$, which called the *entropy* of f. The entropy of f should be an invariant with respect to certain *automorphisms* of X. The complexity of the dynamics of f should be reflected by $h(f)$: the larger $h(f)$ the more complex is its dynamics.

The subject of this short survey paper is mostly concerned with the entropy of a holomorphic $f : X \to X$, where X is a compact Kähler manifold, and the entropy of a rational map of $f : Y \dashrightarrow Y$, where Y is a smooth projective variety. In the holomorphic case the author [12; 13; 14] showed that entropy of f is equal to the logarithm of the spectral radius of the finite dimensional f_* on the total homology group $H_*(X)$ over $\mathbb{R}$.

Mathematics Subject Classification: 28D20, 30D05, 37F, 54H20.

Keywords: holomorphic self maps, rational dominating self maps, dynamic spectral radius, entropy.

Most of the paper is devoted to the rational map $f : Y \dashrightarrow Y$ which can be assumed dominating. In this case we have some partial results and inequalities. We recall three possible definition of the entropy $h_B(f), h(f), h_F(f)$ which are related as follows: $h_B(f) \leq h(f) = h_F(f)$. The analog of the *dynamical homological spectral radius* are given by $\rho_{\text{dyn}}(f_*), e^{\text{lov}(f)}$ and $e^{H(f)}$, where the three quantities can be viewed as the volume growth. It is known that $h_F(f) \leq \text{lov}(f) \leq H(f)$. $H(f)$ is a birational invariant. I.e. let $\hat{Y}$ be a smooth projective variety such that there exists $\iota : Y \dashrightarrow \hat{Y}$ which is a *birational* map. Then $f : Y \dashrightarrow Y$ can be lifted to dominating $\hat{f} := \iota f \iota^{-1} : \hat{Y} \dashrightarrow \hat{Y}$, and $H(f) = H(\hat{f})$. However $h_F(f)$ does not have to be equal to $h_F(\hat{f})$. The main conjecture of this paper are the equalities

$$ h_B(\hat{f}) = h(\hat{f}) = h_F(\hat{f}) = \text{lov}(\hat{f}) = H(\hat{f}) = \rho_{\text{dyn}}(\hat{f}_*), \qquad (1\text{--}1) $$

for some $\hat{f}$ birationally equivalent to f. For polynomial automorphisms of $\mathbb{C}^2$, which are birational maps of $\mathbb{P}^2$, the results of the papers [16; 36; 8] prove the above conjecture for $\hat{f} = f$. Some other examples where this conjecture holds are given in [21; 22].

The pioneering inequality of Gromov $h_F(f) \leq \text{lov}(f)$ [20] uses basic results in entropy theory, Riemannian geometry and complex manifolds. Author's results are using basic results in entropy theory, algebraic geometry and the results of Gromov, Yomdin [38] and Newhouse [31]. From the beginning of 90's the notion of *currents* were introduced in the study of the dynamics of holomorphic and rational maps in several complex variables. See the survey paper [35]. In fact the inequality $\text{lov}(f) \leq H(f)$ proved in [8; 9; 10] and [23], as well as most of the results in are derived [21; 22], are using the theory of currents.

The author believes that in dealing with the notion of the entropy solely, one can cleverly substitute the theory of currents with the right notions of algebraic geometry. All the section of this paper except the last one are not using currents. It seems to the author that to prove the conjecture (1–1) one needs to prove a correct analog of Yomdin's inequality [38].

We now survey briefly the contents of this paper. §2 deals with the entropy of $f : X \to X$, where first X is a compact metric space and f is continuous, and second X is compact Kähler and f is holomorphic. §3 is devoted to the study of three definitions of entropy of a continuous map $f : X \to X$, where X is an arbitrary subset of a compact metric space Y. In §4 we discuss rational dominating maps $f : Y \dashrightarrow Y$, where Y is a smooth projective variety. §5 discusses various notions and results on the entropy of rational dominating maps. In §6 we discuss briefly the recent results, in particular the inequality $\text{lov}(f) \leq H(f)$ which uses currents.

It is impossible to mention all the relevant existing literature, and I apologize to authors of papers not mentioned. It is my pleasure to thank S. Cantat, V. Guedj, J. Propp, N. Sibony and C.-M. Viallet for pointing out related papers.

2. Entropy of continuous and holomorphic maps

The first rigorous definition of the entropy was introduced by Kolmogorov [27]. It assumes that X is a probability space $(X, \mathcal{B}, \mu)$, where f preserves the probability measure μ. It is denoted by $h_\mu(f)$, and is usually referred under the following names: *metric* entropy, *Kolmogorov–Sinai* entropy, or *measure entropy*. $h_\mu(f)$ is an invariant under measure preserving invertible automorphism $A : X \to X$: $h_\mu(f) = h_\mu(A \circ f \circ A^{-1})$.

Assume that X is a compact metric space and $f : X \to X$ a continuous map. Then Adler, Konheim and McAndrew defined the *topological* entropy $h(f)$ [1]. $h(f)$ has a *maximal characterization* in terms of measure entropies f. Let $\mathcal{B}$ be the Borel sigma algebra generated by open set in X. Denote by $\Pi(X)$ the compact space of probability measures on $(X, \mathcal{B})$. Let $\Pi(f) \subseteq \Pi(X)$ be the compact set of all f-invariant probability measures. (Krylov–Bogolyubov theorem implies that $\Pi(f) \neq \varnothing$.) Then the *variational principle* due to Goodwyn, Dinaburg and Goodman [18; 7; 17] states $h(f) = \max_{\mu \in \Pi(f)} h_\mu(f)$. Hence $h(f)$ depends only to the topology induced by the metric on X. In particular, $h(f)$ is invariant under any homeomorphism $A : X \to X$.

The next step is to consider the case where X is a compact smooth manifold and $f : X \to X$ is a differentiable map, say $f \in C^r(X)$, where r is usually at least 1. The most remarkable subclasses of f are strongly hyperbolic maps, and in particular axiom A diffeomorphisms [34]. The dynamics of an Axiom A diffeomorphism on the nonwandering set can be coded as a *subshift of a finite type* (SOFT), hence its entropy is given by the exponential growth of the periodic points of f, i.e., $h(f) = \limsup_{k \to \infty} \frac{\log \mathrm{Fix}\, f^k}{k}$, where $\mathrm{Fix}\, f^k$ the number of periodic points of f of period k.

It is well known in topology that $\mathrm{Fix}\, f^k$ can be estimated below by the Lefschetz number of f^k. Let $H_*(X)$ denote the total homology group of X over $\mathbb{R}$, i.e.,

$$H_*(X) = \bigoplus_{i=0}^{\dim_\mathbb{R} X} H_i(X),$$

the direct sum of the homology groups of X of all dimensions with coefficients in $\mathbb{R}$. Then f induces the linear operator $f_* : H(X) \to H(X)$, where $f_{*,i} : H_i(X) \to H_i(X), i = 0, \ldots, \dim_\mathbb{R} X$. The *Lefschetz number* of f^k is defined as

$$\Lambda(f^k) := \sum_{i=0}^{\dim_\mathbb{R} X} (-1)^i \operatorname{Trace} f_{*,i}^k.$$

Intuitively, $\Lambda(f^k)$ is the algebraic sum of k-periodic points of f, counted with their multiplicities.

Denote by $\rho(f_*)$ and $\rho(f_{*,i})$ the spectral radius of f_* and $f_{*,i}$ respectively. Recall that

$$\rho(f_{*,i}) = \limsup_{k\to\infty} |\operatorname{Trace} f_{*,i}^k|^{1/k}$$

and $\rho(f_*) = \max_{i=0,\dots,\dim_\mathbb{R} X} \rho(f_{*,i})$. Hence

$$\limsup_{k\to\infty} \frac{\log |\Lambda(f^k)|}{k} \le \log \rho(f_*).$$

The arguments in [34] yield that for any f in the subset H of an Axiom A diffeomorphism, (H is defined in [34]), one has the inequality $|\operatorname{Trace} f_{*,i}^k| \le$ Fix f^k for each $= 1, \dots, \dim_\mathbb{R} X$. ($H$ is C^0 dense in $\operatorname{Diff}^r(X)$ [34, Thm 3.1].) Hence for any $f \in H$ one has the inequality [34, Prop 3.3]

$$h(f) \ge \log \rho(f_*). \tag{2--1}$$

It was conjectured in [34] that the above inequality holds for any differentiable f.

Let $\deg f$ be the *topological degree* of $f : X \to X$. Then

$$|\deg f| = \rho(f_{*,\dim_\mathbb{R} X}).$$

Hence $\rho(f_*) \ge |\deg f|$. It was shown by Misiurewicz and Przytycki [30] that if $f \in C^1(X)$ then $h(f) \ge |\deg f|$. However this inequality may fail if $f \in C^0(X)$. The entropy conjecture (2--1) for a smooth f, i.e., $f \in C^\infty(X)$, was proved by Yomdin [38]. Conversely, Newhouse [31] showed that for $f \in C^{1+\varepsilon}(X)$ the *volume growth* of smooth submanifolds of f is an upper bound for $h(f)$. See also a related upper bound in [32].

This paper is devoted to study the entropy of f where X is a complex Kähler manifold and f is either holomorphic map, or X is a projective variety and f is a rational map dominating map. We first discuss the case where f is holomorphic.

Let $\mathbb{P}$ be the complex projective space. Then $f : \mathbb{P} \to \mathbb{P}$ is holomorphic if and only if $f|\mathbb{C}$ is a rational map. Hence $\deg f$ is the cardinality of the set $f^{-1}(z)$ for all but a finite number of $z \in \mathbb{C}$. So $\deg f = \rho(f_*)$ in this case. Lyubich [28] showed that $h(f) = \log \deg f$. Gromov in preprint dated 1977, which appeared as [20], showed that if $f : \mathbb{P}^d \to \mathbb{P}^d$ is holomorphic then $h(f) = \log \deg f$. It is well known in this case $\rho(f_*) = \deg f$.

In [12] the author showed that if X is a complex projective variety and $f : X \rightarrow X$ is holomorphic, then $h(f) = \log \rho(f_*)$. Note that one can view f_* a linear operator on $H_*(X, \mathbb{Z})$, i.e., the total homology group with integer coefficients. Hence f_* can be represented by matrix with integer coefficients. In particular, $\rho(f_*)$ is an algebraic integer, i.e., the entropy is the logarithm of an algebraic integer. (This fact was observed in [3] for certain rational maps.) In [14] the author extended this result to a compact Kähler manifold.

Examples of the dynamics of biholomorphic maps $f : X \rightarrow X$, where X is a compact $K3$ surface which is Kähler but not necessary a projective variety, are given in [6; 29]. See also [9] for higher dimensional examples. In summary, the entropy of a holomorphic self map f of a compact Kähler manifold is determined by the spectral radius of the induced action of f on the total homology of X.

3. Definitions of entropy

In this paper Y will be always a compact matrix space with the metric

$$\text{dist}(\cdot, \cdot) : Y \times Y \rightarrow \mathbb{R}_+.$$

Let $X \subseteq Y$ be a nonempty set, and assume that $f : X \rightarrow X$ is a continuous map with respect the topology induced by the metric dist on X. For $x, y \in X$ and $n \in \mathbb{N}$ let

$$\text{dist}_n(x, y) = \max_{k=0,...,n-1} \text{dist}(f^k(x), f^k(y)).$$

So $\text{dist}_1(x, y) = \text{dist}(x, y)$ and the sequence $\text{dist}_n(x, y), n \in \mathbb{N}$ is nondecreasing. Hence for each $n \in \mathbb{N}$ dist_n is a distance on X. Furthermore, each metric dist_n induces the same topology X as the metric dist. For $\varepsilon > 0$ a set $S \subseteq X$ is called (n, ε) separated if $\text{dist}_n(x, y) \geq \varepsilon$ for any $x, y \in S, x \neq y$. For any set $K \subseteq X$ denote by $N(n, \varepsilon, K) \in \mathbb{N} \cup \{\infty\}$ the maximal cardinality of (n, ε) separated set $S \subseteq K$. Clearly, $N(n_1, \varepsilon, K) \geq N(n_2, \varepsilon, K)$ if $n_1 \geq n_2$, $N(n, \varepsilon_1, K) \geq N(n, \varepsilon_2, K)$ if $0 < \varepsilon_1 \leq \varepsilon_2$, and $N(n, \varepsilon, K_1) \geq N(n, \varepsilon, K_2)$ if $K_1 \supseteq K_2$.

We now discuss a few possible definitions of the entropy of f. Let $K \subseteq X$. Then

$$h(f, K) := \lim_{\varepsilon \searrow 0} \limsup_{n \rightarrow \infty} \frac{\log N(n, \varepsilon, K)}{n}. \tag{3-1}$$

We call $h(f, K)$ the *topological entropy* of $f|K$. (Note that $h(f, K) = \infty$ if $N(n, \varepsilon, K) = \infty$ for some $n \in \mathbb{N}$ and $\varepsilon > 0$.) Equivalently, $h(f, K)$ can be viewed as the *exponential growth* of the maximal number of (n, ε) separated sets (in n).

Clearly $h(f, K_1) \geq h(f, K_2)$ if $X \supseteq K_1 \supseteq K_2$. Then $h(f) := h(f, X)$ is the *topological entropy* of f.

Bowen's definition of the entropy of f, denoted here as $h_B(f)$, is given as follows [37, §7.2]. Let $K \subseteq X$ be a compact set. Then K is a compact set with respect to dist_n. Hence $N(n, \varepsilon, K) \in \mathbb{N}$. Then $h_B(f, X)$ is the supremum of $h(f, K)$ for all compact subsets K of X. I.e.

$$h_B(f, X) = \sup_{K \in X} h(f, K).$$

When no ambiguity arises we let $h_B(f) := h_B(f, X)$. Clearly, if X is compact then $h_B(f) = h(f)$. (It is known that $h(f) \in [0, \infty]$ for a compact X; see [37].)

Since $N(n, \varepsilon, K) \leq N(n, \varepsilon, X)$ for any $K \subseteq X$ it follows that $h(f) \geq h_B(f)$. The following example, pointed out to me by Jim Propp, shows that it is possible that $h(f) > h_B(f)$. Let $Y := \{z \in \mathbb{C}, \, |z| \leq 1\}, X := \{z \in \mathbb{C}, \, |z| < 1\}$ be the closed and the open unit disk respectively in the complex plane. Let $2 \leq p \in \mathbb{N}$ and assume that $f(z) := z^p$. It is well known that $h(f, Y) = \log p$. It is straightforward to show that $h(f, X) = h(f, Y)$. Let $K \subset X$ be a compact set. Let $D(0, r)$ be the closed disk or radius $r < 1$, centered at 0, such that $K \subseteq D(0, r)$. Since $f(D(0, r)) \subseteq D(0, r)$ it follows that $h_B(f, X) \leq h(f, D(0, r)) = 0$.

Our last definition of the entropy of h, denoted by $h_F(f, X)$, or simply $h_F(f)$ is based on the notion of the orbit space. Let $\mathcal{Y} := Y^{\mathbb{N}}$ be the space of the sequences $\mathbf{y} = (y_i)_{i \in \mathbb{N}}$, where each $y_i \in Y$. We introduce a metric on $\mathcal{Y}$:

$$d(\{x_i\}, \{y_i\}) := \sum_{i=1}^{\infty} \frac{\text{dist}(x_i, y_i)}{2^{i-1}}, \quad \{x_i\}_{i \in \mathbb{N}}, \{y_i\}_{i \in \mathbb{N}} \in \mathcal{Y}.$$

Then $\mathcal{Y}$ is a compact metric space, whose diameter is twice the diameter of Y. The *shift* transformation $\sigma : \mathcal{Y} \to \mathcal{Y}$ is given by $\sigma(\{y_i\}_{i \in \mathbb{N}}) = \{y_{i+1}\}_{i \in \mathbb{N}}$. Then $d(\sigma(\mathbf{x}), \sigma(\mathbf{y})) \leq 2d(\mathbf{x}, \mathbf{y})$, that is, σ is a Lipschitz map. Given $x \in X$ then the f-orbit of x, or simply the orbit of x, is the point orb $x := \{f^{i-1}(x)\}_{i \in \mathbb{N}} \in \mathcal{Y}$. Denote by orb $X \subseteq \mathcal{Y}$, the *orbit space*, the set of all f-orbits. Note that $\sigma(\text{orb } x) = \text{orb } f(x)$. Hence $\sigma(\text{orb } X) \subseteq \text{orb } X$. $\sigma|\text{orb } X$, the restriction of σ to the orbit space, is "equivalent" to the map $f : X \to X$. I.e. let $\omega : X \to \text{orb } X$ be given by $\omega(x) := \text{orb } x$. Clearly ω is a homeomorphism. Let $\mathcal{X}$ be the closure of orb X with respect to the metric d defined above. Since $\mathcal{Y}$ is compact, $\mathcal{X}$ is compact. Clearly $\sigma(\mathcal{X}) \subseteq \mathcal{X}$. Then the following diagram is commutative:

$$
\begin{array}{ccc}
X & \xrightarrow{f} & X \\
\omega \downarrow & & \omega \downarrow \\
\mathcal{X} & \xrightarrow{\sigma} & \mathcal{X}
\end{array}
$$

Following [12, §4] we define $h_F(f, X)$ to be equal to the topological entropy of $\sigma|\mathcal{X}$:

$$h_F(f, X) := h(\sigma|\mathcal{X}) = h(\sigma, \mathcal{X}).$$

When no ambiguity arises we let $h_F(f) := h_F(f, X)$. Since the closure of orb X is $\mathcal{X}$, it is not difficult to show that $h_F(f) = h(\sigma, \text{orb } X)$.

Observe first that if X is a compact subset of Y then $h_F(f)$ is the topological entropy $h(f)$ of f. Indeed, since f is continuous and X is compact $\mathcal{X} = \text{orb } X$. Since ω is a homeomorphism, the variational principle implies that $h(f) = h_F(f)$.

We observe next that $h(f) \leq h_F(f)$. Let

$$d_n(\mathbf{x}, \mathbf{y}) := \max_{k=0,\ldots,n-1} d(\sigma^k(\mathbf{x}), \sigma^k(\mathbf{y})).$$

Then $\text{dist}_n(x, y) \leq d_n(\text{orb } x, \text{orb } y)$. Hence $N(n, \varepsilon, X) \leq N(n, \varepsilon, \mathcal{X})$. Hence $h(f) \leq h_F(f)$. The arguments of the proof [22, Lemma 1.1] show that $h(f) = h_F(f)$. (In [22] $h_{\text{top}}^{\text{Bow}}(f)$ is our $h(f)$, and $h_{\text{top}}^{\text{Gr}}(f)$ is the topological entropy with respect to the metric

$$d'(\{x_i\}, \{y_i\}) := \sup_{i \in \mathbb{N}} \frac{\text{dist}(x_i, y_i)}{2^i}.$$

Since d and d' induce the Tychonoff topology on $Y^{\mathbb{N}}$ it follows that $h_{\text{top}}^{\text{Gr}}(f) = h_F(f)$.)

Our discussion of various topological entropies for $f : X \to X$ is very close to the discussion in [25]. The notion of the entropy $h_F(f)$ can be naturally extended to the definition of the entropy of a semigroup acting on X [15]. See [5] for other definition of the entropy of a free semigroup and [11] for an analog of Misiurewicz–Przytycki theorem [30].

4. Rational maps

In this section we use notions and results from algebraic geometry most of which can be found in [19]. Let $\mathbf{z} = (z_0, z_1, \ldots, z_n)$, sometimes denotes as $(z_0 : z_1 : \ldots : z_n)$, be the homogeneous coordinates the n-dimensional complex projective space $\mathbb{P}^n$. Recall that a map $f : \mathbb{P}^n \dashrightarrow \mathbb{P}^n$ is called a rational map if there exists $n + 1$ nonzero coprime homogeneous polynomials $f_0(\mathbf{z}), \ldots, f_n(\mathbf{z})$ of degree $d \in \mathbb{N}$ such that $\mathbf{z} \mapsto f_h(\mathbf{z}) := (f_0(\mathbf{z}), \ldots, f_n(\mathbf{z}))$. Equivalently f lifts to a homogeneous map $f_h : \mathbb{C}^{n+1} \to \mathbb{C}^{n+1}$. The set of singular points of f, denoted by Sing $f \subset \mathbb{P}^n$, sometimes called the *indeterminacy locus* of f, is given by the system $f_0(\mathbf{z}) = \ldots = f_n(\mathbf{z}) = 0$. Sing f is closed subvariety of $\mathbb{P}^n$ of codimension 2 at least. The map f is holomorphic if and only if Sing $f = \varnothing$, i.e., the above system of polynomial equations has only the solution $\mathbf{z} = \mathbf{0}$.

Let Y be an irreducible algebraic variety. It is well known that Y can be embedded as an irreducible subvariety of $\mathbb{P}^n$. For simplicity of notation we will assume that Y is an irreducible variety of $\mathbb{P}^n$. So Y can be viewed as a homogeneous irreducible variety $Y_h \subset \mathbb{C}^{n+1}$, given as the zero set of homogeneous polynomials $p_1(\mathbf{z}) = \ldots = p_m(\mathbf{z}) = 0$. $y \in Y$ is called *smooth* if Y is a complex compact manifold in the neighborhood of y. A nonsmooth $y \in Y$ is called a *singular* point. The set of singular points of Y, denoted by $\mathrm{Sing}\, Y$, is a strict subvariety of Y. Y is called 'smooth if $\mathrm{Sing}\, Y = \varnothing$. Otherwise Y is called *singular*.

Let $f : Y \dashrightarrow Y$ be a rational map. Then one can extend f to $\underline{f} : \mathbb{P}^n \dashrightarrow \mathbb{P}^n$ such that $\mathrm{Sing}\, \underline{f} \cap Y$ is a strict subvariety of Y and $\underline{f}|(Y\backslash\overline{\mathrm{Sing}}\, \underline{f}) = f|(Y\backslash\mathrm{Sing}\, \underline{f})$. $\underline{f}$ is not unique, but the f can be viewed as $\underline{f}|Y$. $\mathrm{Sing}\, f \subset Y$ is the set of the points where f is not holomorphic. $\mathrm{Sing}\, f$ is strict projective variety of X, ($\mathrm{Sing}\, f \subseteq \mathrm{Sing}\, \underline{f}\cap Y$), and each irreducible component of $\mathrm{Sing}\, f$ is at least of codimension 2. The assumption $f : Y \dashrightarrow Y$ means that $\mathbf{w} := f_h(\mathbf{z}) \in Y_h$ for each $\mathbf{z} \in Y_h$. It is known that $Y_1 := \mathrm{Cl}\, f_h(Y_h)$, the closure of $f_h(Y_h)$, is a homogeneous irreducible subvariety of Y. Furthermore either $Y_1 = Y(= Y_0)$, in this case f is called a *dominating* map, or $\dim Y_1 < \dim Y_0$. In the second case the dynamics of $f_0 := f$ is reduced to the dynamics of the rational map $f_1 : Y_1 \dashrightarrow Y_1$. Continuing in the same manner we deduce that there exists a finite number of strictly descending irreducible subvarieties $Y_0 := Y \supsetneq \ldots \supsetneq Y_k$ such that $f_k : Y_k \dashrightarrow Y_k$ is a rational dominating map. (Note that Y_k may be a singular variety.) Thus one needs only to study the dynamics of a rational dominating map $f : Y \dashrightarrow Y$, where Y may be a singular variety.

The next notion is the resolution of singularities of Y and f. An irreducible projective variety Z birationally equivalent to Y if the exists a birational map $\iota : Z \dashrightarrow Y$. Z is called a *blow up* of Y if there exists a birational map $\pi : Z \to Y$ such π is holomorphic. Y is called a blow down of Z. Hironaka's result claims that any irreducible singular variety Y has a smooth blow up Z. Let $f : Y \dashrightarrow Y$ be a rational dominating map. Let Y be a birationally equivalent to Z. Then f lifts to a rationally dominating map $\hat{f} : Z \dashrightarrow Z$. Hence to study the dynamics of f one can assume that $f : Y \dashrightarrow Y$ is rational dominating map and Y is smooth. Hironaka's theorem implies that there exists a smooth blow up Z of Y such that f lifts to a holomorphic map $\tilde{f} : Z \to Y$. Then one has the induced dual linear maps on the homologies and the cohomologies of Y and Z:

$$\tilde{f}_* : \mathrm{H}_*(Z) \to \mathrm{H}_*(Y), \quad \tilde{f}^* : \mathrm{H}^*(Y) \to \mathrm{H}^*(Z).$$

We view the homologies $\mathrm{H}_*(Y), \mathrm{H}_*(Z)$ as homologies with coefficients in $\mathbb{R}$, and hence the cohomologies $\mathrm{H}^*(Y), \mathrm{H}^*(Z)$, which are dual to $\mathrm{H}_*(Y), \mathrm{H}_*(Z)$, as de Rham cohomologies of differential forms. (It is possible to consider these homologies and cohomologies with coefficients in $\mathbb{Z}$ [12].) Recall that the Poincaré duality isomorphism $\eta_Y : H_*(Y) \to H^*(Y)$, which maps a k-cycle to closed $\dim Y - k$ form. ($\eta_Y^* = \eta_Y$.) Then one defines $f^* : H^*(Y) \to H^*(Y)$ and its dual $f_* : H_*(Y) \to H_*(Y)$ as

$$f^* := \eta_Y \pi_* \eta_Z^{-1} \tilde{f}^*, \quad f_* := \tilde{f}_* \eta_Z^{-1} \pi^* \eta_Y.$$

It can be shown that f_*, f^* do not depend on the resolution of f, i.e., on Z. Let $\rho(f_*) = \rho(f^*)$ be the spectral radii of f_*, f^* respectively. (As noted above f_*, f^* can be represented by matrix with integer entries. Hence $\rho(f_*)$ is an algebraic integer.) Then the *dynamical* spectral radius of f_* is defined as

$$\rho_{\mathrm{dyn}}(f_*) = \limsup_{m \to \infty} (\rho((f^m)_*))^{1/m}. \tag{4-1}$$

(Note that $\rho_{\mathrm{dyn}}(f_*)$ is a limit of algebraic integers, so it may not be an algebraic integer.)

Assume that $f : Y \to Y$ is holomorphic. Then f_*, f_* are the standard linear maps on homology and cohomology of Y. So $(f^m)_* = (f_*)^m$, $(f^m)^* = (f^*)^m$ and $\rho_{\mathrm{dyn}}(f_*) = \rho(f_*)$. It was shown by the author that $\log \rho(f_*) = h(f)$ [12]. This equality followed from the observation that $h(f)$ is the *volume growth* induced by f. View Y as a submanifold of $\mathbb{P}^n$, is endowed the induced Fubini–Study Riemannian metric and with the induced Kähler $(1, 1)$ closed form κ. Let $V \subseteq Y$ be any irreducible variety of complex dimension $\dim V \geq 1$. Then the volume of V is given by the Wirtinger formula

$$\mathrm{vol}(V) = \frac{1}{(\dim V)!} \int_V \kappa^{\dim V}(V).$$

Let $L_k \subset \mathbb{P}^n$ be a linear space of codimension k. ($L_0 := \mathbb{P}^n$.) Assume that L_k is in general position. Then $L_k \cap V$ is a variety of dimension $\dim V - k$. For $k < \dim V$ the variety $L_k \cap V$ is irreducible. For $k = \dim V$ the variety $L_k \cap V$ consists of a fixed number of points, independent of a generic L_k, which is called the degree of V, and denoted by $\deg V$. It is well known that $\deg V = \mathrm{vol}(V)$. The homology class of $L_k \cap V$, denoted by $[L_k \cap V]$, is independent of L_k. Since $\mathrm{vol}(L_k \cap V)$ can be expressed in terms of the cup product $\langle [L_k \cap V], [\kappa^{\dim V - k}] \rangle$, or equivalently as $\deg L_k \cap V$, this volume is an *integer*, which is independent of the choice of a generic L_k. Thus the j-th volume growth, of the subvariety $L_{\dim Y - j} \cap Y$ of dimension j, induced by f is given by

$$\beta_j := \limsup_{m \to \infty} \frac{\log \langle (f^m)_*[L_{\dim Y - j} \cap Y], [\kappa^j] \rangle}{m}, \quad j = 1, \ldots, \dim Y,$$

$$H(f) := \max_{j=1,\ldots,\dim Y} \beta_j. \tag{4-2}$$

(See [12, (2)] and [13, (2.8)].) From the equality $\rho(f_*) = \lim_{m \to \infty} \| f_*^m \|^{1/m}$, for any norm $\| \cdot \|$ on $H_*(Y)$, it follows that $H(f) \leq \rho(f_*)$. Newhouse's result [31] claims that $h(f) \leq H(f)$. Combining this inequality with Yomdin's inequality [38] $h(f) \geq \log \rho(f_*)$ we deduced in [12]:

$$H(f) = \log \rho(f) = h(f), \qquad (4\text{--}3)$$

which is a logarithm of an algebraic integer.

Let $K \subset H_*(Y)$ be the cone generated by the homology classes $[V]$ corresponding to all irreducible projective varieties $V \subseteq Y$. Note that $f_*(K) \subseteq K$. Let $H_{*,a}(Y) := K - K \subset H_*(Y)$ be the subspace generated by the homology classes of projective varieties in Y. Then $f_* : H_{*,a}(Y) \to H_{*,a}(Y)$ and denote $f_{*,a} := f_*|H_{*,a}(Y)$. Using the theory of nonnegative operators on finite dimensional cones K, e.g. [4], it follows that $H(f) = \log \rho(f_{*,a})$.

Assume again that $f : Y \dashrightarrow Y$ is rational dominant. Then $f_*(K) \subseteq K$ so $f_* : H_{*,a}(Y) \to H_{*,a}(Y)$ and denote $f_{*,a} := f_*|H_{*,a}*(Y)$. Hence we can define $H(f)$, the volume growth induced by f, as in (4–2) [12; 13]. Similar quantities were considered in [33; 3]. It is plausible to assume that $H(f) = \log \rho_{\mathrm{dyn}}(f_*)$ and we conjecture a more general set of equalities in the next section.

It was shown in [12] that the results of Friedland and Milnor [16] imply the inequalities

$$(f^m)_{*,a} \leq (f_{*,a})^m \quad \text{for all } m \in \mathbb{N}, \qquad (4\text{--}4)$$

for certain polynomial biholomorphisms of $\mathbb{C}^2$, (which are birational maps of $\mathbb{P}^2$.)

It is claimed in [12, p. 367] that if (4–4) holds, the sequence $(\rho((f^m)_{*,a}))^{1/m}$, $m \in \mathbb{N}$ converges. (This is probably wrong. One can show that under the assumption (4–4) for all rational dominant maps $f : Y \dashrightarrow Y$ one has $\rho((f^q)_{*,a})^p \geq \rho((f^{pq})_{*,a})$ for any $p, q \in \mathbb{N}$.) It was also claimed in [12, Lemma 3] that (4–4) holds in general. Unfortunately this result is false, and a counterexample is given in [23, Remark 1.4]. Note that if $f : Y \to Y$ holomorphic then equality in (4–4) holds. Hence all the results of [12] hold for holomorphic maps.

5. Entropy of rational maps

Let $f : Y \dashrightarrow Y$ be a rational dominating map. (We will assume that f is not holomorphic unless stated otherwise.) In order to define the entropy of f we need to find the largest subset $X \subseteq Y \backslash \mathrm{Sing}\, f$ such that $f : X \to X$. Let X_k is the collection of all $x \in Y$ such that $f^j(x) \in Y \backslash \mathrm{Sing}\, f$ for $j = 0, 1, \ldots, k$. Then X_k is open and $Z_k := Y \backslash Y_k$ is a strict subvariety of Y. Clearly $X_k \supseteq X_{k+1}$, $Y_k \subseteq Y_{k+1}$ for $k \in \mathbb{N}$. Then $X := \cap_{k=1}^{\infty} X_k$ is G_δ set. Let κ be the closed

$(1, 1)$-Kähler form on Y. Then $\kappa^{\dim Y}$ is a canonical volume form on Y. Hence $\kappa^{\dim Y}(X) = \kappa^{\dim Y}(Y)$, i.e., X has the full volume.

Since Y is a compact Riemannian manifold, Y is a compact metric space. Thus we can define the three entropies $h(f), h_B(f), h_F(f)$ in §3. So

$$h_B(f) \leq h(f) = h_F(f).$$

Assume that $f : \mathbb{C}^n \to \mathbb{C}^n$ is a polynomial dominating map. Then f lifts to a rational dominating map $f : \mathbb{P}^n \dashrightarrow P^n$, which may be holomorphic. Hence $X \supseteq \mathbb{C}^n$. Assume that f is a proper map of $\mathbb{C}^n$. Recall that one point compactification of $\mathbb{C}^n$, denoted by $\mathbb{C}^n \cup \{\infty\}$, is homeomorphic to the $2n$ sphere S^{2n}. Then f lifts to a continuous map $f_s : S^{2n} \to S^{2n}$. Thus we can define the entropy $h(f_s)$. It is not hard to show that $h(f_s) \leq h_F(f)$.

Let orb $X \subset Y^{\mathbb{N}}$ be the orbit space of f, and let $\mathscr{X}$ be its closure. $\mathscr{X}$ is closely related to the graph construction discussed in [20; 12; 13; 14; 15] and elsewhere. Denote by $\Gamma(f) \subset Y^2$ the closure of the set $\{(x, f(x)), x \in Y \backslash \text{Sing } X\}$ in Y^2. Then $\Gamma(f)$ is an irreducible variety of dimension $\dim Y$ in Y^2. Note that the projection of Γ on the first or second factor of Y in Y^2 is Y. Without a loss of generality we may assume that $\Gamma(f)$ is smooth.

Otherwise let $\pi : Z \to Y$ be a blow up of Y such that $f : Y \dashrightarrow Y$ lifts to a holomorphic map $\tilde{f} : Z \to Y$. Let $\Gamma_1(f) := \{(z, \tilde{f}(z)) : z \in Z\} \subset Z \times Y$. Then $\Gamma_1(f)$ is smooth variety of dimension $\dim Y$. Note that $\hat{\pi} : Z^2 \to Z \times Y$ given by $(z, w) \mapsto (z, \pi(w))$ is a blow up of $Z \times Y$. Lift $\tilde{f}$ to $\hat{f} : Z \dashrightarrow Z$. Then $\Gamma(\hat{f}) \subset Z^2$ is a blow up $\Gamma_1(f)$, hence $\Gamma(\hat{f})$ is smooth.

Let $\Gamma \subset Y^2$ be a closed irreducible smooth variety of dimension $\dim Y$ such that the projection of Γ on the first or second component is Y. Define

$$Y^k(\Gamma) := \{(x_1, \ldots, x_k) \in Y^k, \ (x_i, x_{i+1}) \in \Gamma \text{ for } i = 1, \ldots, k-1\},$$
$$Y^{\mathbb{N}}(\Gamma) := \{(x_1, \ldots, x_k, \ldots) \in Y^{\mathbb{N}}, \ (x_i, x_{i+1}) \in \Gamma \text{ for } i \in \mathbb{N}\}.$$

(here $k = 2, \ldots$). Note that $Y^k(\Gamma)$ is an irreducible variety of dimension $\dim Y$ in Y^k for $k = 2, \ldots$. Note that $Y^{\mathbb{N}}(\Gamma)$ is a σ invariant compact subset of $Y^{\mathbb{N}}$, i.e., $\sigma(Y^{\mathbb{N}}(\Gamma)) \subseteq Y^{\mathbb{N}}(\Gamma)$. Let $h(\Gamma) = h(\sigma | Y^{\mathbb{N}}(\Gamma))$. Y, viewed as a submanifold of $\mathbb{P}^n$, is endowed the induced Fubini–Study Riemannian metric and with the Kähler $(1, 1)$ form κ. Then Y^k has the corresponding induced product Riemannian metric, and Y^k is Kähler, with the $(1, 1)$ form κ_k. Let $\text{vol}(Y^k(\Gamma)) = \kappa_k^{\dim Y}(Y^k(\Gamma))$ be volume of the variety $Y^k(\Gamma)$. Then the volume growth of Γ is given by

$$\text{lov}(\Gamma) := \limsup_{k \to \infty} \frac{\log \text{vol}(Y^k(\Gamma))}{k}. \tag{5-1}$$

The fundamental inequality due to Gromov [20]

$$h(\Gamma) \leq \text{lov}(\Gamma). \tag{5-2}$$

Since the paper of Gromov was not available to the general public until the appearance of [20], the author reproduced Gromov's proof of (5–2) in [13; 14]. Using the above inequality Gromov showed that $h(f) \leq \log \deg f$ for any holomorphic $f : \mathbb{P}^n \to \mathbb{P}^n$.

Let $f : Y \dashrightarrow Y$ be rational dominating. Then $\mathscr{X} = Y^{\mathbb{N}}(\Gamma(f))$. Hence

$$h_F(f) = h(\Gamma(f)). \tag{5-3}$$

If $\Gamma(f)$ is smooth then Gromov's inequality yields that

$$h_F(f) \leq \text{lov}(f) := \text{lov}(\Gamma(f)). \tag{5-4}$$

CONJECTURE 5.1. *Let Y be a smooth projective variety and $f : Y \dashrightarrow Y$ be a rational dominating map. Then there exists a smooth projective variety $\hat{Y}$ and a birational map $\iota : Y \dashrightarrow \hat{Y}$, such that the lifting $\hat{f} : \hat{Y} \dashrightarrow \hat{Y}$ satisfies* (1–1).

We now review briefly certain notions, results and conjectures in [14, S3]. Let $\Gamma \subset Y^2$ be as above, and denote by $\pi_i(\Gamma) \to Y$ the projection of Γ on the i-th component of Y in $Y \times Y$ for $i = 1, 2$. Since $\dim \Gamma = \dim Y$ and $\pi_1(\Gamma) = \pi_2(\Gamma) = Y$, then $\deg \pi_i$ is finite and $\pi_i^{-1}(y)$ consists of exactly $\deg \pi_i$ distinct points for a generic $y \in Y$ for $i = 1, 2$. One can define a linear map $\Gamma_* : H_*(Y) \to H_*(Y)$ given by $\Gamma_* : \pi_1^* \eta_\Gamma^{-1} \pi_2^* \eta_Y$. (This is an analogous definition of f_*, where $f : Y \dashrightarrow Y$ is dominating.) One can show that $\Gamma_*(H_{*,a}(Y)) \subseteq H_{*,a}(Y)$. Let $\Gamma_{*,a} := \Gamma_*|H_{*,a}(Y)$.

$\Gamma \subset Y^2$ is called a *proper* if each π_i is finite to one. Assume that Γ is proper. Then

$$\log \rho(\Gamma_{*,a}) \geq \text{lov}(\Gamma). \tag{5-5}$$

It is conjectured that for a proper Γ

$$\log \rho(\Gamma_{*,a}) = \text{lov}(\Gamma) = h(\Gamma). \tag{5-6}$$

Note that if $f : Y \to Y$ is dominating and holomorphic then $\Gamma(f)$ is proper, $\Gamma_{*,a} = f_*|H_{*,a}(Y)$ and the above conjecture holds.

We close this section with observations and remarks which are not in [14]. Assume that $f : Y \dashrightarrow Y$ be a rational dominating and $Z := \Gamma(f) \subset Y^2$ smooth. Then $\pi_1 : \Gamma(f) \to Y$ is a blow up of Y, and $\pi_2 : \Gamma \to Y$ can be identified with $\tilde{f} : Z \to Y$. It is straightforward to show that $f_* = \Gamma(f)_*$.

It seems to the author that the arguments given in [14, Proof Thm 3.5] imply that (5–5) holds for any smooth variety $\Gamma \subset Y^2$ of dimension $\dim Y$ such that $\pi_1(Y) = \pi_2(Y) = Y$. Suppose that this result is true. Let $f : Y \dashrightarrow Y$ be rational and dominating. Assume that $\Gamma(f) \subset Y^2$ is smooth. Then (5–5) would imply that $\text{lov}(f) \le \log \rho(f_*)$. Applying the same inequality to $(f^k)_*$ and combining it with (5–4) one would able to deduce:

$$h_F(f) \le \text{lov}(f) \le \log \rho_{\text{dyn}}(f_*). \tag{5–7}$$

6. Currents

Many recent advances in complex dynamics in several complex variables were achieved using the notion of a *current*. See for example the survey article [35]. Recall that on an m-dimensional manifold M a current of degree $m - p \ge 0$ is a linear functional on all smooth p-differential forms $\mathcal{D}^p(M)$ with a compact support, where p is a nonnegative integer.

Let $f : Y \dashrightarrow Y$ be a meromorphic dominating self map of a compact Kähler manifold of complex dimension $\dim Y$, with the $(1, 1)$ Kähler form κ. Let $f^*\kappa$ be a pullback of κ. Then $f^*\kappa$ is a current on $Y \backslash \text{Sing } f$. Define the *p-dynamic degree* of f by

$$\lambda_p(f) := \limsup_{k \to \infty} \left(\int_{Y \backslash \text{Sing } f^k} (f^k)^* \kappa^p \wedge \kappa^{\dim Y - p} \right)^{1/k}, \quad p = 1, \dots, \dim Y.$$

It is shown in [8] that the dynamical degrees are invariant with respect to a bimeromorphic map $\iota : Y \dashrightarrow Z$, where Z is a compact Kähler manifold. (See also [23] for the case where Y, Z are projective varieties.) Moreover

$$\text{lov}(f) \le \max_{p=1,\dots,\dim Y} \log \lambda_p(f). \tag{6–1}$$

Assume that Y is a projective variety. It can be shown that the dynamic degree $\lambda_p(f)$ is equal to $e^{\beta_{\dim Y - p}}$ for $p = 1, \dots, \dim Y$, which are defined in (4–2), where $\beta_0 := \beta_{\dim Y}$. Hence

$$H(f) = \max_{p=1,\dots,\dim Y} \log \lambda_p(f), \tag{6–2}$$

where $H(f)$ is defined in (4–2). Thus $H(f)$ can be viewed as the *algebraic entropy* of f [3]. [24, Lemma 4.3] computes $H(f)$ for a large class of automorphisms of $\mathbb{C}^k$, and see also [10; 23]. Combine (5–4) with (6–1) and (6–2) to deduce the inequality $h_F(f) \le H(f)$, which was conjectured in [13, Conjecture 2.9].

Consider the following example $f : \mathbb{C}^2 \to \mathbb{C}^2, (z, w) \mapsto (z^2, w + 1)$ [22, Example 1.4]. Since f is proper we have $f_s : S^4 \to S^4$. Clearly S^4 is the domain of attraction of the fixed point $f_s(\infty) = \infty$. Hence $h(f_s) = 0$. Lift f to $f : \mathbb{P}^2 \dashrightarrow \mathbb{P}^2$. Then f has a singular point $\mathbf{a} := (0, 1, 0)$ and any other point at the line at infinity $(1, w, 0)$ is mapped to a fixed point $\mathbf{b} := (1, 0, 0)$. So $X = \mathbb{P}^2 \backslash \{a\}$, and $\Gamma(f) = \{(\mathbf{z}, f(\mathbf{z})) : \mathbf{z} \in \mathbb{P}^2 \backslash \{a\}\} \cup \{(\mathbf{a}, (z : w : 0)) :$

$(z : w) \in \mathbb{P}\}$, which is equal to the blow up of $\mathbb{P}^2$ at $\mathbf{a}$. On $(\mathbb{P}^2)^{\mathbb{N}}(\Gamma(f))$ σ has two fixed points: $(\mathbf{a}, \mathbf{a}, \ldots), (\mathbf{b}, \mathbf{b}, \ldots)$. The set $\mathscr{X}_0 := ((\mathbf{x}, f(\mathbf{x}), \ldots,) :$ $\mathbf{x} \in A_0 := \{(z, w, 1), |z| \le 1\}\}$ is in the domain of the attraction of $(\mathbf{a}, \mathbf{a}, \ldots)$. The set $(\mathbb{P}^2)^{\mathbb{N}}(\Gamma(f))\backslash(\mathscr{X}_0 \cup \{(\mathbf{a}, \mathbf{a}, \ldots)\}$ is in the domain of the attraction of $(\mathbf{b}, \mathbf{b}, \ldots)$. Hence $h(f) = 0$. Observe that $\hat{f} : (\mathbb{P} \times \mathbb{P}) \to (\mathbb{P} \times \mathbb{P})$, given as $((z : s), (w : t)) \mapsto ((z^2 : s^2), (w + t : t))$, is the lift of f to $(\mathbb{P} \times \mathbb{P})$. $\hat{f}$ is holomorphic and $h(\hat{f}) = H(\hat{f}) = \log 2$. Since $\mathbb{P} \times \mathbb{P}$ is birational to $\mathbb{P}^2$ it follows that $H(f) = \log 2 > h_F(f) = 0$. In particular $h_F(f)$ is not a birational invariant [22]. Note that Conjecture 5.1 is valid for this example. Additional examples in [21; 22] support the Conjecture 5.1.

Assume now that $f : \mathbb{C}^2 \to \mathbb{C}^2$ is a polynomial automorphism, hence f is proper. It is shown in [16] that $h(f_s) = h(f, K)$ for some compact subset of $\mathbb{C}^2$. Furthermore the results of [16] and [36] imply that $h(f, K) = H(f)$. One easily deduce that $H(f) = \rho_{\mathrm{dyn}}(f_*)$. Clearly $h_B(f) \ge h(f, K)$. Then the inequalities $h_F(f) \le \mathrm{lov}(f) \le H(f)$ yield Conjecture 5.1. See [2; 9; 26] for additional results on entropy of certain rational maps.

The inequality (6–1) and its suggested variant (5–7) can be viewed as Newhouse type upper bounds [31] which shows that the volume growth bounds from above the entropy of a rational dominating map. In order to prove Conjecture 5.1 one needs to prove a suitable Yomdin type lower bound for the entropy of f.

References

[1] R. L. Adler, A. G. Konheim and M. H. McAndrew, Topological entropy, *Trans. Amer. Math. Soc.* **114** (1965), 309–311.

[2] E. Bedford and J. Diller, Real and complex dynamics of a family of birational maps of the plane: the golden mean subshift, *Amer. J. Math.* **127** (2005), 595–646.

[3] M. P. Bellon and C.-M. Viallet, Algebraic entropy, *Comm. Math. Phys.* **204** (1999), 425–437.

[4] A. Berman and R. J. Plemmons, *Nonnegative matrices in the mathematical sciences*, Academic Press, New York, 1979.

[5] A. Bufetov, Topological entropy of free semigroup actions and skew-product transformations, *J. Dynam. Control Systems* **5** (1999), 137–143.

[6] S. Cantat, Dynamique des automorphismes des surfaces $K3$, *Acta Math.* **187** (2001), 1–57.

[7] E. I. Dinaburg, A correlation between topological entropy and metric entropy, *Dokl. Akad. Nauk SSSR* **190** (1970), 19–22.

[8] T.-C. Dinh and N. Sibony, Regularization of currents and entropy, *Ann. Sci. École Norm. Sup.* **37** (2004), 959–971.

[9] T. C. Dinh and N. Sibony, Green currents for holomorphic automorphisms of compact Kähler manifolds, *J. Amer. Math. Soc.* **18** (2005), 291–312.

[10] T.-C. Dinh and N. Sibony, Une borne supérieure pour l'entropie topologique d'une application rationnelle, *Ann. of Math. (2)* **161** (2005), 1637–1644.

[11] A. Yu. Fishkin, An analogue of the Misiurewicz–Przytycki theorem for some mappings, (Russian), *Uspekhi Mat. Nauk* **56** (2001), No. 337, 183–184. Translation in *Russian Math. Surveys* **56** (2001), No. 1, 158–159.

[12] S. Friedland, Entropy of polynomial and rational maps, *Ann. of Math. (2)* **133** (1991), 359–368.

[13] S. Friedland, Entropy of rational self-maps of projective varieties, *International Conference on Dynamical Systems and Related Topics*, editor: K. Shiraiwa, *Advanced Series in Dynamical Systems* **9**, 128–140, World Scientific Publishing Co., Singapore, 1991.

[14] S. Friedland, Entropy of algebraic maps, Proceedings of the Conference in Honor of Jean-Pierre Kahane, *J. Fourier Anal. Appl.* (1995), Special Issue, 215–228.

[15] S. Friedland, Entropy of graphs, semigroups and groups, Ergodic theory of Z^d actions (Warwick, 1993–1994), 319–343, *London Math. Soc. Lecture Note Ser.* **228**, Cambridge Univ. Press, Cambridge, 1996.

[16] S. Friedland and J. Milnor, Dynamical properties of plane polynomial automorphisms, *Ergodic Theory Dynam. Systems* **9** (1989), 67–99.

[17] T. N. T. Goodman, Relating topological entropy and measure entropy, *Bull. London Math. Soc.* **3** (1971), 176–180.

[18] L. W. Goodwyn, Topological entropy bounds measure-theoretic entropy, *Proc. Amer. Math. Soc.* **23** (1969), 679–688.

[19] P. Griffiths and J. Harris, *Principles of algebraic geometry*, Wiley Interscience, New York, 1978.

[20] M. Gromov, On the entropy of holomorphic maps, *Enseign. Math.* **49** (2003), 217–235.

[21] V. Guedj, Courants extrémaux et dynamique complexe, *Ann. Sci. École Norm. Sup.* **38** (2005), 407–426.

[22] V. Guedj, Entropie topologique des applications méromorphes, *Ergodic Theory Dynam. Systems* **25** (2005), 1847–1855.

[23] V. Guedj, Ergodic properties of rational mappings with large topological degree, *Ann. of Math. (2)* **161** (2005), 1589–1607.

[24] V. Guedj and N. Sibony, Dynamics of polynomial automorphisms of $\mathbf{C}^k$, *Ark. Mat.* **40** (2002), 207–243.

[25] B. Hasselblatt, Z. Nitecki and J. Propp, Topological entropy for non-uniformly continuous maps, 2005. arXiv:math.DS/0511495.

[26] B. Hasselblatt and J. Propp, Monomial maps and algebraic entropy, 2006. arXiv: mathDS/0604521.

[27] A. N. Kolmogorov, A new metric invariant of transitive dynamical systems and Lebesgue space automorphisms, *Dokl. Acad. Sci. USSR* **119** (1958), 861–864.

[28] M. Yu. Lyubich, Entropy of analytic endomorphisms of the Riemann sphere. *Funktsional. Anal. i Prilozhen.* **15** (1981), 83–84.

[29] C. T. McMullen, Dynamics on $K3$ surfaces: Salem numbers and Siegel disks, *J. Reine Angew. Math.* **545** (2002), 201–233.

[30] M. Misiurewicz and F. Przytycki, Topological entropy and degree of smooth mappings, *Bull. Acad. Polon. Sci. Sér. Sci. Math. Astronom. Phys.* **25** (1977), 573–574.

[31] S. E. Newhouse, Entropy and volume, *Ergodic Theory Dynam. Systems* **8*** (1988), Charles Conley Memorial Issue, 283–299.

[32] F. Przytycki, An upper estimation for topological entropy of diffeomorphisms, *Invent. Math.* **59** (1980), 205–213.

[33] A. Russakovskii and B. Shiffman, Value distribution for sequences of rational mappings and complex dynamics, *Indiana Univ. Math. J.* **46** (1997), 897–932.

[34] M. Shub, Dynamical systems, filtrations and entropy, *Bull. Amer. Math. Soc.* **80** (1974), 27–41.

[35] N. Sibony, Dynamique des applications rationnelles de $\mathbf{P}^k$, *Panor. Synthèses* **8**, 97–185, Soc. Math. France, Paris, 1999.

[36] J. Smillie, The entropy of polynomial diffeomorphisms of C^2, *Ergodic Theory Dynam. Systems* **10** (1990), 823–827.

[37] P. Walters, *An introduction to ergodic theory*, Graduate Texts in Mathematics **79**, Springer, New York, 1982.

[38] Y. Yomdin, Volume growth and entropy, *Israel J. Math.* **57** (1987), 285–300.

SHMUEL FRIEDLAND
DEPARTMENT OF MATHEMATICS, STATISTICS AND COMPUTER SCIENCE
UNIVERSITY OF ILLINOIS AT CHICAGO
CHICAGO, ILLINOIS 60607-7045
UNITED STATES
friedlan@uic.edu

Recent Progress in Dynamics
MSRI Publications
Volume **54**, 2007

Causes of stretching of Birkhoff sums and mixing in flows on surfaces

ANDREY KOCHERGIN

On the Anniversary of Anatole Katok, my Friend and Teacher.

ABSTRACT. We study causes of stretching of Birkhoff sums and study their action in the mixing of various surface flows. In so doing, we succeed in amplifying the result of Khanin and Sinai about mixing in the Arnold's example of flow with nonsingular fixed points on a two-dimensional torus.

1. Introduction

There are three known kinds of mixing flows on two-dimensional surfaces: continuous flows without fixed points on a torus, smooth flows with singular fixed points, and smooth flows with nonsingular fixed points (Arnold's example). However, in the last case mixing arises not on the whole torus but only on an ergodic component.

We suggest a special flow S^t, constructed over a circle rotation or an interval exchange transformation (which we denote by T) and under some positive "roof" function, as an ergodic relative of a Borel measure-preserving flow on a two-dimensional surface. In such special flows the only possible cause of mixing is the difference in the times that various points take to get from the "floor" to the "roof". This can cause, as time passes, a small rectangle to be strongly stretched and almost uniformly distributed along trajectories and hence over the phase space.

The divergence of adjacent points is described via Birkhoff sums of the "roof" function

$$f^r(x) = \sum_{k=0}^{r-1} f(T^k x).$$

This is obvious from the relation $S^t(x, y) = S^{y+t-f^r(x)}(T^r x, 0)$, where (x, y) denotes a point in phase space. Strong and almost uniform distribution of a little rectangle over the phase space is ensured by strong almost uniform stretching of Birkhoff sums for $r \approx t$.

Formally this is described by the next theorem. In order to state it, we introduce, for $x \in \mathbb{T}^1$ and $t > 0$, the notation $\mathcal{R}(t, x)$ for the number of jumps that the point $(x; 0)$ undergoes under the action of S^t over a time t. For any measurable $X \subset \mathbb{T}^1$ we set

$$\mathcal{R}(t, X) = \bigcup_{x \in X} \mathcal{R}(t, x).$$

THEOREM 1 (SUFFICIENT CONDITION FOR MIXING). *Let T be an ergodic circle rotation and suppose $t_0 > 0$. Assume that the following objects are fixed for each $t > t_0$:*

– a finite partial partition ξ_t of the circle $\mathbb{T}^1$ into closed intervals: $\xi_t = \{C\}$ with

$$\lim_{t \to +\infty} \max_{C \in \xi_t} |C| = 0, \qquad \lim_{t \to +\infty} \mu([\xi_t]) = 1,$$

where $[\xi_t]$ denotes the union of elements of ξ_t; and
– positive functions ε and H such that $\varepsilon(t) \to 0$ and $H(t) \to +\infty$ for $t \to +\infty$.

If for each $t > t_0$ for any $C \in \xi_t$ and any $r \in \mathcal{R}(t, [\xi_t])$, the Birkhoff sum $f^r|C$ is $(\varepsilon(t), H(t))$-uniformly distributed, then the special flow constructed over T and under the function f is mixing.

That $f^r|C$ is (ε, H)-uniformly distributed means that the function $f^r|C$ is, in some sense, ε-uniformly distributed in an interval of length no less then H. The exact definition of an ε-uniform distribution varies slightly with the circumstances.

We identify three causes of stretching of Birkhoff sums which we may tentatively call ergodic, resonant and individual. They exert various influences on point divergence in various kinds of mixing flows.

2. Flows without fixed points

It is shown in my paper [5] that if f is of bounded variation, the special flow over the circle rotation with roof function f cannot be mixing. A. Katok [3] generalized this result to special flows over interval exchange transformations.

In this situation one can say that the absence of mixing is a corollary of an ergodic effect, the effect of averaging, which results in the Birkhoff sums f^r, for certain values of r, having relatively small variation on sufficiently large sets.

It is possible to gain mixing for a special flow over a circle rotation and under continuous functions at the expense of a resonance condition on T and f.

The main idea is the following. Let p_n/q_n be the sequence of convergents of the rotation angle ρ of the circle $\mathbb{T}^1 = \mathbb{R}/\mathbb{Z}$. Choose ρ and positive sequences a_n and t_n such that for $n \to +\infty$

$$a_n t_n \nearrow +\infty, \quad a_{n+1} t_n \searrow 0, \quad \frac{a_n q_n}{a_{n+1} q_{n+1}} \to 0, \quad t_n/q_n \to 0.$$

Set

$$u_0(x) = \min(\{x\}, \{1-x\}) - \tfrac{1}{4}, \quad u_n(x) = a_n u_0(q_n x), \qquad (2\text{–}1)$$

$$f = F + \sum_{n=1}^{\infty} u_n, \qquad (2\text{–}2)$$

where F is a positive Lipschitz function on $\mathbb{T}^1$ with unit integral.

THEOREM 2. *The special flow over the rotation of the circle by ρ and under the function f constructed above is mixing.*

For instance, u_0 could be the functions $\sin 2\pi x$ or $\cos 2\pi x$.

The stretching and almost uniform distribution of Birkhoff sums f^r for $r \in (t_n/2, 2t_{n+1})$ is ensured by the term u_n, because it "almost resonates" with the rotation through the angle ρ: the period of u_n is $1/q_n$, and thus $u_n(x + \rho) \approx u_n(x)$, and

$$u_n^r(x) \approx r u_n(x) \quad \text{for } r \leqslant 2t_{n+1} \ll q_{n+1}),$$

which yields vertical stretching of u_n^r in each segment of length $1/q_n$, the requirement $a_n t_n \nearrow +\infty$ guaranteeing strong stretching for $r \in (t_n/2, 2t_{n+1})$ and n sufficiently large, since the amplitude of u_n^r is approximately equal to $r a_n/4$. One can call this effect ergodic.

However, the growth of f^r cannot be ensured indefinitely by a single term. In further iterations, due to the accumulation of errors in the shifts the growth of u_n^r breaks down. For example, if $r \approx q_{n+1}/2$, the shifted function $u_n(T^r x)$ is almost exactly half a period out of phase with $u_n(x)$. Hence the next term u_{n+1} must be charged with the stretching in the next interval $r \in (t_{n+1}/2, 2t_{n+2})$; moreover, for $r \approx t_{n+1}$ the term u_{n+1}^r has to grow enough, and therefore it must be also be taken into account in the estimates.

This fact and the requirement that the function $(u_n^r + u_{n+1}^r)|C$ (where C is an element of the partition) be stretched and almost uniformly distributed for $r \in (t_n/2, 2t_{n+1})$ impose the additional constraints on a_n and q_n given above.

Due to the rapid change of a_n and q_n all other terms can, in effect, be discarded: the preceding ones because their derivatives are relatively small and the following ones because their amplitudes are not yet sufficiently large.

The construction of a mixing special flow over an arbitrary ergodic automorphism and under a continuous function [6] is also based on this two-term model, and the terms in this case are related to the Rokhlin towers.

Using the model above we can construct mixing flows over circle rotations and under the roof functions with additional regularity [9].

THEOREM 3. *For any sufficiently regular modulus of continuity weaker than the Lipschitz condition, there exists a mixing special flow over some circle rotation and under a roof function with this modulus of continuity.*

THEOREM 4. *For any $\gamma \in (0, 1)$ and $\theta > 0$, and any circle rotation through the angle ρ satisfying $q_{n+1} \asymp q_n^{1+\theta}$, there exists a positive function $f \in C^\gamma(\mathbb{T}^1)$ such that the special flow S^t over this circle rotation and under f is mixing with power-rate behavior, that is, there exist an exponent $\beta > 0$, a constant M and a time moment t^* such that*

$$|\mu_2(S^t Q_1 \cap Q_2) - \mu_2(Q_1) \cdot \mu_2(Q_2)| < M t^{-\beta}.$$

for any rectangles Q_1, Q_2 and any $t > t^*$.

For example we'll construct the flow satisfying this theorem.

Let $\theta > 0$ and $0 < \gamma < 1$ be given. Choose ρ such that the sequence q_n of denominators of convergents to ρ satisfies

$$q_{n+1} \asymp q_n^{1+\theta}.$$

Choose a sequence a_n satisfying

$$a_n \asymp q_n^{-\gamma},$$

and then construct f by (2–1) and (2–2). The times for switching from one term to another are given by $t_n = q_n^{\varkappa}$, where $\varkappa$ and the exponent β of the mixing rate are found from some system of inequalities; moreover β depends on θ and γ. For example, if $\theta = 1$ and $\gamma = 1/2$, we may set $\beta = 1/9$, and for $\theta = 0.754$, $\gamma = 0.57$ one may set $\beta \approx 0.118$.

Bassam Fayad cleverly implemented this two-term model in the construction of an analytical mixing special flow over a translation on $\mathbb{T}^2$ [2]. He represents each term in (2–2) as $u_n(x, y) = X_n(x) + Y_n(y)$ and thus arranges the shifts in each direction on $\mathbb{T}^2$ so that two successive terms do not interfere: the terms

$X_n(x)$, $Y_n(y)$, $X_{n+1}(x)$, $Y_{n+1}(y)$ consequently charge with stretching, and functions of different variables change one another; as a result u_{n+1} can be substantially smaller than u_n, and f can be made analytical.

To summarize this section we can say that the mixing in the model above is obtained with a resonant effect which is in a certain sense stronger than the ergodic one. In this case rapid growth is needed for q_n.

The natural question is for which moduli of continuity is it possible to obtain a mixing special flow over a circle rotation by a typical angle. Perhaps it is necessary to construct another model realizing the resonant effect, if possible.

3. Singular fixed points

Another variant of mixing flow on a surface is a smooth flow with singular fixed points. Such a flow is isomorphic to a special flow over an interval exchange with a roof function which is smooth everywhere except the break point of T, which are power singularities of the function.

To describe precisely the effects arising from singular points, we introduce after [7] (with some simplifications) a class of functions, denoted by $\mathcal{F}(a, b)$.

Let $M : (0, 1] \to \mathbb{R}_+$ be a nondecreasing function with $M(1) \geqslant 1$, let and $\omega : (0, 1) \to \mathbb{R}_+$ be a nondecreasing function such that $\lim_{x \to +0} \omega(x) = 0$. We say that $\varphi \in \mathcal{F}_{M,\omega}(a, b)$ if

(1) $\varphi \in C^2(a, b)$,
(2) for any $x, y \in (a, b)$ and any $\theta \in (0, 1)$, if $\theta(x - a) \leqslant y - a \leqslant \dfrac{x - a}{\theta}$ then

$$\frac{\varphi''(x)}{M(\theta)} \leqslant \varphi''(y) \leqslant M(\theta)\varphi''(x),$$

and
(3) for any $x \in (a, b)$,

$$\varphi''(x) \geqslant \frac{1}{(x - a)^2 \omega(x - a)}.$$

Then we set

$$\mathcal{F}(a, b) = \bigcup_{M,\omega} \mathcal{F}_{M,\omega}(a, b).$$

THEOREM 5. *Let T be an ergodic interval exchange of the circle $\mathbb{T}^1$, and let $\bar{x}_1, \ldots, \bar{x}_K$ be a finite set of points containing all the break points of T. Assume that for $x \in \mathbb{T}^1 \setminus \cup_i \bar{x}_i$ we have $f(x) \geqslant c > 0$ and*

$$f(x) = f_0(x) + \sum_{i=1}^{K} (f_i(\{x - \bar{x}_i\}) + g_i(\{\bar{x}_i - x\})),$$

where f_i, $g_i \in \mathcal{F}(0, 1)$, and $f_0 \in C^2(\mathbb{T}^1)$. Then the special flow over T with roof function f is mixing.

(Functions of type $x^\alpha(A + o(1))$ for $\alpha \in (0, 1)$ and $A > 0$ belong to $\mathcal{F}(0, 1)$, so flows with singular fixed points are mixing.)

In this case, for large r, the strong and almost uniform stretching of Birkhoff sums f^r in the interval of continuity (a, b) is provided by two terms having singularities at the points a and b; the other terms do not oppose it. One can say that the mixing in the flow is provided by the *individual* effect of fixed points.

This statement is supported by two following facts (for simplicity we suppose that $a = 0$).

LEMMA 1. *Suppose $\varphi \in \mathcal{F}_{M,\omega}(0, 1)$, $0 < b < 1$, and $0 \leqslant h_j < 1 - b$ for $j = 0, \dots, N$. Then*

$$\sum_{j=0}^{N} \varphi(x + h_j) \in \mathcal{F}_{M,\omega}(0, b).$$

LEMMA 2. *Suppose $\varphi, \psi \in \mathcal{F}_{M,\omega}(0, b)$. If b is small enough then $\varphi(x) + \psi(b - x)|(0, b)$ is almost uniformly distributed in a long enough interval.*

(For exact statements see [7].)

This fact is interesting since it implies that the presence of nonsingular fixed points in the flow on surfaces (or logarithmic singularities of roof function) isn't sufficient for mixing.

4. Functions with logarithmic singularities

We say that a roof function has *logarithmic singularities* if it suffices the next conditions:

(1) f has K singular points $\bar{x}_1, \dots, \bar{x}_K$.

(2) $f \in C^1\left(\mathbb{T}^1 \setminus \bigcup_{i=1}^{K} \bar{x}_i\right)$ and $f(x) \geqslant c > 0$.

(3) For any $i = 1, \dots, K$

$$f'(x) = \frac{1}{\{x - \bar{x}_i\}}(-A_i + o(1)) \quad \text{for} \ x \to \bar{x}_i + 0,$$

$$f'(x) = \frac{1}{\{\bar{x}_i - x\}}(B_i + o(1)) \quad \text{for} \ x \to \bar{x}_i - 0,$$

where A_i, $B_i \geqslant 0$.

We set

$$A = \sum_{i=1}^{K} A_i, \quad B = \sum_{i=1}^{K} B_i.$$

The function f is called *symmetric* if $A = B \neq 0$, *asymmetric* if $A \neq B$, and *strongly asymmetric* if

$$\text{sign}(A_i - B_i) = \text{sign}(A - B) \neq 0 \quad \text{for any } i.$$

For symmetric functions there is the following theorem [8].

THEOREM 6. *If*

$$f(x) = f_0(x) + \sum_{i=1}^{K} \left(A_i \log \frac{1}{\{x - \bar{x}_i\}} + B_i \log \frac{1}{\{\bar{x}_i - x\}} \right),$$

where f_0 has a bounded variation, $A = B$, and ρ allows approximation by rationals with rate $\text{const}/(q^2 \log q)$, then the special flow over the circle rotation by ρ with roof function f is not mixing.

So, the deceleration of points and the stretching of a little rectangle in the neighborhood of a regular fixed point are not sufficient for mixing: the stretching produced by moving on one side of a fixed point is compensated while moving on the other side.

Examples of smooth flows on the two-dimensional torus with nonsingular fixed points appear naturally in Arnold's paper [1]. The phase space of such a flow decomposes into cells bounded by closed separatrices of regular fixed points and filled with periodic orbits, and an ergodic component in which orbits move on one side of a fixed point frequently then on the other. The ergodic component of such a flow is isomorphic to a special flow over a circle rotation and under a roof function with an asymmetric logarithmic singular point. A conjecture about the possibility of mixing in such flows was proposed in [8]. Khanin and Sinai proved it with a certain restriction on the rotation angle.

Due to estimates of the ergodic and resonant effects on the stretching of Birkhoff sums, it is possible to weaken this restriction in the case of an asymmetric function and to prove mixing for any irrational angle in the case of a *strongly asymmetric* function.

Let $\rho = [k_1, \ldots, k_s, \ldots]$ be the expansion ρ in a continued fraction. Let p_s/q_s be the s-th convergent to ρ.

THEOREM 7. [10] *Let f be an asymmetric function with logarithmic singularities and ρ an irrational satisfying*

$$\log k_{n+1} = o \left(\log q_n \right). \tag{$*$}$$

Then the special flow over the circle rotation through the angle ρ under the roof function f is mixing.

In [4], the restriction on ρ is stronger: $k_{n+1} \leqslant \text{const } n^{1+\gamma}$, where $0 < \gamma < 1$. It is easy to show that, if for some $\gamma > 0$, perhaps great than 1, we have $k_{n+1} \leqslant \text{const } n^{1+\gamma}$ for all n, then ρ satisfies (∗).

THEOREM 8. [10] *If f is a strongly asymmetric function with logarithmic singularities and ρ is an arbitrary irrational angle then the special flow over a circle rotation through ρ with roof function f is mixing.*

In a special flow with an asymmetric roof function, a new relationship between the ergodic and resonant effects is detected. To illustrate this we'll describe the main ideas of the proof of the previous two theorems. The full presentation takes about fifty pages.

To estimate the stretching of Birkhoff sums f^r, we estimate their derivatives $(f^r)'$. The idea is that, since

$$\text{v.p.} \int_{\mathbb{T}^1} f'(x)\, d\, x = +\infty \text{ or } -\infty,$$

it would be very nice if "according to the ergodic theorem" almost everywhere $(f^r)' \to +\infty$ or $(f^r)' \to -\infty$, and this would ensure the stretching of Birkhoff sums. Moreover, one would want to give this "proof" for the interval exchange. As we'll see, the ergodic component in the expansion of $(f^r)'$ is really present. But the additional term arising from the frequent return of the orbit to a neighborhood of the singularity can essentially violate the "ergodic theorem for non-summable functions". This additional term is large when ρ is well approximable by rationals, and we call this term "resonant".

For such a flow over a circle rotation one can estimate ergodic, resonant and other terms and prove mixing in the cases stated in the theorems.

Let $u, v : \mathbb{R} \to \mathbb{R}$ be the functions with period 1 defined thus:

$$u(x) = 1/x \qquad \text{if } x \in (0, 1],$$
$$v(x) = 1/(1 - x) \quad \text{if } x \in [0, 1).$$

The point $x_0 = 0$ in $\mathbb{T}^1$ is a singular point of both. One can show, that for every x, except singular points of f^r,

$$(f^r)'(x) = \sum_{i=1}^{K} \left(u^r(x - \bar{x}_i)(-A_i + \alpha_i^-(r, x)) + v^r(x - \bar{x}_i)(B_i + \alpha_i^+(r, x)) \right),$$

$$(4-1)$$

where $|\alpha_i^{\pm}(r, x)| \leqslant \alpha(r)$, $\alpha(r) \to 0$ for $r \to +\infty$.

Let

$$q_s \leqslant r < q_{s+1}, \qquad r = l_s q_s + \ldots + l_0 q_0$$

be the expansion of r by denominators q_n with integer nonnegative coefficients, such that

$$1 \leqslant l_s \leqslant k_{s+1}, \quad 0 \leqslant l_n \leqslant k_{n+1} \text{ for } n = 0, 1, \ldots, s-1,$$
$$l_{n-1} q_{n-1} + \ldots + l_0 q_0 < q_n \text{ for } n = 1, \ldots, s.$$

Then

$$u^r(x) = u^{l_s q_s}(x) + \ldots + u^{l_n q_n}(T^{r_{n+1}} x) + \ldots + u^{l_0 q_0}(T^{r_1} x),$$

where $r_n = l_s q_s + \ldots + l_n q_n$.

We set

$$\Delta_n = \min_{0 \leqslant i < j < q_n} |T^i x_0 - T^j x_0|, \quad \delta_n = q_n \rho - p_n.$$

Let us expand u in two terms:

$$\hat{u}_n(x) = u(x), \quad \check{u}_n(x) = 0 \text{ for } x \in (0, \Delta_n),$$

$$\hat{u}_n(x) = 0, \quad \check{u}_n(x) = u(x) \text{ for } x \notin (0, \Delta_n).$$

Then $u^{l_n q_n}(T^{r_{n+1}} x) = \hat{u}_n^{l_n q_n}(T^{r_{n+1}} x) + \check{u}_n^{l_n q_n}(T^{r_{n+1}} x)$.

One can show that for any x

$$\check{u}_n^{l_n q_n}(x) = l_n q_n \log q_n + P_{l_n q_n}^{-}(T^{r_{n+1}} x), \quad |P_{l_n q_n}^{-}(T^{r_{n+1}} x)| < 4 l_n q_n.$$

The term $l_n q_n \log q_n$ is called the *ergodic* component of $u^{l_n q_n}$. This component does not depend on x and for a given q_n is proportional to l_n.

We'll present the term $\hat{u}_n^{l_n q_n}$ as a sum

$$\hat{u}_n^{l_n q_n}(T^{r_{n+1}} x) = I_n^{-}(x) + Z_n^{-}(x)$$

by the following way. We denote by $x_n^{-}(x)$ the singular point of the function $\hat{u}^{l_n q_n}(T^{r_{n+1}} x)$ nearest to x on its left hand side, if such exists, and set

$$I_n^{-}(x) = u(x - x_n^{-}(x)), \quad Z_n^{-}(x) = \hat{u}_n^{l_n q_n}(T^{r_{n+1}} x) - I_n^{-}(x).$$

If $x_n^{-}(x)$ does not exist, then we set $I_n^{-}(x) = 0$, $Z_n^{-}(x) = 0$. Thus, we obtain the expansion

$$u^{l_n q_n}(T^{r_{n+1}} x) = l_n q_n \log q_n + I_n^{-}(x) + Z_n^{-}(x) + P_n^{-}(T^{r_{n+1}} x).$$

We conditionally call the term $Z_n^{-}(x)$ *resonant*. Its value essentially depends on arrangement of the point x and the singular points of $\hat{u}_n^{l_n q_n}(T^{r_{n+1}} x)$. Its maximal value is approximately $q_{n+1} \log k_{n+1}$, which depends on the precision of the approximation of ρ by p_n/q_n. We take the sum over n of the expansion above and denote

$$e(r) = \sum_{n=1}^{s} l_n q_n \log q_n, \quad Z^-(x) = \sum_{n=0}^{s} Z_n^-(x) \quad P^-(x) = \sum_{n=0}^{s} P_n^-(T^{r_{n+1}}x).$$

It is not difficult to show that

$$\sum_{n=0}^{s} I_n^-(x) = I^-(x) + I_{rem}(x),$$

where $0 \leqslant I^-(x) \leqslant 2/\{x^-(x) - x\}, 0 \leqslant I_{rem}(x) \leqslant 2q_s(\log s + 1)$.

Similar estimates are valid for v^r. Thus we get

$$u^r(x) = e(r) + Z^-(x) + I^-(x) + o(e(r)),$$
$$v^r(x) = e(r) + Z^+(x) + I^+(x) + o(e(r)),$$

where

$$Z^+(x) = \sum_{n=0}^{s} Z_n^+(x), \quad Z^-(x) = \sum_{n=0}^{s} Z_n^-(x).$$

For $|P^\pm(x)|$ and $I^\pm(x)$ we have the estimates

$$|P^\pm(x)| \leqslant 4r, \quad 0 < I^\pm(x) < \frac{2}{|x^\pm(x) - x|},$$

where $x^-(x)$ and $x^+(x)$ are the nearest singular points of u^r and v^r respectively to x on its left and right hand sides respectively.

In the expansion given, all three components are present: ergodic $e(r)$, resonant $Z^\pm$ and individual $I^\pm$. The last becomes inessential after a slight restriction of the set on which it is considered. The ergodic component would be enough for stretching of Birkhoff sums if the resonant terms "wouldn't oppose" or "would help" it. It turns out that in the case $\log k_{n+1} = o(\log q_n)$ the resonant terms are small in comparison with the ergodic term, and in the case of a strongly asymmetric function they go with the ergodic term on a large set.

We consider this assertion more explicitly. Choose a sequence σ_n, $n \in \mathbb{Z}_+$, depending on ρ and satisfying the conditions

$$\sigma_n \searrow 0; \quad \sigma_n > (\log q_n)^{-1/4} \quad (\text{for } n > 1), \quad \sigma_n^2 \log q_n \nearrow +\infty.$$

If $\log k_{n+1} = o(\log q_n)$, it is easy to show that σ_n can satisfy an additional condition $\log k_{n+1} \leqslant \sigma_n^2 \log q_n$.

Fix t large enough and choose m such that $\sqrt{2}q_m \leqslant t < \sqrt{2}q_{m+1}$. Define

$$V_m = \{x : |x - T^{-j}x_0| \geqslant 3\sigma_m/q_m \qquad \text{for } j = 0, 1, \ldots, 2q_m - 1\},$$
$$V'_m = \{x : |x - T^{-j}x_0| \geqslant 3\sigma_m/q_{m+1} \quad \text{for } j = 0, 1, \ldots, 2q_{m+1} - 1\}.$$

Also set

$$m' = m, \ V(t) = V_m, \text{ for } \sqrt{2}q_m \leqslant t < \sqrt{2}\sigma_m q_{m+1},$$
$$m' = m + 1, \ V(t) = V'_m, \text{ for } \sqrt{2}\sigma_m q_{m+1} \leqslant t < \sqrt{2}q_{m+1}.$$

The set $V(t)$ consists of disjoint closed intervals (or isolated points); the number of these intervals is no more than $2q_{m'}$; $\mu\,(V(t)) > 1 - 12\sigma_m$; the length of each interval is no more $2/q_{m'}$.

It is not difficult to show that for any $x \in V(t)$ and any $r \in (t/\sqrt{2}, \sqrt{2}t)$ the set $X(r)$ of singular points of u^r and v^r together with their $\sigma_m/q_{m'}$-neighborhoods does not intersect $V(t)$, and thus

$$I^{\pm}(x) < \frac{2}{\sigma_m^2 \log q_m} e(r).$$

(Note that $\sigma_m^2 \log q_m \to +\infty$ for $m \to +\infty$.)

One more object is necessary to describe the properties of the resonant terms. We decompose the set $X^{(n)} = X^{(n)}(r)$ of singular points of $u^{l_n q_n}(T^{r_{n+1}}x)$ into subsets

$$X_i^{(n)} = \{T^{-r_{n+1}-i-jq_n}, \ j = 0, \ldots, l_n - 1\}, \quad i = 0, \ldots, q_n - 1,$$

which we call *clusters* of rank n. Each cluster consists of l_n points, producing an arithmetic progression with the step δ_n. By $[X_i^{(n)}]$ we denote the minimal segment containing $X_i^{(n)}$, $[X^{(n)}] = \bigcup_i [X_i^{(n)}]$, $\partial[X^{(n)}]$ is the bound of $[X^{(n)}]$. The length of each segment is

$$|X_i^{(n)}| = (l_n - 1)|\delta_n| \approx \frac{l_n/k_{n+1}}{q_n}, \quad \mu([X^{(n)}]) \approx l_n/k_{n+1}.$$

The Segments $[X_i^{(n)}]$ are disjoint, so $\partial[X^{(n)}]$ is the union of the ends of $[X_i^{(n)}]$.

For the set W, we define $U(\varepsilon, W) = \bigcup_{x \in W} U(\varepsilon, x)$, where $U(\varepsilon, x)$ is ε-neighborhood of x.

THEOREM 9 (ABOUT THE MAIN RESONANT TERM). *For m sufficiently large, there are the following possible situations.*

(1) *The main resonant term is absent: for any $s < m$*

$$q_{s+1} \log k_{s+1} \leqslant \sigma_m t \log q_m$$

and additionally $\log k_{m+1} \leqslant \sigma_m^2 \log q_m$ *or* $\sqrt{2}q_m \leqslant t < \sqrt{2}\sigma_m q_{m+1}$. *Then for any* $r \in (t/\sqrt{2}, \sqrt{2}t)$ *and* $x \in V(t)$

$$0 \leqslant \sum_n Z_n^{\pm}(x) < 20\sigma_m e(r).$$

(2) *The main resonant term is of rank m: for any* $s < m$ $q_{s+1} \log k_{s+1} \leqslant \sigma_m t \log q_m$, *and*

$$\log k_{m+1} > \sigma_m^2 \log q_m, \quad \sqrt{2}\sigma_m q_{m+1} \leqslant t < \sqrt{2}q_{m+1}.$$

Then for any x

$$\sum_{n \neq m} Z_n^{\pm}(x) < 16\sigma_m e(r),$$

and for $Z_m^{\pm}(x) = Z_m^{\pm}(r, x)$, *when* $r \in (t/\sqrt{2}, \sqrt{2}t)$ *and*

$$x \in V(t) \setminus U(\sigma_m/q_m, \partial[X^{(m)}(r)]),$$

there is an alternative:
— *if* $x \notin [X^{(m)}(r)]$, *then* $Z_m^{\pm}(x) < \sigma_m e(r)$;
— *if* $x \in [X^{(m)}(r)]$, *then* $q_{m+1} \ln k_{m+1} - \sigma_m e(r) < Z_m^{\pm}(x) < q_{m+1} \ln k_{m+1} + \sigma_m e(r)$.

(3) *The main resonant term is of rank* $s < m$: *there exists* $s < m$ *such that* $q_{s+1} \log k_{s+1} > \sigma_m t \log q_m$. *Then for any* $r \in (t/\sqrt{2}, \sqrt{2}t)$ *and* $x \in V(t)$

$$\sum_{n \neq s} Z_n^{\pm}(x) < \sigma_m e(r);$$

for $Z_s^{\pm}(x) = Z_s^{\pm}(r, x)$, *when* $r \in (t/\sqrt{2}, \sqrt{2}t)$ *and*

$$x \in V(t) \setminus U(\sigma_s/q_s, \partial[X^{(s)}(r)]),$$

there is an alternative:
— *if* $x \notin [X^{(s)}(r)]$, *then* $Z_s^{\pm}(x) < \sigma_m e(r)$;
— *if* $x \in [X^{(s)}(r)]$, *then* $q_{s+1} \log k_{s+1} - \sigma_m e(r) < Z_s^{\pm}(x) < q_{s+1} \log k_{s+1} + \sigma_m e(r)$.

Now we may define the functions $\varepsilon(t) \to 0$, $H(t) \to +\infty$ for $t \to +\infty$, for the sufficient condition of mixing (Theorem 1). Let

$$\alpha_t = \max_{r \in (t/\sqrt{2}, \sqrt{2}t)} \alpha(r),$$

$$\varepsilon_e(t) = \frac{21(A+B)}{|B-A|}\sigma_m + \frac{4K}{|B-A|}\alpha_t, \quad \varepsilon_L(t) = \max_{1 \leqslant i \leqslant K} \frac{2}{|A_i - B_i|}\alpha_t,$$

where K is the number of singular points of f, $\alpha(r)$ is the infinitesimal sequence defined in (4–1).

Next set

$$\varepsilon(t) = 2\max(\varepsilon_e(t), \varepsilon_L(t)), \quad H(t) = \frac{|B - A|\sigma_m^2 \log q_m}{4}.$$

For each of the situations (1)–(3) we'll define a partial partition ξ_t such that each element is a segment and for $t \to +\infty$ they satisfies the following conditions:

(1) $\mu([\xi_t]) \to 1$.
(2) $\max_{C \in \xi_t} |C| \to 0$.
(3) For any element $C \in \xi_t$ there exists a constant $L(C) \geqslant 0$, such that for any $r \in \mathcal{R}(t, [\xi_t])$ and $x \in C$,

$$(f^r)'(x) = (B - A)(e(r) + L(C))(1 + \gamma(r, x))), \quad |\gamma(r, x)| < \varepsilon(t)/2. \quad (4\text{--}2)$$

(4) For any element $C \in \xi_t$ and $r \in (t/\sqrt{2}, \sqrt{2}t)$,

$$|C|e(r) > \frac{\sigma_m^2 \log q_m}{2}.$$

Thus we'll verify the sufficient condition for mixing.

In situation 1 we set

$$\overline{V}(t) = \bigcap_{i=1}^{K}(V(t) + \bar{x}_i), \quad L(C) = 0.$$

Then for any $x \in \overline{V}(t)$ and any i $x - \bar{x}_i \in V(t)$, and substituting the expansions of $u^r(x - \bar{x}_i)$ and $v^r(x - \bar{x}_i)$ to (4–1) it is easy to make sure that for any $r \in (t/\sqrt{2}, \sqrt{2}t)$ the relation (4–2) is satisfied.

As elements of the partition ξ_t we take those connected components of $\overline{V}(t)$ whose length is at least $\sigma_m/q_{m'}$.

If $\log k_{n+1} = o(\log q_n)$, then for each t situation 1 is realized.

In situations 2 and 3, the principle of construction of ξ_t is the same, but it is necessary to slightly narrow the set $\overline{V}(t)$. We show how to do it in the situation 3, for situation 2 it is necessary to replace the index s by the index m everywhere.
Let $\bar{r} = \min \mathcal{R}(t, \overline{V}(t))$. Set

$$\widetilde{V}(t) = \left(V(t) \setminus U(\sigma_m/q_s, \partial[X^{(s)}(\bar{r})])\right) \cap (V_s - \bar{r}\rho),$$

$$\overline{\overline{V}}(t) = \bigcap_{i=1}^{K}(\widetilde{V}(t) + \bar{x}_i).$$

As elements of the partition ξ_t we take those connected components of $\overline{\overline{V}}(t)$ whose length is at least $\sigma_m/q_{m'}$.

For $C \in \xi_t$ set

$$L(C) = \left(\sum_{i=1}^{K} \frac{B_i - A_i}{B - A} \chi_i(x) \right) q_{s+1} \log k_{s+1},$$

where χ_i is the indicator function of the set $[X^{(s)}] + \bar{x}_i$, $x \in C$. This definition does not depend on the choice of representative $x \in C$, as it follows the construction of ξ_t. Also the inequality $L(C) \geqslant 0$ follows from the construction and the condition that $\text{sign}(B_i - A_i) = \text{sign}(B - A) \neq 0$.

Note, that the partition ξ_t depends on some fixed $\bar{r}$ in the situations 2 and 3, and thus (4–2) is valid only for r closed to $\bar{r}$, whereas the sufficient condition of mixing requires it for any $r \in \mathcal{R}(t, [\xi_t])$. By estimating the oscillation of f^r in the set $\overline{V}(t)$ it is possible to prove that the range $\mathcal{R}(t, [\xi_t])$ is not too large, and (4–2) is valid for the whole range.

5. Some problems

(1) It is not known whether the restriction $(*)$ on the angle in the theorem 7 is appreciable. For angles which don't satisfy $(*)$, it is possible to construct an asymmetric function f with logarithmic singularities, such that for an unbounded set of moments t and corresponding r, the oscillation of f^r on each element $C \in \xi_t : C \subset ([X^{(m)}]) + \bar{x}_i$ is small since $e(r) + L(C) = o(q_{m+1})$ in the expansion (4–2), but the oscillation of f^r on each set

$$[X_i^{(m)}] \setminus U(\sigma_m, \partial[X_i^{(m)}]) \cap [\xi_t]$$

is large enough; also the distribution of $f^r | ([X_i^{(m)}] \setminus U(\sigma_m, \partial[X_i^{(m)}]))$ is not almost uniform, it is almost discrete. Thus we cannot use the sufficient condition for mixing given above, and yet cannot prove the absence of mixing.

(2) It is not known whether the theorem 6 for symmetric function with logarithmic singularities is right for angles satisfying $k_{n+1} = o(\log q_n)$ (or the same $\log k_{n+1} = o(\log \log q_n)$).

Using techniques from Fourier analysis, M. Lemańczyk has slightly extended the class of functions considered but not the class of angles.

(3) It would be interesting to clarify how for mixing special flows the modulus of continuity of the roof function relates to the speed of rational approximation of the angle of the rotation in the base. Maybe there exist models other than the model of "two terms" described above, in which more terms are simultaneously involved in the stretching of Birkhoff sums.

(4) It is also not known what the maximum rate of mixing is for special flows over ergodic rotations and under continuous roof functions. This question

is interesting in connection with the existence of such flows with Lebesgue spectrum.

(5) Is the presence of only one singular fixed point, even in the presence of other nonsingular fixed points, sufficient for mixing of an ergodic flow on a surface ? It seems the answer should be positive but it has not yet been proved.

(6) Is the mixing in the above flows mixing of all orders ?

References

[1] V. I. Arnold. Topological and ergodic properties of closed 1-forms with incommensurable periods. (Russian). *Funct. Anal. i Prilozhen.* **25**:2 (1991), 1–12, 96. English transl. in *Funct. Anal. Appl.* **25**:2 (1991), 81–90.

[2] B. R. Fayad. Reparamétrage des flots irrationnels sur le tore. Thèse, L'École Polytechniqe, Paris, 2000.

[3] A. B. Katok. Interval exchange transformations and some special flows are not mixing. *Israel. J. Math.* **35** (1980), 301–310.

[4] K. M. Khanin and Y. G. Sinai. Mixing of some classes of special flows over rotations of the circle. *Funct. Anal. i Prilozh.* **26** (1992), 155–169.

[5] A. V. Kochergin. On the absence of mixing in special flows over a rotation of the circle and in flows on a two-dimensional torus. *Doklady Akad. Nauk SSSR* **205** (1972), 515–518. English transl. in *Soviet Math. Dokl.* **13** (1972), 949–952.

[6] A. V. Kochergin. The time change in flows and mixing. *Izv. Akad. Nauk SSSR Ser. Mat.* **37** (1973). English transl. in *Math. USSR Izvestija* **7** (1973), 1272–1294.

[7] A. V. Kochergin. On mixing in special flows over a shifting of segments and in smooth flows on surfaces. *Mat. Sb.* **96 (138)**:3 (1975), 471–502. English transl. in *Mat. USSR-Sb.* **25**:3 (1975), 441–469.

[8] A. V. Kochergin. Nondegenerate saddles and the absence of mixing. *Mat. Zametki* **19**:3 (1976), 453–468. English transl. in *Math. Notes* **19**:3 (1976), 277–286.

[9] A. V. Kochergin. A mixing special flow over a circle rotation with almost Lipschitz function. *Mat. Sb.* **193**:3 (2002), 51–78. English transl. in *Sb. Mat.* **193**:3 (2002), 359–385.

[10] A. V. Kochergin. Nondegenerate fixed points and mixing in flows on two-dimensional torus. *Mat. Sb.* **194**:8 (2003), 83–112. English transl. in *Sb. Math.* **194**:8 (2003), 1195–1224.

[11] M. Lemańczyk. Sur l'absence de mélange pour des flots spéciaux au dessus d'une rotation irrationelle. *Colloq. Math.* **84–85** (2000), 29–41.

ANDREY KOCHERGIN
DEPARTMENT OF ECONOMICS
MOSCOW STATE UNIVERSITY
LENINSKIE GORY
MOSCOW
RUSSIA
  avk@econ.msu.ru

Recent Progress in Dynamics
MSRI Publications
Volume **54**, 2007

Solenoid functions for hyperbolic sets on surfaces

ALBERTO A. PINTO AND DAVID A. RAND

ABSTRACT. We describe a construction of a moduli space of *solenoid functions* for the C^{1+}-conjugacy classes of hyperbolic dynamical systems f on surfaces with hyperbolic basic sets Λ_f. We explain that if the holonomies are sufficiently smooth then the diffeomorphism f is *rigid* in the sense that it is C^{1+} conjugate to a hyperbolic affine model. We present a moduli space of *measure solenoid functions* for all Lipschitz conjugacy classes of C^{1+}-hyperbolic dynamical systems f which have a invariant measure that is absolutely continuous with respect to Hausdorff measure. We extend *Livšic and Sinai's eigenvalue formula* for Anosov diffeomorphisms which preserve an absolutely continuous measure to hyperbolic basic sets on surfaces which possess an invariant measure absolutely continuous with respect to Hausdorff measure.

CONTENTS

1. Introduction

We say that (f, Λ) is a C^{1+} *hyperbolic diffeomorphism* if it has the following properties:

(i) $f : M \to M$ is a $C^{1+\alpha}$ diffeomorphism of a compact surface M with respect to a $C^{1+\alpha}$ structure $\mathscr{C}_f$ on M, for some $\alpha > 0$.

(ii) Λ is a hyperbolic invariant subset of M such that $f|\Lambda$ is topologically transitive and Λ has a local product structure.

We allow both the case where $\Lambda = M$ and the case where Λ is a proper subset of M. If $\Lambda = M$ then f is Anosov and M is a torus [16; 33]. Examples where Λ is a proper subset of M include the Smale horseshoes and the codimension one attractors such as the Plykin and derived-Anosov attractors.

THEOREM 1.1 (EXPLOSION OF SMOOTHNESS). *Let f and g be any two C^{1+} hyperbolic diffeomorphisms with basic sets Λ_f and Λ_g, respectively. If f and g are topologically conjugate and the conjugacy has a derivative at a point with nonzero determinant, then f and g are C^{1+} conjugate.*

See definitions of topological and C^{1+} conjugacies in Section 2.3. A weaker version of this theorem was first proved by D. Sullivan [47] and E. de Faria [8] for expanding circle maps. Theorem 1.1 follows from [13] using the results presented in [1] and in [13] which apply to Markov maps on train tracks and to nonuniformly hyperbolic diffeomorphisms.

For every C^{1+} hyperbolic diffeomorphism f we denote by $\delta_{f,s}$ the Hausdorff dimension of the stable-local leaves of f intersected with Λ, and we denote by $\delta_{f,u}$ the Hausdorff dimension of the unstable-local leaves of f intersected with Λ. Let $\mathscr{P}$ be the set of all periodic points of Λ under f. For every $x \in \mathscr{P}$, let us denote by $\lambda_{f,s}(x)$ and $\lambda_{f,u}(x)$ the stable and unstable eigenvalues of the periodic orbit containing x. A. Livšic and Ya. Sinai [25] proved that an Anosov diffeomorphism f admits an f-invariant measure that is absolutely continuous with respect to the Lebesgue measure on M if, and only if, $\lambda_{f,s}(x)\lambda_{f,u}(x) = 1$ for every periodic point $x \in \mathscr{P}$. In Theorem 1.1 of [42], it is proved the following extension of Livšic and Sinai's Theorem to C^{1+} hyperbolic diffeomorphisms with hyperbolic sets on surfaces such as Smale horseshoes and codimension one attractors.

THEOREM 1.2 (LIVŠIC AND SINAI'S EXTENDED FORMULA). *A C^{1+} hyperbolic diffeomorphism f admits an f-invariant measure that is absolutely continuous with respect to the Hausdorff measure on Λ if, and only if, for every periodic point $x \in \mathscr{P}$,*

$$\lambda_{f,s}(x)^{\delta_{f,s}}\lambda_{f,u}(x)^{\delta_{f,u}} = 1 \ .$$

By the *flexibility* of a given topological model of hyperbolic dynamics we mean the extent of different smooth realizations of this model. Thus a typical result provides a moduli space to parameterise these realizations. To be effective it is important that these moduli spaces should be easily characterised. For example, for C^2 Anosov diffeomorphisms of the torus that preserve a smooth invariant measure, the eigenvalue spectrum is a complete invariant of smooth conjugacies as shown by De la Llave, Marco and Moriyon [26; 27; 30; 31]. However, for hyperbolic systems on surface other than Anosov systems the eigenvalue spectra are only a complete invariant of Lipschitz conjugacy (see [42]). In [39], the notions of a *HR-structure* and of a *solenoid function* are used to construct the moduli space.

Consider affine structures on the stable and unstable lamination in Λ. These are defined in terms of a pair of *ratio functions* r^s and r^u (see Section 3.1). If r^s and r^u are Hölder continuous and invariant under f then we call the associated structure a HR-structure (HR for Hölder-ratios). Theorem 5.1 in [39] gives a one-to-one correspondence between HR-structures and C^{1+} conjugacy classes of $f|\Lambda$ (see Theorem 3.1). The main step in the proof of this and related results is to show that, given a HR-structure, there is a canonical construction of a representative in the corresponding conjugacy class. By Theorem 5.3 in [39], this representative has the following maximum smoothness property: the holonomies for the representative are as smooth as those of any diffeomorphism that is C^{1+} conjugate to it. In particular, if there is an affine diffeomorphism with this HR-structure, then this representative is the affine diffeomorphism. In Section 3.9, we present the definition of *stable and unstable solenoid functions* and we introduce the set $\mathcal{PS}(f)$ of all pairs of solenoid functions. To each HR-structure one can associate a pair (σ^s, σ^u) of solenoid functions corresponding to the stable and unstable laminations of Λ, where the solenoid functions σ^s and σ^u are the restrictions of the ratio functions r^s and r^u, respectively, to a set determined by a Markov partition of f. Theorem 6.1 in [39] says that there is a one-to-one correspondence between Hölder solenoid function pairs and HR-structures (see Theorem 3.4). Since these solenoid function pairs form a nice space with a simply characterised completion they provide a good moduli space. For example, in the classical case of Smale horseshoes the moduli space is the set of all pairs of positive Hölder continuous functions with domain $\{0, 1\}^{\mathbb{N}}$.

Let $\mathcal{T}(f, \Lambda)$ be the set of all C^{1+} hyperbolic diffeomorphisms (g, Λ_g) such that (g, Λ_g) and (f, Λ) are topologically conjugate (See definitions of topological and C^{1+} conjugacies in Section 2.3).

THEOREM 1.3 (FLEXIBILITY). *The natural map $c : \mathcal{T}(f, \Lambda) \to \mathcal{PS}(f)$ which associates a pair of solenoid functions to each C^{1+} conjugacy class is a bijection.*

The solenoid functions were first introduced in [36; 39] inspired by the scaling functions introduced by M. Feigenbaum [10; 11] and D. Sullivan [48]. The completion of the image of c is the set of pairs of continuous solenoid functions which is a closed subset of a Banach space. They correspond to f-invariant affine structures on the stable and unstable laminations for which the holonomies are uniformly asymptotically affine (uaa) as defined in [47].

In [41], the moduli space of solenoid functions is used to study the existence of *rigidity* for diffeomorphisms on surfaces. In dynamics, rigidity occurs when simple topological and analytical conditions on the model system imply that there is no flexibility and so a unique smooth realization. One can paraphrase this by saying that the moduli space for such systems is a singleton. For example, a famous result of this type due to Arnol'd, Herman and Yoccoz [3; 20; 51] is that a sufficiently smooth diffeomorphism of the circle with an irrational rotation number satisfying the usual Diophantine condition is C^{1+} conjugate to a rigid rotation. The rigidity depends upon both the analytical hypothesis concerning the smoothness and the topological condition given by the rotation number and if either are relaxed then it fails. The analytical part of the rigidity hypotheses for hyperbolic surface dynamics will be a condition on the smoothness of the holonomies along stable and unstable manifolds.

THEOREM 1.4 (RIGIDITY). *If f is a C^r diffeomorphism with a hyperbolic basic set Λ and the holonomies of f are C^r with uniformly bounded C^r norm and with $r-1$ greater than the Hausdorff dimension along the stable and unstable leaves intersected with Λ then f is C^{1+} conjugated to a hyperbolic affine model.*

See the definition of a *hyperbolic affine model* in Section 4.2. Theorem 4.1 contains a slightly stronger version of Theorem 1.4 using the notion of a HD^l *complete set of holonomies*. Both theorems are proved in [41]. In these theorems we allow both the case where $\Lambda = M$ (so that f is Anosov and $M \cong \mathbb{T}^2$ [16; 33]) and the case where Λ is a proper subset. In the case of the Smale horseshoe f, as presented in Figure 8, the hyperbolic affine maps $\hat{f}$ topologically conjugate to f, up to affine conjugacy, form a two-dimension set homeomorphic to $\mathbb{R}^+ \times \mathbb{R}^+$. In the case of hyperbolic attractors with $HD^s < 1$, there are no affine maps as proved in [14]. Hence, Theorem 1.4 implies that the stable holonomies can never be smoother than $C^{1+\alpha}$ with α greater than the Hausdorff dimension along the stable leaves intersected with Λ (see [14]). This result is linked with J. Harrison's conjecture of the nonexistence of $C^{1+\alpha}$ diffeomorphisms of the circle with $\alpha > HD$, where HD is the Hausdorff dimension of its nonwandering domain. A weaker version of this conjecture was proved by A. Norton [35] using box dimension instead of Hausdorff dimension. In the case of Anosov diffeomorphisms of the torus, the hyperbolic affine model is a hyperbolic toral automorphism and is unique up to affine conjugacy [15; 16; 29; 33]. In general, the topological conjugacy between such a diffeomorphism and the corresponding hyperbolic affine model is only Hölder continuous and need not be any smoother. This is the case if there is a periodic orbit of f whose eigenvalues differ from those of the hyperbolic affine model. For Anosov diffeomorphisms f of the torus there are the following results, all of the form that if a C^k f has

C^r foliations then f is C^s-*rigid*, i.e. f is C^s-conjugate to the corresponding hyperbolic affine model:

(i) Area-preserving Anosov maps f with $r = \infty$ are C^∞-rigid (Avez [4]).

(ii) C^k area-preserving Anosov maps f with $r = 1 + o(t \mid \log t \mid)$ are C^{k-3}-rigid (Hurder and Katok [22]).

(iii) C^1 area-preserving Anosov maps f with $r \geq 2$ are C^r-rigid (Flaminio and Katok [17]).

(iv) C^k Anosov maps f $(k \geq 2)$ with $r \geq 1 + \text{Lipshitz}$ are C^k-rigid (Ghys [18]).

The moduli space of solenoid functions is used in [42] to construct classes of smooth hyperbolic diffeomorphisms with an invariant measure μ absolutely continuous with respect to the Hausdorff measure. It is interesting to note that when we consider the C^{1+} hyperbolic diffeomorphisms realising a particular topological model then the stable and unstable ratio functions are indendent in the following sense. If r^s is a stable ratio function for some hyperbolic diffeomorphism and r^u is the unstable ratio function for some other hyperbolic diffeomorphism then there is a hyperbolic diffeomorphism that has the pair (r^s, r^u) as its HR structure. The same is no longer true if we ask the C^{1+} hyperbolic realizations to have an invariant measure μ absolutely continuous with respect to the Hausdorff measure. For $\iota \in \{s, u\}$, let us denote by ι' the element of $\{s, u\}$ which is not ι.

THEOREM 1.5 (MEASURE RIGIDITY FOR ANOSOV DIFFEOMORPHISMS). *For $\iota \in \{s, u\}$, given an ι-solenoid function σ_ι there is a unique ι'-solenoid function such that the C^{1+} Anosov diffeomorphisms determined by the pair (σ_s, σ_u) satisfy the property of having an invariant measure μ absolutely continuous with respect to Lebesgue measure of their hyperbolic sets.*

In the case of Smale horseshoes, the ι'-solenoid function is not anymore unique but belongs to a well-characterized set which is the δ-solenoid equivalence class of the Gibbs measure determined by the ι-solenoid function (see Section 5.6).

THEOREM 1.6 (MEASURE FLEXIBILITY FOR SMALE HORSESHOES). *For $\iota \in \{s, u\}$, given an ι-solenoid function σ_ι there is an infinite dimensional space of solenoid functions $\sigma_{\iota'}$ (but not all) such that the C^{1+} hyperbolic Smale horseshoes determined by the pairs (σ_s, σ_u) have the property of having an invariant measure μ absolutely continuous with respect to the Hausdorff measure of their hyperbolic sets.*

Codimension one attractors partly inherit the properties of Anosov diffeomorphisms and partly those of Smale horseshoes because locally they are a product of lines with Cantor sets embedded in lines.

THEOREM 1.7. (i) (MEASURE FLEXIBILITY FOR CODIMENSION ONE AT-
TRACTORS) *Given an u-solenoid function σ_u there is an infinite dimensional
space of s-solenoid functions σ_s (but not all) such that the C^{1+} hyperbolic
codimension one attractors determined by the pairs (σ_s, σ_u) have the property of having an invariant measure μ absolutely continuous with respect to
the Hausdorff measure of their hyperbolic sets.*

(ii) (MEASURE RIGIDITY FOR CODIMENSION ONE ATTRACTORS) *Given an
s-solenoid function σ_s there is a unique u-solenoid function σ_u such that the
C^{1+} hyperbolic codimension one attractors determined by the pair (σ_s, σ_u)
have the property of having an invariant measure μ absolutely continuous
with respect to the Hausdorff measure of their hyperbolic sets.*

Theorem 5.9 contains a stronger version of Theorems 1.5, 1.6 and 1.7, and it is
proved in Lemmas 8.17 and 8.18 in [42].

Since (f, Λ) is a C^{1+} hyperbolic diffeomorphism it admits a Markov partition $\mathcal{R} = \{R_1, \ldots, R_k\}$. This implies the existence of a two-sided subshift of
finite type Θ in the symbol space $\{1, \ldots, k\}^{\mathbb{Z}}$, and an inclusion $i : \Theta \to \Lambda$ such
that (a) $f \circ i = i \circ \tau$ and (b) $i(\Theta_j) = R_j$ for every $j = 1, \ldots, k$. For every
$g \in \mathcal{T}(f, \Lambda)$, the inclusion $i_g = h_{f,g} \circ i : \Theta \to \Lambda_g$ is such that $g \circ i_g = i_g \circ \tau$.
We call such a map $i_g : \Theta \to \Lambda_g$ a *marking* of (g, Λ_g).

DEFINITION 1.1. If $(g, \Lambda_g) \in \mathcal{T}(f, \Lambda)$ is a C^{1+} hyperbolic diffeomorphism as
above and ν is a Gibbs measure on Θ then we say that (g, Λ_g, ν) is a Hausdorff
realisation if $(i_g)_* \nu$ is absolutely continuous with respect to the Hausdorff measure on Λ_g. If this is the case then we will often just say that ν is a Hausdorff
realisation for (g, Λ_g).

We note that the Hausdorff measure on Λ_g exists and is unique, and if a Hausdorff realisation exists then it is unique. However, a Hausdorff realisation need
not exist.

DEFINITION 1.2. Let $\mathcal{T}_{f,\Lambda}(\delta_s, \delta_u)$ be the set of all C^{1+} hyperbolic diffeomorphisms (g, Λ_g) in $\mathcal{T}(f, \Lambda)$ such that (i) $\delta_{g,s} = \delta_s$ and $\delta_{g,u} = \delta_u$; (ii) there is a
g-invariant measure μ_g on Λ_g which is absolutely continuous with respect to
the Hausdorff measure on Λ_g. We denote by $[\nu] \subset \mathcal{T}_{f,\Lambda}(\delta_s, \delta_u)$ the subset of
all C^{1+}-realisations of a Gibbs measure ν in $\mathcal{T}_{f,\Lambda}(\delta_s, \delta_u)$.

De la Llave, Marco and Moriyon [26; 27; 30; 31] have shown that the set of
stable and unstable eigenvalues of all periodic points is a complete invariant of
the C^{1+} conjugacy classes of Anosov diffeomorphisms. We extend their result
to the sets $[\nu] \subset \mathcal{T}_{f,\Lambda}(\delta_s, \delta_u)$.

THEOREM 1.8 (EIGENVALUE SPECTRA). (i) *Any two elements of the set $[\nu] \subset$
$\mathcal{T}_{f,\Lambda}(\delta_s, \delta_u)$ have the same set of stable and unstable eigenvalues and these*

sets are a complete invariant of [v] *in the sense that if* $g_1, g_2 \in \mathcal{T}_{f,\Lambda}(\delta_s, \delta_u)$
have the same eigenvalues if, and only if, they are in the same subset [v].

(ii) *The map* $v \rightarrow [v] \subset \mathcal{T}_{f,\Lambda}(\delta_s, \delta_u)$ *gives a* $1 - 1$ *correspondence between*
  C^{1+}-*Hausdorff realisable Gibbs measures* v *and Lipschitz conjugacy classes*
  in $\mathcal{T}_{f,\Lambda}(\delta_s, \delta_u)$.

Theorem 1.8 is proved in [42], where it is also proved that the set of stable and
unstable eigenvalues of all periodic orbits of a C^{1+} hyperbolic diffeomorphism
$g \in \mathcal{T}(f, \Lambda)$ is a complete invariant of each Lipschitz conjugacy class. Furthermore, for Anosov diffeomorphisms every Lipschitz conjugacy class is a C^{1+}
conjugacy class.

REMARK 1.9. We have restricted our discussion to Gibbs measures because it
follows from Theorem 1.8 that, if $g \in \mathcal{T}_{f,\Lambda}(\delta_s, \delta_u)$ has a g-invariant measure
μ_g which is absolutely continuous with respect to the Hausdorff measure then
μ_g is a C^{1+}-Hausdorff realisation of a Gibbs measure v so that $\mu_g = (i_g)_* v$.

If f is a Smale horseshoe then every Gibbs measure v is C^{1+}-Hausdorff realisable by a hyperbolic diffeomorphism contained in $\mathcal{T}_{f,\Lambda}(\delta_s, \delta_u)$ (see [42]).
However, this is not the case for Anosov diffeomorphisms and codimension
one attractors. E. Cawley [6] characterised all C^{1+}-Hausdorff realisable Gibbs
measures as Anosov diffeomorphisms using cohomology classes on the torus.
In [42], it is used measure solenoid functions to classify all C^{1+}-Hausdorff
realisable Gibbs measures, in an integrated way, of all C^{1+} hyperbolic diffeomorphisms on surfaces. In Section 5.3, the stable and unstable measure solenoid
functions are easily built from the Gibbs measures, and, in Section 5.6, we define
the infinite dimensional metric space $\mathcal{SOL}^\iota$.

THEOREM 1.10 (MEASURE SOLENOID FUNCTIONS). *Let* f *be an Anosov
diffeomorphism or a codimension one attractor. The following statements are
equivalent:*

(i) *The Gibbs measure* v *is* C^{1+}-*Hausdorff realisable by a hyperbolic diffeomorphism contained in* $\mathcal{T}_{f,\Lambda}(\delta_s, \delta_u)$.

(ii) *The* ι-*measure solenoid function* $\sigma_{v,\iota} : \mathrm{msol}^s \rightarrow \mathbb{R}^+$ *has a nonvanishing
  Hölder continuous extension to the closure of* msol^s *belonging to* $\mathcal{SOL}^\iota$.

We present a more detailed version of this theorem in Theorems 5.5 and 5.8.
These theorems are proved in [42].

By Theorems 1.8 and 1.10, for ι equal to s and u, we obtain that the map
$v \rightarrow \sigma_{v,\iota}$ gives a one-to-one correspondence between the sets [v] contained in
$\mathcal{T}_{f,\Lambda}(\delta_s, \delta_u)$ and the space of measure solenoid functions $\sigma_{g,\iota}$ whose continuous
extension is contained in $\mathcal{SOL}^\iota$.

COROLLARY 1.11 (MODULI SPACE). *The set $\mathcal{SOL}^{\iota}$ is a moduli space parame-*
terizing all Lipschitz conjugacy classes [v] *of C^{1+} hyperbolic diffeomorphisms*
contained in $\mathcal{T}_{f,\Lambda}(\delta_s, \delta_u)$.

2. Hyperbolic diffeomorphisms

In this section, we present some basic facts on hyperbolic dynamics, that we
include for clarity of the exposition.

2.1. Stable and unstable superscripts. Throughout the paper we will use the
following notation: we use ι to denote an element of the set $\{s, u\}$ of the stable
and unstable superscripts and ι' to denote the element of $\{s, u\}$ that is not ι.
In the main discussion we will often refer to objects which are qualified by ι
such as, for example, an ι-leaf: This means a leaf which is a leaf of the stable
lamination if $\iota = s$, or a leaf of the unstable lamination if $\iota = u$. In general the
meaning should be quite clear.

We define the map $f_{\iota} = f$ if $\iota = u$ or $f_{\iota} = f^{-1}$ if $\iota = s$.

2.2. Leaf segments. Let d be a metric on M. For $\iota \in \{s, u\}$, if $x \in \Lambda$ we denote
the local ι-manifolds through x by

$$W^{\iota}(x, \varepsilon) = \left\{ y \in M : d(f_{\iota}^{-n}(x), f_{\iota}^{-n}(y)) \leq \varepsilon, \text{ for all } n \geq 0 \right\}.$$

By the Stable Manifold Theorem (see [21]), these sets are respectively contained
in the stable and unstable immersed manifolds

$$W^{\iota}(x) = \bigcup_{n \geq 0} f_{\iota}^{n} \left(W^{\iota} \left(f_{\iota}^{-n}(x), \varepsilon_0 \right) \right)$$

which are the image of a $C^{1+\gamma}$ immersion $\kappa_{\iota,x} : \mathbb{R} \to M$. An *open* (resp.
closed) full ι-leaf segment I is defined as a subset of $W^{\iota}(x)$ of the form $\kappa_{\iota,x}(I_1)$
where I_1 is an open (resp. closed) subinterval (nonempty) in $\mathbb{R}$. An *open (resp.
closed) ι-leaf segment* is the intersection with Λ of an open (resp. closed) full
ι-leaf segment such that the intersection contains at least two distinct points.
If the intersection is exactly two points we call this closed ι-leaf segment an
ι-leaf gap. A *full ι-leaf segment* is either an open or closed full ι-leaf segment.
An *ι-leaf segment* is either an open or closed ι-leaf segment. The *endpoints*
of a full ι-leaf segment are the points $\kappa_{\iota,x}(u)$ and $\kappa_{\iota,x}(v)$ where u and v are
the endpoints of I_1. The *endpoints* of an ι-leaf segment I are the points of
the minimal closed full ι-leaf segment containing I. The *interior* of an ι-leaf
segment I is the complement of its boundary. In particular, an ι-leaf segment I
has empty interior if, and only if, it is an ι-leaf gap. A map $c : I \to \mathbb{R}$ is an *ι-leaf
chart* of an ι-leaf segment I if has an extension $c_E : I_E \to \mathbb{R}$ to a full ι-leaf

segment I_E with the following properties: $I \subset I_E$ and c_E is a homeomorphism onto its image.

2.3. Topological and smooth conjugacies. Let (f, Λ) be a C^{1+} hyperbolic diffeomorphism. Somewhat unusually we also desire to highlight the C^{1+} structure on M in which f is a diffeomorphism. By a C^{1+} *structure on M* we mean a maximal set of charts with open domains in M such that the union of their domains cover M and whenever U is an open subset contained in the domains of any two of these charts i and j then the overlap map $j \circ i^{-1} : i(U) \to j(U)$ is $C^{1+\alpha}$, where $\alpha > 0$ depends on i, j and U. We note that by compactness of M, given such a C^{1+} structure on M, there is an atlas consisting of a finite set of these charts which cover M and for which the overlap maps are $C^{1+\alpha}$ compatible and uniformly bounded in the $C^{1+\alpha}$ norm, where $\alpha > 0$ just depends upon the atlas. We denote by $\mathscr{C}_f$ the C^{1+} structure on M in which f is a diffeomorphism. Usually one is not concerned with this as, given two such structures, there is a homeomorphism of M sending one onto the other and thus, from this point of view, all such structures can be identified. For our discussion it will be important to maintain the identity of the different smooth structures on M.

We say that a map $h : \Lambda_f \to \Lambda_g$ is a *topological conjugacy* between two C^{1+} hyperbolic diffeomorphisms (f, Λ_f) and (g, Λ_g) if there is a homeomorphism $h : \Lambda_f \to \Lambda_g$ with the following properties:

(i) $g \circ h(x) = h \circ f(x)$ for every $x \in \Lambda_f$.
(ii) The pull-back of the ι-leaf segments of g by h are ι-leaf segments of f.

We say that a topological conjugacy $h : \Lambda_f \to \Lambda_g$ is a *Lipschitz conjugacy* if h has a bi-Lipschitz homeomorphic extension to an open neighborhood of Λ_f in the surface M (with respect to the C^{1+} structures $\mathscr{C}_f$ and $\mathscr{C}_g$, respectively).

Similarly, we say that a topological conjugacy $h : \Lambda_f \to \Lambda_g$ is a C^{1+} *conjugacy* if h has a $C^{1+\alpha}$ diffeomorphic extension to an open neighborhood of Λ_f in the surface M, for some $\alpha > 0$.

Our approach is to fix a C^{1+} hyperbolic diffeomorphism (f, Λ) and consider C^{1+} hyperbolic diffeomorphism (g_1, Λ_{g_1}) topologically conjugate to (f, Λ). The topological conjugacy $h : \Lambda \to \Lambda_{g_1}$ between f and g_1 extends to a homeomorphism H defined on a neighborhood of Λ. Then, we obtain the new C^{1+}-realization (g_2, Λ_{g_2}) of f defined as follows: (i) the map $g_2 = H^{-1} \circ g_1 \circ H$; (ii) the basic set is $\Lambda_{g_2} = H^{-1}|\Lambda_{g_1}$; (iii) the C^{1+} structure $\mathscr{C}_{g_2}$ is given by the pull-back $(H)_* \mathscr{C}_{g_1}$ of the C^{1+} structure $\mathscr{C}_{g_1}$. From (i) and (ii), we get that $\Lambda_{g_2} = \Lambda$ and $g_2|\Lambda = f$. From (iii), we get that g_2 is C^{1+} conjugated to g_1. Hence, to study the conjugacy classes of C^{1+} hyperbolic diffeomorphisms (f, Λ) of f, we can just consider the C^{1+} hyperbolic diffeomorphisms (g, Λ_g)

with $\Lambda_g = \Lambda$ and $g|\Lambda_g = f|\Lambda$, which we will do from now on for simplicity of our exposition.

2.4. Rectangles. Since Λ is a hyperbolic invariant set of a diffeomorphism $f : M \to M$, for $0 < \varepsilon < \varepsilon_0$ there is $\delta = \delta(\varepsilon) > 0$ such that, for all points $w, z \in \Lambda$ with $d(w, z) < \delta$, $W^u(w, \varepsilon)$ and $W^s(z, \varepsilon)$ intersect in a unique point that we denote by $[w, z]$. Since we assume that the hyperbolic set has a *local product structure*, we have that $[w, z] \in \Lambda$. Furthermore, the following properties are satisfied: (i) $[w, z]$ varies continuously with $w, z \in \Lambda$; (ii) the bracket map is continuous on a δ-uniform neighborhood of the diagonal in $\Lambda \times \Lambda$; and (iii) whenever both sides are defined $f([w, z]) = [f(w), f(z)]$. Note that the bracket map does not really depend on δ provided it is sufficiently small.

Let us underline that it is a standing hypothesis that all the hyperbolic sets considered here have such a local product structure.

A *rectangle* R is a subset of Λ which is (i) closed under the bracket i.e. $x, y \in R \Longrightarrow [x, y] \in R$, and (ii) proper i.e. is the closure of its interior in Λ. This definition imposes that a rectangle has always to be proper which is more restrictive than the usual one which only insists on the closure condition.

If ℓ^s and ℓ^u are respectively stable and unstable leaf segments intersecting in a single point then we denote by $[\ell^s, \ell^u]$ the set consisting of all points of the form $[w, z]$ with $w \in \ell^s$ and $z \in \ell^u$. We note that if the stable and unstable leaf segments ℓ and ℓ' are closed then the set $[\ell, \ell']$ is a rectangle. Conversely in this 2-dimensional situations, any rectangle R has a product structure in the following sense: for each $x \in R$ there are closed stable and unstable leaf segments of Λ, $\ell^s(x, R) \subset W^s(x)$ and $\ell^u(x, R) \subset W^u(x)$ such that $R = [\ell^s(x, R), \ell^u(x, R)]$. The leaf segments $\ell^s(x, R)$ and $\ell^u(x, R)$ are called *stable and unstable spanning leaf segments* for R (see Figure 1). For $\iota \in \{s, u\}$, we denote by $\partial\ell^\iota(x, R)$ the set consisting of the endpoints of $\ell^\iota(x, R)$, and we denote by int $\ell^\iota(x, R)$ the set $\ell^\iota(x, R) \setminus \partial\ell^\iota(x, R)$. The *interior of* R is given by int $R = [\text{int } \ell^s(x, R), \text{int } \ell^u(x, R)]$, and the *boundary of* R is given by $\partial R = [\partial\ell^s(x, R), \ell^u(x, R)] \cup [\ell^s(x, R), \partial\ell^u(x, R)]$.

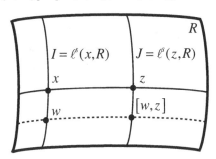

Figure 1. A rectangle.

2.5. Markov partitions. By a *Markov partition of f* we understand a collection $\mathcal{R} = \{R_1, \ldots, R_k\}$ of rectangles such that (i) $\Lambda \subset \bigcup_{i=1}^{k} R_i$; (ii) $R_i \cap R_j = \partial R_i \cap \partial R_j$ for all i and j; (iii) if $x \in \iint R_i$ and $fx \in \iint R_j$ then

(a) $f(\ell^s(x, R_i)) \subset \ell^s(fx, R_j)$ and $f^{-1}(\ell^u(fx, R_j)) \subset \ell^u(x, R_i)$
(b) $f(\ell^u(x, R_i)) \cap R_j = \ell^u(fx, R_j)$ and $f^{-1}(\ell^s(fx, R_j)) \cap R_i = \ell^s(x, R_i)$.

The last condition means that $f(R_i)$ goes across R_j just once. In fact, it follows from condition (a) providing the rectangles R_j are chosen sufficiently small (see [28]). The rectangles making up the Markov partition are called *Markov rectangles*.

We note that there is a Markov partition $\mathcal{R}$ of f with the following *disjointness property* (see [5; 34; 46]):

(i) if $0 < \delta_{f,s} < 1$ and $0 < \delta_{f,u} < 1$ then the stable and unstable leaf boundaries of any two Markov rectangles do not intersect.
(ii) if $0 < \delta_{f,\iota} < 1$ and $\delta_{f,\iota'} = 1$ then the ι'-leaf boundaries of any two Markov rectangles do not intersect except, possibly, at their endpoints.

If $\delta_{f,s} = \delta_{f,u} = 1$, the disjointness property does not apply and so we consider that it is trivially satisfied for every Markov partition. For simplicity of our exposition, we will just consider Markov partitions satisfying the disjointness property.

2.6. Leaf *n*-cylinders and leaf *n*-gaps. For $\iota = s$ or u, an *ι-leaf primary cylinder of a Markov rectangle R* is a spanning ι-leaf segment of R. For $n \geq 1$, an *ι-leaf n-cylinder of R* is an ι-leaf segment I such that

(i) $f_\iota^n I$ is an ι-leaf primary cylinder of a Markov rectangle M;
(ii) $f_\iota^n \left(\ell^{\iota'}(x, R) \right) \subset M$ for every $x \in I$.

For $n \geq 2$, an *ι-leaf n-gap G of R* is an ι-leaf gap $\{x, y\}$ in a Markov rectangle R such that n is the smallest integer such that both leaves $f_\iota^{n-1} \ell^{\iota'}(x, R)$ and $f_\iota^{n-1} \ell^{\iota'}(y, R)$ are contained in ι'-boundaries of Markov rectangles; An *ι-leaf primary gap G* is the image $f_\iota G'$ by f_ι of an ι-leaf 2-gap G'.

We note that an ι-leaf segment I of a Markov rectangle R can be simultaneously an n_1-cylinder, $(n_1 + 1)$-cylinder, $\ldots$, n_2-cylinder of R if $f^{n_1}(I)$, $f^{n_1+1}(I)$, $\ldots$, $f^{n_2}(I)$ are all spanning ι-leaf segments. Furthermore, if I is an ι-leaf segment contained in the common boundary of two Markov rectangles R_i and R_j then I can be an n_1-cylinder of R_i and an n_2-cylinder of R_j with n_1 distinct of n_2. If $G = \{x, y\}$ is an ι-gap of R contained in the interior of R then there is a unique n such that G is an n-gap. However, if $G = \{x, y\}$ is contained in the common boundary of two Markov rectangles R_i and R_j then G can be an n_1-gap of R_i and an n_2-gap of R_j with n_1 distinct of n_2. Since the number

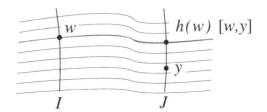

Figure 2. A basic stable holonomy from I to J.

of Markov rectangles $R_1, \ldots, R_k$ is finite, there is $C \geq 1$ such that, in all the above cases for cylinders and gaps we have $|n_2 - n_1| \leq C$.

We say that a leaf segment K is the i-th *mother* of an n-cylinder or an n-gap J of R if $J \subset K$ and K is a leaf $(n-i)$-cylinder of R. We denote K by $m^i J$.

By the properties of a Markov partition, the smallest full ι-leaf $\hat{K}$ containing a leaf n-cylinder K of a Markov rectangle R is equal to the union of all smallest full ι-leaves containing either a leaf $(n+j)$-cylinder or a leaf $(n+i)$-gap of R, with $i \in \{1, \ldots, j\}$, contained in K.

2.7. Metric on Λ. We say that a rectangle R is an (n_s, n_u)-*rectangle* if there is $x \in R$ such that, for $\iota = s$ and u, the spanning leaf segments $\ell^\iota(x, R)$ are either an ι-leaf n_ι-cylinder or the union of two such cylinders with a common endpoint.

The reason for allowing the possibility of the spanning leaf segments being inside two touching cylinders is to allow us to regard geometrically very small rectangles intersecting a common boundary of two Markov rectangles to be small in the sense of having n_s and n_u large.

If $x, y \in \Lambda$ and $x \neq y$ then $d_\Lambda(x, y) = 2^{-n}$ where n is the biggest integer such that both x and y are contained in an (n_s, n_u)-rectangle with $n_s \geq n$ and $n_u \geq n$. Similarly if I and J are ι-leaf segments then $d_\Lambda(I, J) = 2^{-n_{\iota'}}$ where $n_\iota = 1$ and $n_{\iota'}$ is the biggest integer such that both I and J are contained in an (n_s, n_u)-rectangle.

2.8. Basic holonomies. Suppose that x and y are two points inside any rectangle R of Λ. Let $\ell(x, R)$ and $\ell(y, R)$ be two stable leaf segments respectively containing x and y and inside R. Then we define $\theta : \ell(x, R) \to \ell(y, R)$ by $\theta(w) = [w, y]$. Such maps are called the *basic stable holonomies* (see Figure 2). They generate the pseudo-group of all stable holonomies. Similarly we define the basic unstable holonomies.

By Theorem 2.1 in [40], the holonomy $\theta : \ell^\iota(x, R) \to \ell^\iota(y, R)$ has a $C^{1+\alpha}$ extension to the leaves containing $\ell^\iota(x, R)$ and $\ell^\iota(y, R)$, for some $\alpha > 0$.

2.9. Foliated lamination atlas. In this section when we refer to a C^r object r is allowed to take the values $k + \alpha$ where k is a positive integer and $0 < \alpha \leq 1$. Two ι-leaf charts i and j are C^r compatible if whenever U is an open subset of an ι-leaf segment contained in the domains of i and j then $j \circ i^{-1} : i(U) \to j(U)$ extends to a C^r diffeomorphism of the real line. Such maps are called *chart overlap maps*. A *bounded C^r ι-lamination atlas* $\mathscr{A}^\iota$ is a set of such charts which (a) cover Λ, (b) are pairwise C^r compatible, and (c) the chart overlap maps are uniformly bounded in the C^r norm.

Let $\mathscr{A}^\iota$ be a bounded $C^{1+\alpha}$ ι-lamination atlas, with $0 < \alpha \leq 1$. If $i : I \to \mathbb{R}$ is a chart in $\mathscr{A}^\iota$ defined on the leaf segment I and K is a leaf segment in I then we define $|K|_i$ to be the length of the minimal closed interval containing $i(K)$. Since the atlas is bounded, if $j : J \to \mathbb{R}$ is another chart in $\mathscr{A}^\iota$ defined on the leaf segment J which contains K then the ratio between the lengths $|K|_i$ and $|K|_j$ is universally bounded away from 0 and ∞. If $K' \subset I \cap J$ is another such segment then we can define the ratio $r_i(K : K') = |K|_i / |K'|_i$. Although this ratio depends upon i, the ratio is exponentially determined in the sense that if T is the smallest segment containing both K and K' then

$$r_j \left(K : K' \right) \in \left(1 \pm \mathcal{O} \left(|T|_i^\alpha \right) \right) r_i \left(K : K' \right) .$$

This follows from the $C^{1+\alpha}$ smoothness of the overlap maps and Taylor's Theorem.

A C^r lamination atlas $\mathscr{A}^\iota$ has *bounded geometry* (i) if f is a C^r diffeomorphism with C^r norm uniformly bounded in this atlas; (ii) if for all pairs I_1, I_2 of ι-leaf n-cylinders or ι-leaf n-gaps with a common point, we have that $r_i(I_1 : I_2)$ is uniformly bounded away from 0 and ∞ with the bounds being independent of i, I_1, I_2 and n; and (iii) for all endpoints x and y of an ι-leaf n-cylinder or ι-leaf n-gap I, we have that $|I|_i \leq \mathcal{O}\left((d_\Lambda(x, y))^\beta \right)$ and $d_\Lambda(x, y) \leq \mathcal{O}\left(|I|_i^\beta \right)$, for some $0 < \beta < 1$, independent of i, I and n.

A C^r bounded lamination atlas $\mathscr{A}^\iota$ is C^r *foliated* (i) if $\mathscr{A}^\iota$ has bounded geometry; and (ii) if the basic holonomies are C^r and have a C^r norm uniformly bounded in this atlas, except possibly for the dependence upon the rectangles defining the basic holonomy. A bounded lamination atlas $\mathscr{A}^\iota$ is C^{1+} *foliated* if $\mathscr{A}^\iota$ is C^r foliated for some $r > 1$.

2.10. Foliated atlas $\mathscr{A}^\iota(g, \rho)$. Let $g \in \mathcal{T}(f, \Lambda)$ and $\rho = \rho_g$ be a C^{1+} Riemannian metric in the manifold containing Λ. The *ι-lamination atlas $\mathscr{A}^\iota(g, \rho)$ determined by ρ* is the set of all maps $e : I \to \mathbb{R}$ where $I = \Lambda \cap \hat{I}$ with $\hat{I}$ a full ι-leaf segment, such that e extends to an isometry between the induced Riemannian metric on $\hat{I}$ and the Euclidean metric on the reals. We call the maps $e \in \mathscr{A}^\iota(\rho)$ the *ι-lamination charts*. If I is an ι-leaf segment (or a full ι-leaf segment) then by $|I|_\rho$ we mean the length in the Riemannian metric ρ of the minimal full ι-leaf containing I. By Theorem 2.2 in [40], the lamination atlas $\mathscr{A}^\iota(g, \rho)$ is C^{1+} foliated for $\iota = \{s, u\}$.

3. Flexibility

In this section, we construct the stable and unstable solenoid functions, and we show an equivalence between C^{1+} hyperbolic diffeomorphisms and pairs of stable and unstable solenoid functions.

3.1. HR-Hölder ratios.
A *HR-structure* associates an affine structure to each stable and unstable leaf segment in such a way that these vary Hölder continuously with the leaf and are invariant under f.

An affine structure on a stable or unstable leaf is equivalent to a *ratio function* $r(I : J)$ which can be thought of as prescribing the ratio of the size of two leaf segments I and J in the same stable or unstable leaf. A *ratio function* $r(I : J)$ is positive (we recall that each leaf segment has at least two distinct points) and continuous in the endpoints of I and J. Moreover,

$$r(I : J) = r(J : I)^{-1} \text{ and } r(I_1 \cup I_2 : K) = r(I_1 : K) + r(I_2 : K) \quad (3\text{--}1)$$

provided I_1 and I_2 intersect at most in one of their endpoints.

We say that r is an *ι-ratio function* if (i) for all ι-leaf segments K, $r(I : J)$ defines a ratio function on K, where I and J are ι-leaf segments contained in K; (ii) r is invariant under f, i.e. $r(I : J) = r(fI : fJ)$ for all ι-leaf segments; and (iii) for every basic ι-holonomy $\theta : I \to J$ between the leaf segment I and the leaf segment J defined with respect to a rectangle R and for every ι-leaf segment $I_0 \subset I$ and every ι-leaf segment or gap $I_1 \subset I$,

$$\left| \log \frac{r(\theta I_0 : \theta I_1)}{r(I_0 : I_1)} \right| \leq \mathbb{O}\left((d_\Lambda(I, J))^\varepsilon \right) \quad (3\text{--}2)$$

where $\varepsilon \in (0, 1)$ depends upon r and the constant of proportionality also depends upon R, but not on the segments considered.

A *HR-structure* on Λ invariant by f is a pair (r_s, r_u) consisting of a stable and an unstable ratio function.

3.2. Realised ratio functions.
Let $(g, \Lambda) \in \mathcal{T}(f, \Lambda)$ and let $\mathscr{A}(g, \rho)$ be an ι-lamination atlas which is C^{1+} foliated. Let $|I| = |I|_\rho$ for every ι-leaf segment I. By hyperbolicity of g on Λ, there are $0 < \nu < 1$ and $C > 0$ such that for all ι-leaf segments I and all $m \geq 0$ we get $|g_{\iota'}^m I| \leq C\nu^m |I|$. Thus, using the mean value theorem and the fact that g_ι is C^r, for all short leaf segments K and all leaf segments I and J contained in it, the ι-realised ratio function $r_{\iota,g}$ given by

$$r_{\iota,g}(I : J) = \lim_{n \to \infty} \frac{|g_{\iota'}^n I|}{|g_{\iota'}^n J|}$$

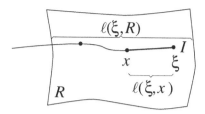

Figure 3. The embedding $e : I \to \mathbb{R}$.

is well-defined, where $\alpha = \min\{1, r - 1\}$. This construction gives the HR-structure on Λ determined by g, and so also invariant by f. By [39], we get the following equivalence:

THEOREM 3.1. *The map* $g \to (r_{s,g}, r_{u,g})$ *determines a one-to-one correspondence between* C^{1+} *conjugacy classes in* $\mathcal{T}(f, \Lambda)$ *and HR-structures on* Λ *invariant by* f.

3.3. Foliated atlas $\mathcal{A}(r)$. Given an ι-ratio function r, we define the embeddings $e : I \to \mathbb{R}$ by

$$e(x) = r(\ell(\xi, x), \ell(\xi, R)) \tag{3-3}$$

where ξ is an endpoint of the ι-leaf segment I and R is a Markov rectangle containing ξ (see Figure 3). For this definition it is not necessary that R contains I. We denote the set of all these embeddings e by $\mathcal{A}(r)$.

The embeddings e of $\mathcal{A}(r)$ have overlap maps with affine extensions. Therefore, the atlas $\mathcal{A}(r)$ extends to a $C^{1+\alpha}$ lamination structure $\mathcal{L}(r)$. By Proposition 4.2 in [40], we obtain that $\mathcal{A}(r)$ is a C^{1+} foliated atlas.

Let $g \in \mathcal{T}(f, \Lambda)$ and $\mathcal{A}(g, \rho)$ a C^{1+} foliated ι-lamination atlas determined by a Riemmanian metric ρ. Combining Proposition 2.5 and Proposition 3.5 of [39], we get that the overlap map $e_1 \circ e_2^{-1}$ between a chart $e_1 \in \mathcal{A}(g, \rho)$ and a chart $e_2 \in \mathcal{A}(r_{\iota,g})$ has a C^{1+} diffeomorphic extension to the reals. Therefore, the atlasses $\mathcal{A}(g, \rho)$ and $\mathcal{A}(r_{\iota,g})$ determine the same C^{1+} foliated ι-lamination. In particular, for all short leaf segments K and all leaf segments I and J contained in it, we obtain that

$$r_{\iota,g}(I : J) = \lim_{n \to \infty} \frac{|g_{\iota'}^n I|_\rho}{|g_{\iota'}^n J|_\rho} = \lim_{n \to \infty} \frac{|g_{\iota'}^n I|_{i_n}}{|g_{\iota'}^n J|_{i_n}}$$

where i_n is any chart in $\mathcal{A}(r_{\iota,g})$ containing the segment $g_{\iota'}^n K$ in its domain.

3.4. Realised solenoid functions. For $\iota = s$ and u, let sol^ι denote the set of all ordered pairs (I, J) of ι-leaf segments with the following properties:

(i) The intersection of I and J consists of a single endpoint.
(ii) If $\delta_{\iota,f} = 1$ then I and J are primary ι-leaf cylinders.

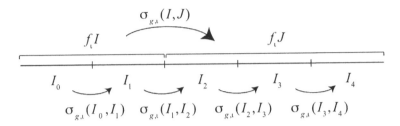

Figure 4. The f-matching condition for ι-leaf segments.

(iii) If $0 < \delta_{\iota,f} < 1$ then $f_{\iota'} I$ is an ι-leaf 2-cylinder of a Markov rectangle R and $f_{\iota'} J$ is an ι-leaf 2-gap also of the same Markov rectangle R.

(See section 2.4 for the definitions of leaf cylinders and gaps). Pairs (I, J) where both are primary cylinders are called *leaf-leaf pairs*. Pairs (I, J) where J is a gap are called *leaf-gap pairs* and in this case we refer to J as a *primary gap*. The set sol^ι has a very nice topological structure. If $\delta_{\iota',f} = 1$ then the set sol^ι is isomorphic to a finite union of intervals, and if $\delta_{\iota',f} < 1$ then the set sol^ι is isomorphic to an embedded Cantor set on the real line. We define a pseudo-metric $d_{\mathrm{sol}^\iota} : \mathrm{sol}^\iota \times \mathrm{sol}^\iota \to \mathbb{R}^+$ on the set sol^ι by

$$d_{\mathrm{sol}^\iota}\left((I, J), (I', J')\right) = \max\left\{d_\Lambda\left(I, I'\right), d_\Lambda\left(J, J'\right)\right\}.$$

Let $g \in \mathcal{T}(f, \Lambda)$. For $\iota = s$ and u, we call the restriction of an ι-ratio function $r_{\iota,g}$ to sol^ι a *realised solenoid function* $\sigma_{\iota,g}$. By construction, for $\iota = s$ and u, the restriction of an ι-ratio function to sol^ι gives an Hölder continuous function satisfying the matching condition, the boundary condition and the cylinder-gap condition as we pass to describe.

3.5. Hölder continuity of solenoid functions. This means that for $t = (I, J)$ and $t' = (I', J')$ in sol^ι, $|\sigma_\iota(t) - \sigma_\iota(t')| \le \mathcal{O}\left((d_{\mathrm{sol}^\iota}(t, t'))^\alpha\right)$. The Hölder continuity of $\sigma_{g,\iota}$ and the compactness of its domain imply that $\sigma_{g,\iota}$ is bounded away from zero and infinity.

3.6. Matching condition. Let $(I, J) \in \mathrm{sol}^\iota$ be a pair of primary cylinders and suppose that we have pairs

$$(I_0, I_1), (I_1, I_2), \ldots, (I_{n-2}, I_{n-1}) \in \mathrm{sol}^\iota$$

of primary cylinders such that $f_\iota I = \bigcup_{j=0}^{k-1} I_j$ and $f_\iota J = \bigcup_{j=k}^{n-1} I_j$. Then

$$\frac{|f_\iota I|}{|f_\iota J|} = \frac{\sum_{j=0}^{k-1} |I_j|}{\sum_{j=k}^{n-1} |I_j|} = \frac{1 + \sum_{j=1}^{k-1} \prod_{i=1}^{j} |I_i|/|I_{i-1}|}{\sum_{j=k}^{n-1} \prod_{i=1}^{j} |I_i|/|I_{i-1}|}.$$

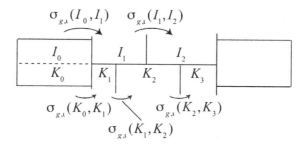

Figure 5. The boundary condition for ι-leaf segments.

Hence, noting that $g|\Lambda = f|\Lambda$, the realised solenoid function $\sigma_{\iota,g}$ must satisfy the *matching condition* (see Figure 4) for all such leaf segments:

$$\sigma_{\iota,g}(I:J) = \frac{1 + \sum_{j=1}^{k-1} \prod_{i=1}^{j} \sigma_{\iota,g}(I_i : I_{i-1})}{\sum_{j=k}^{n-1} \prod_{i=1}^{j} \sigma_{\iota,g}(I_i : I_{i-1})}. \tag{3-4}$$

3.7. Boundary condition. If the stable and unstable leaf segments have Hausdorff dimension equal to 1, then leaf segments I in the boundaries of Markov rectangles can sometimes be written as the union of primary cylinders in more than one way. This gives rise to the existence of a boundary condition that the realised solenoid functions have to satisfy.

If J is another leaf segment adjacent to the leaf segment I then the value of $|I|/|J|$ must be the same whichever decomposition we use. If we write $J = I_0 = K_0$ and I as $\bigcup_{i=1}^{m} I_i$ and $\bigcup_{j=1}^{n} K_j$ where the I_i and K_j are primary cylinders with $I_i \neq K_j$ for all i and j, then the above two ratios are

$$\sum_{i=1}^{m} \prod_{j=1}^{i} \frac{|I_j|}{|I_{j-1}|} = \frac{|I|}{|J|} = \sum_{i=1}^{n} \prod_{j=1}^{i} \frac{|K_j|}{|K_{j-1}|}.$$

Thus, noting that $g|\Lambda = f|\Lambda$, a realised solenoid function $\sigma_{\iota,g}$ must satisfy the following *boundary condition* (see Figure 5) for all such leaf segments:

$$\sum_{i=1}^{m} \prod_{j=1}^{i} \sigma_{\iota,g}\left(I_j : I_{j-1}\right) = \sum_{i=1}^{n} \prod_{j=1}^{i} \sigma_{\iota,g}\left(K_j : K_{j-1}\right). \tag{3-5}$$

3.8. Cylinder-gap condition. If the ι-leaf segments have Hausdorff dimension less than one and the ι'-leaf segments have Hausdorff dimension equal to 1, then a primary cylinder I in the ι-boundary of a Markov rectangle can also be written as the union of gaps and cylinders of other Markov rectangles. This gives rise to the existence of a cylinder-gap condition that the ι-realised solenoid functions have to satisfy.

Before defining the cylinder-gap condition, we will introduce the scaling function that will be useful to express the cylinder-gap condition, and also, in Definition 3.2, the bounded equivalence classes of solenoid functions and, in Definition 5.3, the δ-bounded solenoid equivalence classes of a Gibbs measure.

Let scl^ι be the set of all pairs (K, J) of ι-leaf segments with the following properties:

(i) K is a leaf n_1-cylinder or an n_1-gap segment for some $n_1 > 1$;
(ii) J is a leaf n_2-cylinder or an n_2-gap segment for some $n_2 > 1$;
(iii) $m^{n_1-1}K$ and $m^{n_2-1}J$ are the same primary cylinder.

LEMMA 3.2. *Every function $\sigma_\iota : \mathrm{sol}^\iota \to \mathbb{R}^+$ has a canonical extension s_ι to scl^ι. Furthermore, if σ_ι is the restriction of a ratio function $r_\iota|\mathrm{sol}^\iota$ to sol^ι then*

$$s_\iota = r_\iota|\mathrm{scl}^\iota.$$

The above map $s_\iota : \mathrm{scl}^\iota \to \mathbb{R}^+$ is the *scaling function* determined by the solenoid function $\sigma_\iota : \mathrm{sol}^\iota \to \mathbb{R}^+$. Lemma 3.2 is proved in Section 3.8 in [42].

Let (I, K) be a leaf-gap pair such that the primary cylinder I is the ι-boundary of a Markov rectangle R_1. Then the primary cylinder I intersects another Markov rectangle R_2 giving rise to the existence of a cylinder-gap condition that the realised solenoid functions have to satisfy as we pass to explain. Take the smallest $l \geq 0$ such that $f_{\iota'}^l I \cup f_{\iota'}^l K$ is contained in the intersection of the boundaries of two Markov rectangles M_1 and M_2. Let M_1 be the Markov rectangle with the property that $M_1 \cap f_{\iota'}^l R_1$ is a rectangle with nonempty interior (and so $M_2 \cap f_{\iota'}^l R_2$ also has nonempty interior). Then, for some positive n, there are distinct n-cylinder and gap leaf segments $J_1, \ldots, J_m$ contained in a primary cylinder of M_2 such that $f_{\iota'}^l K = J_m$ and the smallest full ι-leaf segment containing $f_{\iota'}^l I$ is equal to the union $\bigcup_{i=1}^{m-1} \hat{J}_i$, where $\hat{J}_i$ is the smallest full ι-leaf segment containing J_i. Hence,

$$\frac{|f_{\iota'}^l I|}{|f_{\iota'}^l K|} = \sum_{i=1}^{m-1} \frac{|J_i|}{|J_m|}.$$

Hence, noting that $g|\Lambda = f|\Lambda$, a realised solenoid function $\sigma_{g,\iota}$ must satisfy the *cylinder-gap condition* for all such leaf segments:

$$\sigma_{g,\iota}(I, K) = \sum_{i=1}^{m-1} s_{g,\iota}(J_i, J_m),$$

where $s_{g,\iota}$ is the scaling function determined by the solenoid function $\sigma_{g,\iota}$. See Figure 6.

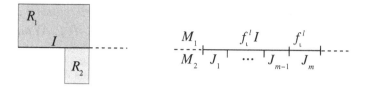

Figure 6. The cylinder-gap condition for ι-leaf segments.

3.9. Solenoid functions. Now, we are ready to present the definition of an ι-solenoid function.

DEFINITION 3.1. A Hölder continuous function $\sigma_\iota : \mathrm{sol}^\iota \to \mathbb{R}^+$ is an ι-*solenoid function* if it satisfies the matching condition, the boundary condition and the cylinder-gap condition.

We denote by $\mathcal{PS}(f)$ the set of pairs (σ_s, σ_u) of stable and unstable solenoid functions.

REMARK 3.3. Let $\sigma_\iota : \mathrm{sol}^\iota \to \mathbb{R}^+$ be an ι-solenoid function. The matching, the boundary and the cylinder-gap conditions are trivially satisfied except in the following cases:

(i) The matching condition if $\delta_{\iota, f} = 1$.
(ii) The boundary condition if $\delta_{s, f} = \delta_{u, f} = 1$.
(iii) The cylinder-gap condition if $\delta_{\iota, f} < 1$ and $\delta_{\iota', f} = 1$.

THEOREM 3.4. *The map* $r_\iota \to r_\iota | \mathrm{sol}^\iota$ *gives a one-to-one correspondence between* ι-*ratio functions and* ι-*solenoid functions.*

PROOF. Every ι-ratio function restricted to the set sol^ι determines an ι-solenoid function $r_\iota | \mathrm{sol}^\iota$. Now we prove the converse. Since the solenoid functions are continuous and their domains are compact they are bounded away from 0 and ∞. By this boundedness and the f-matching condition of the solenoid functions and by iterating the domains sol^s and sol^u of the solenoid functions backward and forward by f, we determine the ratio functions r^s and r^u at very small (and large) scales, such that f leaves the ratios invariant. Then, using the boundedness again, we extend the ratio functions to all pairs of small adjacent leaf segments by continuity. By the boundary condition and the cylinder-gap condition of the solenoid functions, the ratio functions are well determined at the boundaries of the Markov rectangles. Using the Hölder continuity of the solenoid function, we deduce inequality (3–2). □

The set $\mathcal{PS}(f)$ of all pairs (σ_s, σ_u) has a natural metric. Combining Theorem 3.1 with Theorem 3.4, we obtain that the set $\mathcal{PS}(f)$ forms a moduli space for the C^{1+} conjugacy classes of C^{1+} hyperbolic diffeomorphisms $g \in \mathcal{T}(f, \Lambda)$:

COROLLARY 3.5. *The map $g \to (r_{s,g}|\text{sol}^s, r_{u,g}|\text{sol}^u)$ determines a one-to-one correspondence between C^{1+} conjugacy classes of $g \in \mathcal{T}(f, \Lambda)$ and pairs of solenoid functions in $\mathcal{PS}(f)$.*

DEFINITION 3.2. We say that any two ι-solenoid functions $\sigma_1 : \text{sol}^\iota \to \mathbb{R}^+$ and $\sigma_2 : \text{sol}^\iota \to \mathbb{R}^+$ are in the same *bounded equivalence class* if the corresponding scaling functions $s_1 : \text{scl}^\iota \to \mathbb{R}^+$ and $s_2 : \text{scl}^\iota \to \mathbb{R}^+$ satisfy the following property: There is $C > 0$ such that

$$\left| \log s_1(J, m^i J) - \log s_2(J, m^i J) \right| < C \tag{3-6}$$

for every ι-leaf $(i+1)$-cylinder or $(i+1)$-gap J.

In Lemma 8.8 in [42], it is proved that two C^{1+} hyperbolic diffeomorphisms g_1 and g_2 are Lipschitz conjugate if, and only if, the solenoid functions $\sigma_{g_1,\iota}$ and $\sigma_{g_2,\iota}$ are in the same bounded equivalence class for ι equal to s and u.

4. Rigidity

If the holonomies are sufficiently smooth then the system is essentially affine. To see that, rather than consider all holonomies, it is enough to consider a C^{1,HD^ι} complete set of holonomies.

4.1. Complete sets of holonomies.
Before introducing the notion of a C^{1,α^ι} complete set of holonomies, we define the $C^{1,\alpha}$ regularities, with $0 < \alpha \leq 1$, for diffeomorphisms.

DEFINITION 4.1. Let $\theta : I \subset \mathbb{R} \to J \subset \mathbb{R}$ be a diffeomorphism. For $0 < \alpha < 1$, the diffeomorphism θ is $C^{1,\alpha}$ if, for all points $x, y \in I$,

$$|\theta'(y) - \theta'(x)| \leq \chi_{\theta,\alpha}(|y - x|) \tag{4-1}$$

where the positive function $\chi_{\theta,\alpha}(t)$ is $o(t^\alpha)$ i.e. $\lim_{t \to 0} \chi_{\theta,\alpha}(t)/t^\alpha = 0$.
The map $\theta : I \to J$ is $C^{1,1}$ if, for all points $x, y \in I$,

$$\left| \log \theta'(x) + \log \theta'(y) - 2 \log \theta'\left(\frac{x+y}{2} \right) \right| \leq \chi_{\theta,1}(|y - x|) \tag{4-2}$$

where the positive function $\chi_{\theta,1}(t)$ is $o(t)$, i.e. $\lim_{t \to 0} \chi_{\theta,1}(t)/t = 0$. For $0 < \alpha \leq 1$, the functions $\chi_{\theta,\alpha}$ are called the α-*modulus of continuity* of θ.

In particular, for every $\beta > \alpha > 0$, a $C^{1+\beta}$ diffeomorphism is $C^{1,\alpha}$, and, for every $\gamma > 0$, a $C^{2+\gamma}$ diffeomorphism is $C^{1,1}$. We note that the regularity $C^{1,1}$ (also denoted by $C^{1+\text{zygmund}}$) of a diffeomorphism θ used in this paper is stronger than the regularity $C^{1+\text{Zygmund}}$ (see [32]). The importance of these $C^{1,\alpha}$ smoothness classes for a diffeomorphism $\theta : I \to J$ follows from the fact that if $0 < \alpha < 1$ then the map θ will distort ratios of lengths of short intervals

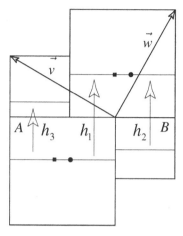

Figure 7. The complete set of holonomies $\mathcal{H} = \{h_1, h_2, h_3, h_1^{-1}, h_2^{-1}, h_3^{-1}\}$ for the Anosov map $g : \mathbf{R}^2 \setminus (\mathbf{Z}\vec{v} \times \mathbf{Z}\vec{w}) \to \mathbf{R}^2 \setminus (\mathbf{Z}\vec{v} \times \mathbf{Z}\vec{w})$ defined by $g(x, y) = (x + y, y)$ and with Markov partition $\mathcal{M} = \{A, B\}$.

in an interval $K \subset I$ by an amount that is $o(|I|^\alpha)$, and if $\alpha = 1$ the map θ will distort the cross-ratios of quadruples of points in an interval $K \subset I$ by an amount that is $o(|I|)$ (see Appendix in [41]).

Suppose that R_i and R_j are Markov rectangles, $x \in R_i$ and $y \in R_j$. We say that x and y are ι-*holonomically related* if there is an ι'-leaf segment $\ell^{\iota'}(x, y)$ such that $\partial \ell^{\iota'}(x, y) = \{x, y\}$, and there are two distinct spanning ι'-leaf segments $\ell^{\iota'}(x, R_i)$ and $\ell^{\iota'}(y, R_j)$ such that their union contains $\ell^{\iota'}(x, y)$.

For every Markov rectangle $R_i \in \mathcal{R}$, let x_i be a chosen point in R_i. Let $\mathcal{I}^\iota = \{I_i = \ell^\iota(x_i, R_i) : R_i \in \mathcal{R}\}$. A *complete set of ι-holonomies* $\mathcal{H}^\iota = \{h_\beta\}$ with respect to $\mathcal{I}^\iota$ consists of a minimal set of basic holonomies with the following property: if $x \in I_i$ is holonomically related to $y \in I_j$, where $I_i, I_j \in \mathcal{I}^\iota$, then for some β either h_β or h_β^{-1} is the holonomy from a neighborhood of x in I_i to I_j which sends x to y (see Figure 7). We call $\mathcal{I}^\iota$ the *domain of the complete set of ι-holonomies* $\mathcal{H}^\iota$. For each $\hat{E}_\iota$-leaf segment I_i in the domain $\mathcal{I}^\iota$ of the complete set of holonomies $\mathcal{H}^\iota$, let $\hat{I}_i$ be a full ι-leaf segment such that $I_i = \hat{I}_i \cap \Lambda$, and let $u_i : \hat{I}_i \to \mathbb{R}$ be a C^r ι-leaf chart of the submanifold structure of $\hat{I}_i$ given by the Stable Manifold Theorem (for instance, we can consider the charts $u_i \in \mathcal{A}(\rho)$ as defined in Section 2.10).

DEFINITION 4.2. A complete set of holonomies $\mathcal{H}^\iota$ is C^{1,α^ι} if for every holonomy $h_\beta : I \to J$ in $\mathcal{H}^\iota$, the map $u_j \circ h_\beta \circ u_i^{-1}$ and its inverse have C^{1,α^ι} diffeomorphic extensions to $\mathbb{R}$ such that the modulus of continuity does not depend upon $h_\beta \in \mathcal{H}^\iota$.

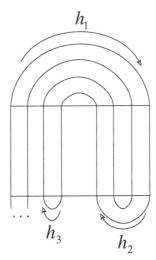

Figure 8. The cardinality of the complete set of holonomies $\mathcal{H} =$ $\{h_1, h_2, h_3, \ldots\}$ is not necessarily finite.

For many systems such as Anosov diffeomorphisms and codimension one attractors there is only a finite number of holonomies in a complete set. In this case the uniformity hypothesis in the modulus of continuity of Definition 4.2 is redundant. However, for Smale horseshoes this is not the case (see Figure 8).

4.2. Hyperbolic affine models. A *hyperbolic affine model for f on Λ* is an *atlas $\mathcal{A}$* with the following properties (see Figure 9):

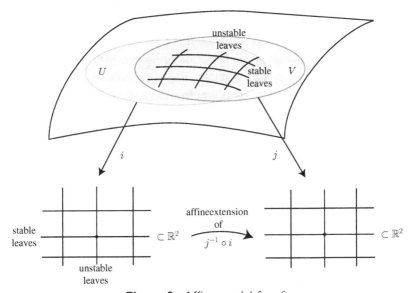

Figure 9. Affine model for f.

(i) the union of the domains of the charts of $\mathcal{A}$ cover an open set of M containing Λ;

(ii) any two charts $i : U \to \mathbb{R}^2$ and $j : V \to \mathbb{R}^2$ in $\mathcal{A}$ have overlap maps $j \circ i^{-1} : i(U \cap V) \to \mathbb{R}^2$ with affine extensions to $\mathbb{R}^2$;

(iii) f is affine with respect to the charts in $\mathcal{A}$;

(iv) Λ is a basic hyperbolic set;

(v) the images of the stable and unstable local leaves under the charts in $\mathcal{A}$ are contained in horizontal and vertical lines; and

(vi) the basic holonomies have affine extensions to the stable and unstable leaves with respect to the charts in $\mathcal{A}$.

THEOREM 4.1. *Let HD^s and HD^u be respectively the Hausdorff dimension of the intersection with Λ of the stable and unstable leaves of f. If f is C^r with $r - 1 > \max\{HD^s, HD^u\}$, and there is a complete set of holonomies for f in which the stable holonomies are C^{1,HD^s}, and the unstable holonomies are C^{1,HD^u}, then the map f on Λ is C^{1+} conjugate to a hyperbolic affine model.*

Theorem 4.1 follows from Theorem 1 in [41]. In assuming that f is C^r with $r - 1 > \max\{HD^s, HD^u\}$ in the previous theorem, we actually only use the fact that f is C^{1,HD^ι} along ι-leaves for $\iota \in \{s, u\}$.

5. Hausdorff measures

We now introduce the notion of stable and unstable measure solenoid functions. We will use the measure solenoid functions to determine which Gibbs measures are C^{1+}-realisable by C^{1+} hyperbolic diffeomorphisms. We define the δ-bounded solenoid equivalence classes of Gibbs measures which allow us to construct all hyperbolic diffeomorphisms with an f-invariant measure absolutely continuous with respect to the Hausdorff measure.

5.1. Gibbs measures. Let us give the definition of an infinite two-sided subshift of finite type $\Theta = \Theta(A)$. The elements of Θ are all infinite two-sided words $w = \ldots w_{-1} w_0 w_1 \ldots$ in the symbols $1, \ldots, k$ such that $A_{w_i w_{i+1}} = 1$, for all $i \in \mathbb{Z}$. Here $A = (A_{ij})$ is any matrix with entries 0 and 1 such that A^n has all entries positive for some $n \geq 1$. We write $w \overset{n_1, n_2}{\sim} w'$ if $w_j = w'_j$ for every $j = -n_1, \ldots, n_2$. The metric d on Θ is given by $d(w, w') = 2^{-n}$ if $n \geq 0$ is the largest such that $w \overset{n,n}{\sim} w'$. Together with this metric Θ is a compact metric space. The two-sided shift map $\tau : \Theta \to \Theta$ is the mapping which sends $w = \ldots w_{-1} w_0 w_1 \ldots$ to $v = \ldots v_{-1} v_0 v_1 \ldots$ where $v_j = w_{j+1}$ for every $j \in \mathbb{Z}$. An (n_1, n_2)-cylinder $\Theta_{w_{-n_1} \ldots w_{n_2}}$, where $w \in \Theta$, consists of all those words w' in Θ such that $w \overset{n_1, n_2}{\sim} w'$. Let Θ^u be the set of all words $w_0 w_1 \ldots$ which extend to words $\ldots w_0 w_1 \ldots$ in Θ, and, similarly, let Θ^s be the set of all words

$\dots w_{-1} w_0$ which extend to words $\dots w_{-1} w_0 \dots$ in Θ. Then $\pi_u : \Theta \to \Theta^u$ and $\pi_s : \Theta \to \Theta^s$ are the natural projection given, respectively, by

$$\pi_u(\dots w_{-1} w_0 w_1 \dots) = w_0 w_1 \dots \quad \text{and} \quad \pi_s(\dots w_{-1} w_0 w_1 \dots) = \dots w_{-1} w_0 \,.$$

An n-cylinder $\Theta^u_{w_0 \dots w_{n-1}}$ is equal to $\pi_u(\Theta_{w_0 \dots w_{n-1}})$, and likewise an n-cylinder $\Theta^s_{w_{-(n-1)} \dots w_0}$ is equal to $\pi_s(\Theta_{w_{-(n-1)} \dots w_0})$. Let $\tau_u : \Theta^u \to \Theta^u$ and $\tau_s : \Theta^s \to \Theta^s$ be the corresponding one-sided shifts.

DEFINITION 5.1. For $\iota = s$ and u, we say that $s_\iota : \Theta^\iota \to \mathbb{R}^+$ is an ι-measure scaling function if s_ι is a Hölder continuous function, and for every $\xi \in \Theta^\iota$

$$\sum_{\tau_\iota \eta = \xi} s_\iota(\eta) = 1 \,,$$

where the sum is upon all $\xi \in \Theta^\iota$ such that $\tau_\iota \eta = \xi$.

For $\iota \in \{s, u\}$, a τ-invariant measure ν on Θ determines a unique τ_ι-invariant measure $\nu_\iota = (\pi_\iota)_* \nu$ on Θ^ι. We note that a τ_ι-invariant measure ν_ι on Θ^ι has a unique τ-invariant natural extension to an invariant measure ν on Θ such that $\nu(\Theta_{w_{n_1} \dots w_{n_2}}) = \nu_\iota(\Theta^\iota_{w_{n_1} \dots w_{n_2}})$.

DEFINITION 5.2. A τ-invariant measure ν on Θ is a *Gibbs measure*:

(i) if the function $s_{\nu,u} : \Theta^u \to \mathbb{R}^+$ given by

$$s_{\nu,u}(w_0 w_1 \dots) = \lim_{n \to \infty} \frac{\nu(\Theta_{w_0 \dots w_n})}{\nu(\Theta_{w_1 \dots w_n})} \,,$$

is well-defined and it is an u-measure scaling function; and

(ii) if the function $s_{\nu,s} : \Theta^s \to \mathbb{R}^+$ given by

$$s_{\nu,s}(\dots w_1 w_0) = \lim_{n \to \infty} \frac{\nu(\Theta_{w_n \dots w_0})}{\nu(\Theta_{w_n \dots w_1})} \,,$$

is well-defined and it is an s-measure scaling function.

The following theorem follows from Corollary 2 in [38].

THEOREM 5.1 (MODULI SPACE FOR GIBBS MEASURES). *If $s_\iota : \Theta^\iota \to \mathbb{R}^+$ is an ι-measure scaling function, for $\iota = s$ or u, then there is a unique τ-invariant Gibbs measure ν such that $s_{\nu,\iota} = s_\iota$.*

5.2. Hausdorff realisations of Gibbs measures for Smale horseshoes.

The properties of the Markov partition $\mathcal{R} = \{R_1, \dots, R_k\}$ of f imply the existence of a unique two-sided subshift τ of finite type $\Theta = \Theta_A$ and a continuous surjection $i : \Theta \to \Lambda$ such that (a) $f \circ i = i \circ \tau$ and (b) $i(\Theta_j) = R_j$ for every $j = 1, \dots, k$. We call such a map $i : \Theta \to \Lambda$ a *marking of a C^{1+} hyperbolic diffeomorphism*

(f, Λ). Since f admits more than one Markov partition, a C^{1+} hyperbolic diffeomorphism (f, Λ) admits always a marking which is not unique.

Recall, from the Introduction, that a Gibbs measure ν on Θ is C^{1+}-*Hausdorff realisable by a hyperbolic diffeomorphism* $g \in \mathcal{T}(f, \Lambda)$ if, for every chart $c : U \to \mathbb{R}^2$ in the C^{1+} structure $\mathscr{C}_g$ of g, the pushforward $(c \circ i)_* \nu$ of ν is absolutely continuous (in fact, equivalent) with respect to the Hausdorff measure on $c(U \cap \Lambda)$.

THEOREM 5.2 (SMALE HORSESHOES). *Let* (f, Λ) *be a Smale horseshoe. Every Gibbs measure* ν *is* C^{1+}-*Hausdorff realisable by a hyperbolic diffeomorphism contained in* $\mathcal{T}_{f,\Lambda}(\delta_s, \delta_u)$.

However, there are Gibbs measures that are not C^{1+}-Hausdorff realisable by Anosov diffeomorphisms and codimension one attractors due to the fact that the Markov rectangles have common boundaries.

Theorem 5.2 follows from Theorem 1.6 in [42].

5.3. Hausdorff realisations of Gibbs measures for Anosov diffeomorphisms.
We will use stable and unstable measure solenoid functions to present a classification of Gibbs measures C^{1+}-Hausdorff realisable by Anosov diffeomorphisms and codimension one attractors.

Let Msol^{ι} be the set of all pairs (I, J) with the following properties: (a) If $\delta_{\iota} = 1$ then $\mathrm{Msol}^{\iota} = \mathrm{sol}^{\iota}$. (b) If $\delta_{\iota} < 1$ then $f_{\iota'} I$ and $f_{\iota'} J$ are ι-leaf 2-cylinders of a Markov rectangle R such that $f_{\iota'} I \cup f_{\iota'} J$ is an ι-leaf segment, i.e. there is a unique ι-leaf 2-gap between them. Let msol^{ι} be the set of all pairs $(I, J) \in \mathrm{Msol}^{\iota}$ such that the leaf segments I and J are not contained in an ι-global leaf containing an ι-boundary of a Markov rectangle. By construction, the set msol^{ι} is dense in Msol^{ι}, and for every pair $(C, D) \subset \mathrm{msol}^{\iota}$ there is a unique $\psi \in \Theta^{\iota'}$ and a unique $\xi \in \Theta^{\iota'}$ such that $i(\pi_{\iota'}^{-1}(\psi)) = C$ and $i(\pi_{\iota'}^{-1}(\xi)) = D$. Hence, we will denote the elements of msol^{ι} by $(\psi_{\Lambda}, \xi_{\Lambda})$, where $\psi_{\Lambda} = i(\pi_{\iota'}^{-1}(\psi))$ and $\xi_{\Lambda} = i(\pi_{\iota'}^{-1}(\xi))$.

LEMMA 5.3. *Let* ν *be a Gibbs measure on* Θ. *The* s-*measure solenoid function* $\sigma_{\nu,s} : \mathrm{msol}^s \to \mathbb{R}^+$ *of* ν *given by*

$$\sigma_{\nu,s}(\psi_{\Lambda}, \xi_{\Lambda}) = \lim_{n \to \infty} \frac{\nu(\Theta_{\psi_0 \dots \psi_n})}{\nu(\Theta_{\xi_0 \dots \xi_n})}$$

is well-defined. The u-*measure solenoid function* $\sigma_{\nu,u} : \mathrm{msol}^u \to \mathbb{R}^+$ *of* ν *given by*

$$\sigma_{\nu,u}(\psi_{\Lambda}, \xi_{\Lambda}) = \lim_{n \to \infty} \frac{\nu(\Theta_{\psi_n \dots \psi_0})}{\nu(\Theta_{\xi_n \dots \xi_0})}$$

is well-defined.

Lemma 5.3 follows from Lemma 5.4 in [42].

LEMMA 5.4. *Let* $\delta_{f,\iota} = 1$. *If an ι-measure solenoid function* $\sigma_{v,\iota} : \text{msol}^{\iota} \rightarrow$ $\mathbb{R}^+$ *has a continuous extension to* sol^{ι} *then its extension satisfies the matching condition.*

PROOF. Let $(J_0, J_1) \in \text{sol}^{\iota}$ be a pair of primary cylinders and suppose that we have pairs

$$(I_0, I_1), (I_1, I_2), \dots, (I_{n-2}, I_{n-1}) \in \text{sol}^{\iota}$$

of primary cylinders such that $f_\iota J_0 = \bigcup_{j=0}^{k-1} I_j$ and $f_\iota J_1 = \bigcup_{j=k}^{n-1} I_j$. Since the set msol^{ι} is dense in sol^{ι} there are pairs $(J_0^l, J_1^l) \in \text{msol}^{\iota}$ and pairs (I_j^l, I_{j+1}^l) with the following properties:

(i) $f_\iota J_0^l = \bigcup_{j=0}^{k-1} I_j^l$ and $f_\iota J_1^l = \bigcup_{j=k}^{n-1} I_j^i$.
(ii) The pair (J_0^l, J_1^l) converges to (J_0, J_1) when i tends to infinity.

Therefore, for every $j = 0, \dots, n-2$ the pair (I_j^l, I_{j+1}^l) converges to (I_j, I_{j+1}) when i tends to infinity. Since v is a τ-invariant measure, we get that the matching condition

$$\sigma_{v,\iota}(J_0^l : J_1^l) = \frac{1 + \sum_{j=1}^{k-1} \prod_{i=1}^{j} \sigma_{v,\iota}(I_j^l : I_{i-1}^l)}{\sum_{j=k}^{n-1} \prod_{i=1}^{j} \sigma_{v,\iota}(I_j^l : I_{i-1}^l)}$$

is satisfied for every $l \geq 1$. Since the extension of $\sigma_{v,\iota} : \text{msol}^{\iota} \rightarrow \mathbb{R}^+$ to the set sol^{ι} is continuous, we get that the matching condition also holds for the pairs (J_0, J_1) and $(I_0, I_1), \dots, (I_{n-2}, I_{n-1})$. □

THEOREM 5.5 (ANOSOV DIFFEOMORPHISMS). *Suppose that f is a C^{1+} Anosov diffeomorphism of the torus Λ. Fix a Gibbs measure v on Θ. Then the following statements are equivalent:*

(i) *The set v, $[v] \subset \mathcal{T}_{f,\Lambda}(1, 1)$ is nonempty and is precisely the set of $g \in \mathcal{T}_{f,\Lambda}(1, 1)$ such that (g, Λ_g, v) is a C^{1+} Hausdorff realisation. In this case $\mu = (i_g)_* v$ is absolutely continuous with respect to Lesbegue measure.*
(ii) *The stable measure solenoid function $\sigma_{v,s} : \text{msol}^s \rightarrow \mathbb{R}^+$ has a nonvanishing Hölder continuous extension to the closure of msol^s satisfying the boundary condition.*
(iii) *The unstable measure solenoid function $\sigma_{v,u} : \text{msol}^u \rightarrow \mathbb{R}^+$ has a nonvanishing Hölder continuous extension to the closure of msol^s satisfying the boundary condition.*

Theorem 5.5 follows from Theorem 1.4 in [42].

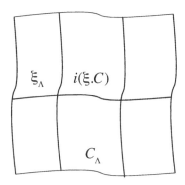

Figure 10. An ι-admissible pair (ξ, C) where $\xi_\Lambda = i(\pi_{\iota'}^{-1}\xi)$ and $C_\Lambda = i(\pi_\iota^{-1}C)$.

5.4. Extended measure scaling functions.

To present a classification of Gibbs measures C^{1+}-Hausdorff realisable by codimension one attractors, we have to define the cylinder-cylinder condition. We will express the cylinder-cylinder condition, in Section 5.5, using the extended measure scaling functions which are also useful to present, in Section 5.6, the δ_ι-bounded solenoid equivalence class of a Gibbs measure.

Throughout the paper, if $\xi \in \Theta^{\iota'}$, we denote by ξ_Λ the leaf primary cylinder segment $i(\pi_{\iota'}^{-1}\xi) \subset \Lambda$. Similarly, if C is an n_ι-cylinder of Θ^ι then we denote by C_Λ the $(1, n_\iota)$-rectangle $i(\pi_\iota^{-1}C) \subset \Lambda$.

Given $\xi \in \Theta^{\iota'}$ and an n-cylinder C of Θ^ι, we say that the pair (ξ, C) is ι-*admissible* if the set

$$\xi.C = \pi_\iota^{-1}C \cap \pi_{\iota'}^{-1}\xi$$

is nonempty (see Figure 10). The set of all ι-admissible pairs (ξ, C) is the ι-*measure scaling set* msc^ι. We construct the *extended ι-measure scaling function* $\rho : msc^\iota \to \mathbb{R}^+$ as follows: If C is a 1-cylinder then we define $\rho_\xi(C) = 1$. If C is an n-cylinder ($\Theta^u_{w_0 \dots w_{(n-1)}}$ or $\Theta^s_{w_{-(n-1)} \dots w_0}$), with $n \geq 2$, then we define

$$\rho_\xi(C) = \prod_{j=1}^{n-1} s_{\nu,\iota}(\pi_{\iota'}\tau_\iota^{n-j}(\xi.m_\iota^{j-1}C))$$

(see Figure 11), where (a) $s_{\nu,\iota}$ is the ι-measure scaling function of the Gibbs measure ν and (b) $m_u^{j-1}\Theta^u_{w_0 \dots w_{(n-1)}} = \Theta^u_{w_0 \dots w_{(n-j)}}$ and $m_s^{j-1}\Theta^s_{w_{-(n-1)} \dots w_0} = \Theta^s_{w_{-(n-j)} \dots w_0}$ (see Section 5.1).

Recall that a τ-invariant measure ν on Θ determines a unique τ_u-invariant measure $\nu_u = (\pi_u)_*\nu$ on Θ^u and a unique τ_s-invariant measure $\nu_s = (\pi_s)_*\nu$ on Θ^s. The following result follows from Theorem 1 in [38].

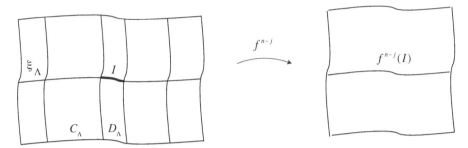

Figure 11. The $(n - j + 1)$-cylinder leaf segment $I = \xi_\Lambda \cap D_\Lambda$ and the primary leaf segment $f^{n-j}(I) = i(\pi_{\iota'}\tau_\iota^{n-j}(\xi.D))$, where $D = m_\iota^{j-1}C$.

THEOREM 5.6 (RATIO DECOMPOSITION OF A GIBBS MEASURE). *Let* ρ : $\mathrm{msc}_\iota \to \mathbb{R}^+$ *be an extended ι-measure scaling function and v the corresponding τ-invariant Gibbs measure. If C is an n-cylinder of Θ^ι then*

$$v_\iota(C) = \int_{\xi \in M} \rho_\xi(C) v_{\iota'}(d\xi),$$

where $M = \pi_{\iota'} \circ \pi_\iota^{-1} C$ *is a 1-cylinder of* $\Theta^{\iota'}$. *The ratios* $v_\iota(C)/\rho_\xi(C)$ *are uniformly bounded away from* 0 *and* ∞.

We note that $\rho_\xi(C)$ is the measure of $\xi.C$ with respect to the probability conditional measure of v on ξ.

5.5. Hausdorff realisations of Gibbs measures for Codimension one attractors.

We introduce the cylinder-cylinder condition that we will use to classify all Gibbs measures that are C^{1+}-Hausdorff realizable by codimension one attractors.

Similarly to the cylinder-gap condition given in Section 3.8 for a solenoid function, we are going to construct the cylinder-cylinder condition for a given measure solenoid function $\sigma_{v,\iota}$. Let $\delta_\iota < 1$ and $\delta_{\iota'} = 1$. Let $(I, J) \in \mathrm{Msol}^\iota$ be such that the ι-leaf segment $f_{\iota'}I \cup f_{\iota'}J$ is contained in an ι-boundary K of a Markov rectangle R_1. Then $f_{\iota'}I \cup f_{\iota'}J$ intersects another Markov rectangle R_2. Take the smallest $k \geq 0$ such that $f_{\iota'}^k I \cup f_{\iota'}^k J$ is contained in the intersection of the boundaries of two Markov rectangles M_1 and M_2. Let M_1 be the Markov rectangle with the property that $M_1 \cap f_{\iota'}^k R_1$ is a rectangle with nonempty interior, and so $M_2 \cap f_{\iota'}^k R_2$ has also nonempty interior. Then, for some positive n, there are distinct ι-leaf n-cylinders $J_1, \ldots, J_m$ contained in a primary cylinder L of M_2 such that $f_{\iota'}^k I = \bigcup_{i=1}^{p-1} J_i$ and $f_{\iota'}^k J = \bigcup_{i=p}^m J_i$. Let $\eta \in \Theta^{\iota'}$ be such that $\eta_\Lambda = L$ and, for every $i = 1, \ldots, m$, let D_i be a cylinder of Θ^ι such that $i(\eta.D_i) = J_i$. Let $\xi \in \Theta^{\iota'}$ be such that $\xi_\Lambda = K$ and C_1 and C_2 cylinders of Θ^ι such that $i(\xi.C_1) = f_{\iota'}I$ and $i(\xi.C_2) = f_{\iota'}J$. We say that the measure solenoid function $\sigma_{v,\iota}$ of the Gibbs measure v satisfies the *cylinder-cylinder condition*

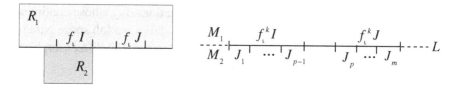

Figure 12. The cylinder-cylinder condition for ι-leaf segments.

(see Figure 12) if, for all such leaf segments,

$$\frac{\rho_\xi(C_2)}{\rho_\xi(C_1)} = \frac{\sum_{i=p}^m \rho_\eta(D_i)}{\sum_{i=1}^{p-1} \rho_\eta(D_i)}$$

where ρ is the measure scaling function determined by ν.

REMARK 5.7. A function $\sigma : \mathrm{msol}^\iota \to \mathbb{R}^+$ that has an Hölder continuous extension to Msol^ι determines an extended scaling function, and so we can check if the function σ satisfies or not the cylinder-cylinder condition.

THEOREM 5.8 (CODIMENSION ONE ATTRACTORS). *Suppose that f is a C^{1+} surface diffeomorphism and Λ is a codimension one hyperbolic attractor. Fix a Gibbs measure ν on Θ. Then the following statements are equivalent:*

(i) *For all $0 < \delta_s < 1$, $[\nu] \subset \mathcal{T}_{f,\Lambda}(\delta_s, 1)$ is nonempty and is precisely the set of $g \in \mathcal{T}_{f,\Lambda}(\delta_s, 1)$ such that (g, Λ_g, ν) is a C^{1+} Hausdorff realisation. In this case $\mu = (i_g)_*\nu$ is absolutely continuous with respect to the Hausdorff measure on Λ_g.*

(ii) *The stable measure solenoid function $\sigma_{\nu,s} : \mathrm{msol}^s \to \mathbb{R}^+$ has a nonvanishing Hölder continuous extension to the closure of msol^s satisfying the cylinder-cylinder condition.*

(iii) *The unstable measure solenoid function $\sigma_{\nu,u} : \mathrm{msol}^u \to \mathbb{R}^+$ has a nonvanishing Hölder continuous extension to the closure of msol^u.*

Theorem 5.8 follows from Theorem 1.5 in [42].

5.6. The moduli space for hyperbolic realizations of Gibbs measures. Let $\mathcal{SOL}^\iota$ be the space of all Hölder continuous functions $\sigma_\iota : \mathrm{Msol}^\iota \to \mathbb{R}^+$ with the following properties:

(i) If $HD^\iota = 1$ then σ_ι is an ι-solenoid function.
(ii) If $HD^\iota < 1$ and $HD^{\iota'} = 1$ then σ_ι satisfies the cylinder-cylinder condition.
(iii) If $HD^\iota < 1$ and $HD^{\iota'} < 1$ then σ_ι does not have to satisfy any extra property.

We recall that HD^ι is the Hausdorff dimension of the ι-leaf segments intersected with Λ.

By Theorems 5.2, 5.5 and 5.8, for ι equal to s and u, we obtain that the map $\nu \to \sigma_{\nu,\iota}$ gives a one-to-one correspondence between the sets $[\nu]$ contained in

$\mathcal{T}_{f,\Lambda}(\delta_s, \delta_u)$ and the space of measure solenoid functions $\sigma_{g,\iota}$ whose continuous extension is contained in $\mathcal{SOL}^\iota$. Hence, the set $\mathcal{SOL}^\iota$ is a moduli space parameterizing all Lipschitz conjugacy classes $[v]$ of C^{1+} hyperbolic diffeomorphisms contained in $\mathcal{T}_{f,\Lambda}(\delta_s, \delta_u)$ (see Corollary 1.11).

DEFINITION 5.3. The δ_ι-*bounded solenoid equivalence class* of a Gibbs measure v is the set of all solenoid functions σ_ι with the following properties: There is $K = K(\sigma_\iota) > 0$ such that for every pair $(\xi, C) \in \mathrm{msc}_\iota$

$$\left| \delta_\iota \log s_\iota(C_\Lambda \cap \xi_\Lambda : \xi_\Lambda) - \log \rho_\xi(C) \right| < K,$$

where (i) ρ is the ι-measure scaling function of v and (ii) s_ι is the scaling function determined by σ_ι.

Let $\sigma_{1,\iota}$ and $\sigma_{2,\iota}$ be two solenoid functions in the same δ_ι-bounded equivalence class of a Gibbs measure v. Using that $\sigma_{1,\iota}$ and $\sigma_{2,\iota}$ are bounded away from zero, we obtain that the corresponding scaling functions also satisfy inequality (3–6) for all pairs $(J, m^i J)$ where J is an ι-leaf $(i+1)$-gap. Hence, the solenoid functions $\sigma_{1,\iota}$ and $\sigma_{2,\iota}$ are in the same bounded equivalence class (see Definition 3.2).

THEOREM 5.9. (i) *There is a natural map* $g \to (\sigma_s(g), \sigma_u(g))$ *which gives a one-to-one correspondence between* C^{1+} *conjugacy classes of* C^{1+} *hyperbolic diffeomorphisms* $g \in [v]$ *and pairs* $(\sigma_s(g), \sigma_u(g))$ *of stable and unstable solenoid functions such that, for* ι *equal to s and u,* $\sigma_\iota(g)$ *is contained in the* δ_ι*-bounded solenoid equivalence class of* v.

(ii) *Given an* ι*-solenoid function* σ_ι *and* $0 < \delta_{\iota'} \le 1$, *there is a unique Gibbs measure* v *and a unique* $\delta_{\iota'}$*-bounded equivalence class of* v *consisting of* ι'*-solenoid functions* $\sigma_{\iota'}$ *such that the* C^{1+} *conjugacy class of hyperbolic diffeomorphisms* $g \in \mathcal{T}_{f,\Lambda}(\delta_s, \delta_u)$ *determined by the pair* (σ_s, σ_u) *have an invariant measure* $\mu = (i_g)_* v$ *absolutely continuous with respect to the Hausdorff measure.*

Theorem 5.9 follows from combining Lemmas 8.17 and 8.18 in [42].

Acknowledgments

We are grateful to Dennis Sullivan and Flávio Ferreira for a number of very fruitful and useful discussions on this work and also for their friendship and encouragement.

We thank the Calouste Gulbenkian Foundation, PRODYN-ESF, POCTI and POSI by FCT and the Portuguese Ministry of Science and Technology, and the Centro de Matemática da Universidade do Porto for their financial support of

A. A. Pinto, and the Wolfson Foundation and the UK Engineering and Physical Sciences Research Council for their financial support of D. A. Rand.

Part of this research was started during a visit by the authors to the IHES and CUNY and continued at the Isaac Newton Institute, IMPA, MSRI and SUNY. We thank them for their hospitality.

References

[1] J. F. Alves, V. Pinheiro and A. A. Pinto, Explosion of smoothness for multimodal maps. In preparation.

[2] D. V. Anosov, Tangent fields of transversal foliations in U-systems. (Russian). *Mat. Zametki* **2** (1967), 539–548. English transl. in *Math. Notes Acad. Sci. USSR* **2**:5 (1967), 818–823.

[3] V. Arnol'd, Small denominators, I: Mapping of a circle into itself. (Russian). *Investijia Akad. Nauk Math* **25**:1 (1961), 21–96. English transl. in *Transl. AMS (2)* **46** (1965), 213–284.

[4] A. Avez, Anosov diffeomorphims. *Proceedings of the Topological Dynamics Symposium* (Fort Collins, CO, 1967), 17–51, W. Gottschalk and J. Auslander (eds.). Benjamin, New York, 1968.

[5] R. Bowen, *Equilibrium states and the ergodic theory of Anosov diffeomorphisms.* Lecture Notes in Mathematics **470**. Springer, Berlin, 1975.

[6] E. Cawley, The Teichmüller space of an Anosov diffeomorphism of T^2. *Invent. Math.* **112** (1993), 351–376.

[7] K. J. Falconer, *The geometry of fractal sets.* Cambridge Tracts in Mathematics **85**. Cambridge University Press, Cambridge, 1986.

[8] E. de Faria, Quasisymmetric distortion and rigidity of expanding endomorphisms of S^1. *Proc. Amer. Math. Soc.* **124** (1996), 1949–1957.

[9] H. Federer, *Geometric measure theory.* Springer, New York, 1969.

[10] M. J. Feigenbaum, Presentation functions, fixed points, and a theory of scaling function dynamics. *J. Stat. Phys.* **52** (1988), 527–569.

[11] M. J. Feigenbaum, Presentation functions, and scaling function theory for circle maps. *Nonlinearity* **1** (1988), 577–602.

[12] F. F. Ferreira and A. A. Pinto, Explosion of smoothness from a point to everywhere for conjugacies between Markov families. *Dyn. Syst.* **16**:2 (2001), 193–212.

[13] F. F. Ferreira and A. A. Pinto, Explosion of smoothness from a point to everywhere for conjugacies between diffeomorphisms on surfaces. *Ergodic Theory Dynam. Systems* **23**:2 (2003), 509–517.

[14] F. F. Ferreira, A. A. Pinto and D. A. Rand, Non-existence of affine models for attractors on surfaces. Submitted.

[15] J. Franks, Anosov diffeomorphisms. *Global Analysis*, S. Smale (ed.), 61–93. American Mathematical Society, Providence, RI, 1970.

[16] J. Franks, Anosov diffeomorphisms on tori. *Trans. Amer. Math. Soc.* **145** (1969), 117–124.

[17] L. Flaminio and A. Katok, Rigidity of symplectic Anosov diffeomorphisms on low dimensional tori. *Ergodic Theory Dynam. Systems* **11** (1991), 427–440.

[18] E. Ghys, Rigidité différentiable des groupes fuchsiens. *Publ. Math. IHES* **78** (1993), 163–185.

[19] B. Hasselblatt, Regularity of the Anosov splitting and of horospheric foliations. *Ergodic Theory Dynam. Systems* **14** (1994), 645–666.

[20] M. R. Herman, Sur la conjugaison différentiable des difféomorphismes du cercle á des rotations. *Publ. Math. IHES* **49** (1979), 5–233.

[21] M. Hirsch and C. Pugh, Stable manifolds and hyperbolic sets. *Global analysis* (Berkeley, CA, 1968). Proc. Symp. Pure Math. **14**, 133–164. American Mathematical Society, Providence, RI, 1970.

[22] S. Hurder and A. Katok, Differentiability, rigidity and Godbillon–Vey classes for Anosov flows. *Publ. Math. IHES* **72** (1990), 5–61.

[23] J. L. Journé, A regularity lemma for functions of several variables. *Rev. Mat. Iberoamericana* **4** (1988), 187–193.

[24] A. Katok and B. Hasselblatt, *Introduction to the modern theory of dynamical systems*. Encyclopedia of Mathematics and its Applications **54**. Cambridge University Press, Cambridge, 1995.

[25] A. N. Livšic and J. G. Sinai, Invariant measures that are compatible with smoothness for transitive C-systems. (Russian). *Dokl. Akad. Nauk SSSR* **207** (1972), 1039–1041.

[26] R. Llave, J. M. Marco and R. Moriyon, Canonical perturbation theory of Anosov systems and regularity results for the Livsic cohomology equations. *Annals of Math.* **123** (1986), 537–612.

[27] R. Llave, Invariants for smooth conjugacy of hyperbolic dynamical systems, II. *Comm. Math. Phys.* **109** (1987), 369–378.

[28] R. Mañé, *Ergodic theory and differentiable dynamics*. Springer, Berlin, 1987.

[29] A. Manning, There are no new Anosov diffeomorphisms on tori. *Amer. J. Math.* **96** (1974), 422.

[30] J. M. Marco and R. Moriyon, Invariants for smooth conjugacy of hyperbolic dynamical systems, I. *Comm. Math. Phys.* **109** (1987), 681–689.

[31] J. M. Marco and R. Moriyon, Invariants for smooth conjugacy of hyperbolic dynamical systems, III. *Comm. Math. Phys.* **112** (1987), 317–333.

[32] W. de Melo and S. van Strien, *One-dimensional dynamics*. Ergebnisse der Mathematik und ihrer Grenzgebiete (3) **25**. Springer, Berlin, 1993.

[33] S. Newhouse, On codimension one Anosov diffeomorphisms. *Amer. J. Math.* **92** (1970), 671–762.

[34] S. Newhouse and J. Palis, Hyperbolic nonwandering sets on two-dimensional manifolds. *Dynamical systems: Proceedings of a Symposium* (Salvador, 1971), M. Peixoto (ed.), 293–301. Academic Press, New York, 1973.

[35] A. Norton, Denjoy's theorem with exponents. *Proc. Amer. Math. Soc.* **127**:10 (1999), 3111–3118.

[36] A. A. Pinto and D. A. Rand, Classifying C^{1+} structures on hyperbolical fractals, I: The moduli space of solenoid functions for Markov maps on train-tracks. *Ergodic Theory Dynam. Systems* **15**:4 (1995), 697–734.

[37] A. A. Pinto and D. A. Rand, Classifying C^{1+} structures on hyperbolical fractals, II: Embedded trees. *Ergodic Theory Dynam. Systems* **15**:5 (1995), 969–992.

[38] A. A. Pinto and D. A. Rand, Existence, uniqueness and ratio decomposition for Gibbs states via duality. *Ergodic Theory Dynam. Systems* **21**:2 (2001), 533–543.

[39] A. A. Pinto and D. A. Rand, Teichmüller spaces and HR structures for hyperbolic surface dynamics. *Ergodic Theory Dynam. Systems* **22**:6 (2002), 1905–1931.

[40] A. A. Pinto and D. A. Rand, Smoothness of holonomies for codimension 1 hyperbolic dynamics. *Bull. London Math. Soc.* **34** (2002), 341–352.

[41] A. A. Pinto and D. A. Rand, Rigidity of hyperbolic sets on surfaces. *J. London Math Soc.* **2** (2004), 1–22.

[42] A. A. Pinto and D. A. Rand, Geometric measures for hyperbolic surface dynamics. Submitted.

[43] A. A. Pinto and D. Sullivan, Assymptotic geometry applied to dynamical systems. Stony Brook preprint, 2004.

[44] D. A. Rand, Universality and renormalisation in dynamical systems. *New directions in dynamical systems*, T. Bedford and J. Swift (eds.), 1–56. Cambridge University Press, Cambridge, 1988.

[45] M. Shub, *Global stability of dynamical systems.* Springer, New York, 1987.

[46] Y. G. Sinai, Markov partitions and U-diffeomorphisms. *Funkcional. Anal. i Priložen.* **2**:1, 64–89. (Russian). English transl. in *Funct. Anal. Appl.* **2** (1968), 61–82.

[47] D. Sullivan, Bounds, quadratic differentials, and renormalization conjectures. *American Mathematical Society centennial publications, II* (Providence, RI, 1988), 417–466. American Mathematical Society, Providence, RI, 1992.

[48] D. Sullivan, Differentiable structures on fractal-like sets determined by intrinsic scaling functions on dual Cantor sets. *The mathematical heritage of Hermann Weyl* (Durham, NC, 1987), 15–23. Proc. Sympos. Pure Math. **48**. American Mathematical Society, Providence, RI, 1988.

[49] D. Sullivan, Linking the universalities of Milnor–Thurston, Feigenbaum and Ahlfors–Bers. *Topological methods in modern mathematics*, L. Goldberg and A. Phillips (eds.), 543–563. Publish or Perish, Boston, 1993.

[50] R. F. Williams, Expanding attractors. *Publ. Math. IHES* **43** (1974), 169–203.

[51] J. C. Yoccoz, Conjugaison différentiable des difféomorphismes du cercle dont le nombre de rotation vérifie une condition diophantienne. *Ann. Sci. École. Norm. Sup. (4)* **17**:3 (1984), 333–359.

ALBERTO A. PINTO
FACULDADE DE CIÊNCIAS
UNIVERSIDADE DO PORTO 4000 PORTO
PORTUGAL
aapinto@fc.up.pt

DAVID A. RAND
MATHEMATICS INSTITUTE, UNIVERSITY OF WARWICK
COVENTRY CV4 7AL
UNITED KINGDOM
dar@maths.warwick.ac.uk

Recent Progress in Dynamics
MSRI Publications
Volume **54**, 2007

Random walks derived from billiards

RENATO FERES

To Anatoly Katok, on his 60th birthday

ABSTRACT. We introduce a class of random dynamical systems derived from billiard maps, which we call *random billiards*, and study certain random walks on the real line obtained from them. The interplay between the billiard geometry and the stochastic properties of the random billiard is investigated. Our main results are concerned with the spectrum of the random billiard's Markov operator. We also describe some basic properties of diffusion limits under appropriate scaling.

1. Introduction

This work is motivated by the following problem about gas kinetics. Suppose that a short pulse of inert gas at very low pressure is released from a point inside a long but finite cylindrical channel. The time at which gas molecules escape the channel through its open ends is then measured by some device such as a mass spectrometer. The inner surface of the cylinder is not perfectly flat due to its molecular structure, imagined as a periodic relief. It is not altogether unreasonable to think that the interaction between the fast moving (inert) gas molecules and the surface is essentially elastic, and that any thermal effects can be disregarded on first approximation. (See [ACM] for a more detailed physical justification of this assumption.) We thus think of the gas-surface interaction as billiard-like. (M. Knudsen, in his classical theoretical and experimental studies on the kinetic theory of gases begun around 1907, used a tennis ball metaphor [Kn, p. 26].) The assumption of low pressure simply means that the collisions among gas molecules are in sufficiently small numbers to be disregarded and only collisions between gas molecules and the channel inner surface are taken into account. The problem is now this: from the time of escape data, possibly for a range of values of channel length, we wish to extract information about the microgeometry of the channel surface.

A mathematical formulation of this problem is proposed below in terms of random dynamical systems derived from deterministic billiards. For each choice of billiard geometry, the interaction between gas molecules and channel inner surface is encoded in a Markov (scattering) operator, P. The study of P is the main focus of the paper, but we also consider to a lesser extent the random flight inside the channel derived from P, and make a number of general remarks about the asymptotic behavior of the time of escape for long channels. More details about the relationship between geometric properties of the billiard cells and diffusion characteristics of (limits of) the random walk will be explored in a future paper. (For a numerical study of this relationship, see [FY2].) The definitions given here were introduced in essentially nonmathematical form in our [FY1].

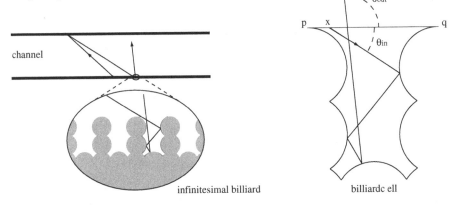

Figure 1. Channel with microgeometry, showing the initial steps of a billiard particle flight (left) and the trajectory inside a billiard cell (right). When the billiard cell is regarded as infinitesimal in the sense defined in the text, the initial point x on the open side $\overline{pq}$ becomes a random variable uniformly distributed over the interval from p to q. The angle θ_{out} then becomes a random function of θ_{in}. The transition probabilities of the events $\theta_{\text{in}} \mapsto \theta_{\text{out}}$ are described by a Markov operator P canonically specified by the cell's shape.

The wished-for general theory, of which the present paper is only the first step, can be seen in a nutshell with the help of a simple example. (The mainly numerical study [FY2] shows a more detailed overview, with many more examples. See also Section 7.) Before describing the example, let us define the time-of-escape function, $f(x)$, a little more precisely. Consider the quantity $\tau(L, r, v)$, defined as the mean time of escape of a pulse of billiard particles released from the middle point of a channel of length $2L$ and radius r, where

v is the scalar velocity, assumed equal for all particles. It is convenient to work with a dimensionless quantity $f(x)$, where $x = L/r$, given by

$$\tau(L, r, v) = \frac{r}{v} f(x).$$

That τ indeed has such a functional form is due to the elementary observation that for any positive number c, the mean time of escape satisfies the relations $\tau(cL, cr, cv) = \tau(L, r, v)$ and $\tau(L, r, cv) = \tau(L, r, v)/c$. Notice that $f(x)$ depends exclusively on the microgeometry of the channel surface encoded by P. The central problem of the theory is to extract information about this microgeometry from $f(x)$.

Let the microgeometry be that of Figure 2. There are two geometric parameters: h and b. It will be clear from the definition of P that it only depends on the shape of the billiard cell up to homothety. So the only geometric parameter we can hope to recover from this time-of-escape experiment is the ratio b/h.

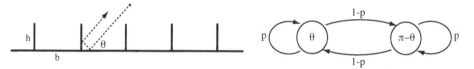

Figure 2. Transition probabilities for this "fence microgeometry" are given by the above Markov diagram. Standard random walk on $\mathbb{R}$ ($p = 1/2$) corresponds to the projection of the random flight in a channel of radius $r = 1/2$, $\theta = \pi/4$, and $b/h = 4$.

The transition probabilities, represented by p and $1 - p$ in the diagram on the right-hand side of Figure 2, are given as follows. Let $k \in \mathbb{Z}$ denote the integer part of $2h/b \tan \theta$, and $s \in [0, 1)$ the fractional part, so that

$$\frac{2h}{b \tan \theta} = k + s.$$

Then, if k is even, $p = 1 - s$, and if k is odd, $p = s$. This is easily obtained by inspecting the unfolding of the billiard cell as shown in Figure 3.

The Markov operator completely determines the microgeometry, up to scale, for this (one-parameter) family of geometries. In other words, the quantity h/b is known if we know the transition probabilities for all values of θ. The standard random walk on a line (which has probability $1/2$ of jumping either forward or backward by a fixed length) corresponds to the horizontal projection of a random flight in a 2-dimensional channel of radius $r = 1/2$, $v = 1$, and box dimensions $\theta = \pi/4$, and $b/h = 4$.

It can also be shown that the parameter b/h is recovered from the time-of-escape function $f(x)$. For concreteness, let us suppose that the initial direction

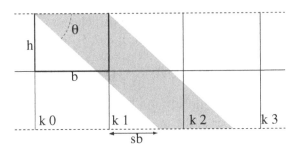

Figure 3. Unfolding of a billiard cell of Figure 2, used to determine the transition probabilities.

is given by the angle $\theta = \pi/4$ and the lengths of the base, b, and height, h, of the rectangular cell satisfy $b > 2h$. Then (see Section 7), for large values of x (i.e., large ratio L/r), we have the asymptotic expression

$$\frac{f(x)}{x^2} \sim \frac{1}{\sqrt{2}}\left(\frac{b}{2h} - 1\right)^{-1}.$$

Therefore, the single geometric parameter of this family of microgeometries, namely the ratio b/h, can be recovered from $f(x)$. This follows from a central limit theorem for Markov chains, which can also be used to show that for long channels the probability density, $u(t, z)$, of finding a billiard particle at time t and position z along the axis of the channel satisfies a standard diffusion equation

$$\frac{\partial u}{\partial t} = D\frac{\partial^2 u}{\partial z^2},$$

where $D = (1/\sqrt{2})(b/2h - 1)rv$ is the diffusion constant, r is the channel radius, and v is the constant scalar velocity of the billiard particles.

The main problem is thus to understand how much geometric information about the billiard cell is contained in $f(x)$. Another, relatively simpler (though not simple), problem is to understand how much geometric information is contained in the spectrum of the operator P. As will be seen later in the paper, P will have discrete real spectrum for a large class of billiard geometries. We should note that the above example is not entirely representative of most microgeometries in some important ways. Most importantly, the correct asymptotic expression of $f(x)$ for large x will typically be $D^{-1}x^2/\ln x$, rather than $D^{-1}x^2$, where D is a diffusion constant. This is briefly discussed in Section 7.

The paper is organized as follows. The first section introduces the idea of *channel microgeometry* and the way a point particle interacts with it. This naturally leads to the definition of a *random billiard map*, generating a random dynamical system on a set V. Here V stands for the interval $[0, \pi]$ in the two-dimensional version of the problem, or the unit hemisphere in the three-dimensional version. The microgeometry specifies a Markov chain with state space V whose transition probabilities operator, P, replaces the ordinary reflection law of deterministic billiards. The Markov chain gives at each moment in time the velocity component of a random flight inside the channel.

The general properties of the random billiard Markov chain are then investigated. It is shown that this random process has a canonical stationary measure, μ, obtained from the Liouville measure of the associated deterministic billiard. A simple ergodicity criterion for μ is provided.

It is shown next that P is a bounded self-adjoint operator on $L^2(V, \mu)$. Self-adjointness is a consequence of time reversibility of the deterministic system, which also implies reversibility of the Markov chain. Under more stringent assumptions (essentially that the underlying deterministic billiard is a dispersing, or Sinai, billiard) it is shown that P is a Hilbert–Schmidt operator. This is one of the central results of the paper.

We then turn to the associated random flight. It is shown by means of an appropriate central limit theorem how the random flight gives rise to Brownian motion, with variance that depends on the spectrum of P, hence on the microgeometry of the channel. This is accomplished most easily in cases when the stationary measure is nonergodic and does not contain the direction of the axis of the channel in its support. The general (ergodic) case, which we consider only briefly, mainly through one example, requires a more careful analysis as the jumps of the random walk have infinite variance. The probability distribution of these jumps is, nevertheless, in the domain of attraction of a normal distribution and ordinary diffusion can still be obtained under an appropriate scaling limit. Explicit values for the diffusion constant are derived for a few simple examples. Finally, explicit formulas for the mean time and mean number of collisions inside a billiard cell are given in terms of parameters associated to the shape of the cell.

Acknowledgment. This paper has benefited from conversations with M. Nicol and D. Szász. I wish to express my thanks to them.

2. Main definitions

The central concepts of the paper—the definitions of a *random billiard*, the *microgeometry* of a channel, the associated Markov operator and random flight in a channel—are introduced here.

2.1. Reflection off a wall with infinitesimal structure. A surface (or curve) *microgeometry*, and how a point particle interacts with it, is the first idea that needs to be explained. The precise description of the particle-surface interaction will involve the notion of a *random billiard*, defined in the next subsection. Although not logically needed, the following remark should provide a conceptual justification for the definitions. The key point to clarify is how to define the reflection law for a billiard particle bouncing off a wall that has periodic geometric features at an infinitesimal scale.

Consider a family of piecewise differentiable surfaces, or curves, S_a, which approximates (in Hausdorff metric) another piecewise differentiable surface S as indicated by Figure 4. Each S_a can be thought of as superposing to S a periodic geometry scaled down in size, with scale parameter a, and slightly deformed to account for the curvature of S. A procedure that yields a well-defined limit for the billiard reflection as $a \to 0$ is the following. As a first step, we replace the incoming velocity v with a random variable V_ε whose probability distribution has smooth density and is sharply concentrated around v, approaching v as $\varepsilon \to 0$. The distribution of reflected velocities is then obtained, resulting in a random variable with probability measure $\mu_{a,\varepsilon}^v$. The reflection law for the incident velocity, v, is now defined by the limit (in the weak* topology)

$$\mu_v := \lim_{\varepsilon \to 0} \lim_{a \to 0} \mu_{a,\varepsilon}^v.$$

It can be shown that the limit exits and the resulting measure is uniquely determined by the scaled down geometry. It is not difficult to obtain, using elementary facts about oscillatory integrals, that the limit has a very explicit form. In dimension two (writing $\mu_v = \mu_\theta$) it is given by

$$\mu_\theta(h) = \int_0^1 h(\Psi_x(\theta))\, dx,$$

where $\theta \in (0, \pi)$ is the incidence angle measured with respect to the tangent line to S at the point of collision, $x \in [0, 1]$ parametrizes a point on the open

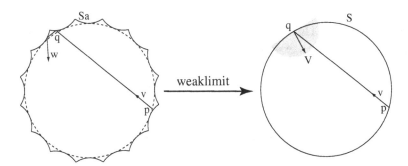

Figure 4. S is fixed and S_a varies so that bumps scale down to zero in size. As sets, $S_a \to S$, but reflection law becomes probabilistic, represented by a *scattering operator*.

side of one of the "microscopic" indentations of S_a inside of which the particle executes a billiard ball motion (see Figure 4), $\Psi_x(\theta)$ is the angle that the particle makes with S as it leaves the indentation, and $\mu_\theta(h)$ denotes the integral of an arbitrary continuous function h on $[0, \pi]$ with measure μ_θ. (See [FY1] for more details about the above limit.) In dimension 3, dx is Lebesgue measure over the unit square.

The key observation here is that the direction of reflection for an incoming velocity v should be regarded as a random variable. Its probability law, μ_v, is obtained by assuming that the point on the open side of the indentation through which the particle passes is a random variable uniformly distributed over the area (or length) of that open side. The indentations are the billiard cells of the random billiard, as defined next.

2.2. Random billiards. The billiard table with an open side that goes into the definition of a random billiard should be thought as representing the individual indentations of the surface S_a for a very small a. Consider an ordinary billiard system consisting of a billiard table B. The boundary of B is the union of a finite number of smooth curves, called the *sides* of the table. One of the sides is distinguished, and for our purposes it will always be a segment of line (or the 2-torus in dimension 3; see below), which will be called the *open side*. This is the segment pq of the billiard cell on the right-hand side of Figure 1.

Choose a number $s \in [0, 1]$ at random with uniform probability and set the initial point of the billiard trajectory to be $x = p + s(q - p)$. Denote by $\Psi_x : [0, \pi] \to [0, \pi]$ the angle component of the first return map of the billiard trajectory back to the open side. Referring to Figure 1 (right-hand side), we have $\theta_{\text{out}} = \Psi_x(\theta_{\text{in}})$. We often identify the open side with the interval $[0, 1]$. The first return map to pq of the ordinary billiard is a map T from the rectangle $[0, 1] \times [0, \pi]$ to itself such that $T(x, \theta) = (y, \Psi_x(\theta))$. The random billiard is then the random dynamical system $(\{\Psi_x : x \in [0, 1]\}, \mathcal{B}, \lambda)$, where λ is Lebesgue measure on the unit interval and $\mathcal{B}$ is the Borel σ-algebra.

Figure 5 provides one interpretation for the iteration of a random billiard map. The table B is doubled over its flat side; a trajectory of the random billiard is

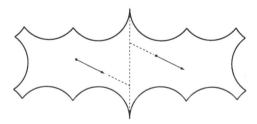

Figure 5. Doubling of billiard table with a random map.

an ordinary billiard trajectory until it crosses the separation line between the two copies of B (the dashed line in Figure 5), at which moment it jumps to a randomly chosen point on the line, keeping the velocity unchanged.

As an example of random billiard map, consider the (doubled up) triangular table shown in Figure 6. Let $\alpha < \pi/6$. Define maps $T_i : [0, \pi] \to \mathbb{R}$ by

$$T_1(\theta) = \theta + 2\alpha$$
$$T_2(\theta) = -\theta + 2\pi - 4\alpha$$
$$T_3(\theta) = \theta - 2\alpha$$
$$T_4(\theta) = -\theta + 4\alpha.$$

The random billiard map $T : [0, \pi] \to [0, \pi]$ is given by $T(\theta) = T_i(\theta)$ with probability $p_i(\theta)$. To specify p_i, first define the function

$$u_\alpha(\theta) = \frac{1}{2}\left(1 + \frac{\tan \alpha}{\tan \theta}\right).$$

Now define

$$p_1(\theta) = \begin{cases} 1 & \theta \in [0, \alpha) \\ u_\alpha(\theta) & \theta \in [\alpha, \pi - 3\alpha) \\ 2\cos(2\alpha)u_{2\alpha}(\theta) & \theta \in [\pi - 3\alpha, \pi - 2\alpha) \\ 0 & \theta \in [\pi - 2\alpha, \pi] \end{cases}$$

$$p_2(\theta) = \begin{cases} 0 & \theta \in [0, \pi - 3\alpha) \\ u_\alpha(\theta) - 2\cos(2\alpha)u_{2\alpha}(\theta) & \theta \in [\pi - 3\alpha, \pi - 2\alpha) \\ u_\alpha(\theta) & \theta \in [\pi - 2\alpha, \pi - \alpha) \\ 0 & \theta \in [\pi - \alpha, \pi] \end{cases}$$

$$p_3(\theta) = \begin{cases} 0 & \theta \in [0, 2\alpha) \\ 2\cos(2\alpha)u_{2\alpha}(-\theta) & \theta \in [2\alpha, 3\alpha) \\ u_\alpha(-\theta) & \theta \in [3\alpha, \pi - \alpha) \\ 1 & \theta \in [\pi - \alpha, \pi] \end{cases}$$

$$p_4(\theta) = \begin{cases} 0 & \theta \in [0, \alpha) \\ u_\alpha(-\theta) & \theta \in [\alpha, 2\alpha) \\ u_\alpha(-\theta) - 2\cos(2\alpha)u_{2\alpha}(-\theta) & \theta \in [2\alpha, 3\alpha) \\ 0 & \theta \in [3\alpha, \pi]. \end{cases}$$

Figure 6 shows the graph of the random map T.

If $\alpha = p\pi/q$, for p, q positive integers, the maps T_i generate a dihedral group, D_m (of order $2m$), where $m = q$ if q is odd, and $m = q/2$ if q is even. The random dynamical system can be regarded as a random walk on an orbit $D_m \cdot \theta$, for $\theta \in [0, \pi]$. The generators of the random walk are $T_1, T_2, T_3 = T_1^{m-1}, T_4 = T_2 \circ T_1^{m-4}$, chosen with probabilities $p_i(\theta')$, for θ' on the orbit of θ. Notice that any pair $(\mathcal{P}, s)$, where $\mathcal{P}$ is a rational polygon in the plane (specified up to

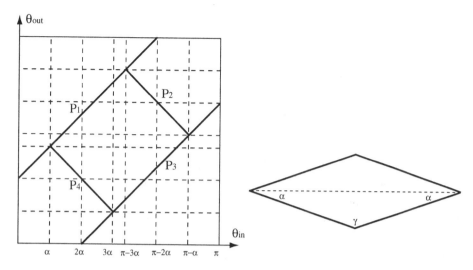

Figure 6. Random billiard map (left) for the shallow triangular table (lower half of the figure on the right).

a similarity transformation) and s is a choice of side of $\mathcal{P}$, determines such a position dependent random walk on a dihedral group.

2.3. Dimension 3. In dimension 3, the boundary surface of the billiard table will be a piecewise smooth surface without boundary, contained in $\mathbb{T}^2 \times [0, a]$, that separates the top and bottom tori; that is, any geodesic segment in $\mathbb{T}^2 \times [0, a]$ that starts in $\mathbb{T}^2 \times \{a\}$ and ends in $\mathbb{T}^2 \times \{0\}$ must intersect the surface. The top torus, denoted simply by $\mathbb{T}^2$, assumes the same role as the distinguished side of the two dimensional billiard. A trajectory of the three dimensional open billiard might look like the one shown in Figure 7. Incoming and outgoing velocities are parametrized by the upper half (unit) hemisphere,

$$S^+ := \{v = (v_1, v_2, v_3) : v_3 > 0, |v| = 1\}.$$

The random dynamical system is now a random map on S^+, denoted $\Psi_x : S^+ \to S^+$, for $x \in \mathbb{T}^2$. The torus is given the normalized Lebesgue measure. Notice that any choice of a (greater than the height of the surface itself) will produce the same random dynamical system.

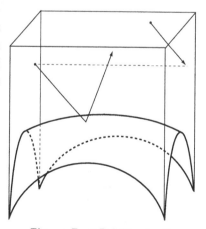

Figure 7. 3-D billiard cell.

The random billiard can be expressed as an ordinary (deterministic) dynamical system on an extended state space, as follows. In dimension 2, let $\Omega = [0, 1]^{\mathbb{N}}$ and $\nu = \lambda^{\mathbb{N}}$, where λ is the Lebesgue measure on the unit interval. Define the shift map $\sigma : \Omega \to \Omega$ by $\sigma(\omega)_i := \omega_{i+1}$ and $\Psi_\omega := \Psi_{\omega_1}$. This gives a map $\Psi : \Omega \times [0, \pi] \to \Omega \times [0, \pi]$. The n-th iterate of Ψ is

$$\Psi^n(\omega, \theta) = (\Psi_{\sigma^n(\omega)} \circ \cdots \circ \Psi_{\sigma(\omega)} \circ \Psi_\omega)(\theta).$$

The three-dimensional case is similarly defined.

We summarize here some notation about billiards that will be used consistently throughout the paper. Let $(X, \mathcal{B}, \lambda)$ denote the measure space $[0, 1]$ or $\mathbb{T}^2$ with the normalized Lebesgue measure and the Borel σ-algebra. Let V denote, respectively, $[0, \pi]$ or S^+, and $\Psi_x : V \to V$, the random billiard maps. At times it will be convenient to write $F_v(x) := \Psi_x(v)$. In what follows, μ will denote (unless explicitly stated otherwise) the probability measure on V defined by $d\mu(\theta) = \frac{1}{2} \sin \theta d\theta$ if $V = [0, \pi]$, and $d\mu(v) = v \cdot n \, dA(v)$, for $V = S^+$, where dA is the area element of S^+. In spherical coordinates, $d\mu(\theta, \phi) = (1/2\pi) \sin(2\theta) d\theta d\phi$, for $0 \le \theta \le \pi/2$ and $0 \le \phi \le 2\pi$.

Let $E = X \times V$ denote the phase space of the open side of the (deterministic) billiard table, both in dimension 2 and dimension 3. The projection to second component will be written $\pi_2 : E \to V$. The Liouville measure of the deterministic billiard gives, after restriction and normalization, the measure $\nu = \lambda \otimes \mu$ on E. Notice that ν is invariant under the first return map $T : E \to E$.

The random system can also be studied as a Markov chain, which is the point of view we mostly take. This will be described shortly, after we explain the connection between the random billiard and random flights in a channel. For detailed definitions and basic properties about random dynamical systems, we refer the reader to [Ar] and [Re].

2.4. Channels with microgeometry and random flights. By a *channel* we mean either a pair of parallel lines (dimension 2) or a cylinder (dimension 3). It may be of finite or infinite length. A channel with *microgeometry* is a channel with a choice of random billiard. In dimension 3, we could more generally define a surface with microgeometry to be a triple (S, ξ, Ψ), where S is a piecewise smooth embedded surface in $\mathbb{R}^3$, ξ is a piecewise smooth orthonormal framing of S, and Ψ is a random billiard. For cylindrical channels, the framing is always the one represented by the vectors e_1, e_2, n of Figure 8. The normal vector, n, is taken so as to point into the region enclosed by the surface and e_2 is parallel to the axis of the cylinder.

To a channel with microgeometry it is associated a random flight in the obvious way: a particle with a given initial velocity moves along a straight line in uniform motion until it hits a point on the channel surface. It then reflects at an

angle specified by the random billiard. Vectors at the point of collision are identified with vectors at the open side of a billiard cell by means of the orthonormal
frame. In other words, we assume that the random billiard is "attached" to the
frame in this sense: defining $C : (v_1, v_2, v_3) \mapsto (v_1, v_2, -v_3)$, a particle that
hits the surface at p with velocity $v = v_1 e_1(p) + v_2 e_2(p) + v_3 n(p)$ follows
the direction specified by $w = w_1 e_1(p) + w_2 e_2(p) + w_3 n(p)$ after reflection,
where $(w_1, w_2, w_3) = C\Psi_x(v_1, v_2, v_3)$, for a randomly chosen x.

Geometric parameters of the channel (its radius and length) are not commen-
surate with those of the billiard cell. Consequently, any geometric characteristic
of the cell that can affect the behavior of the random flight must be invariant
under homothety, such as length ratios and angles.

It is, perhaps, not entirely obvious that iterations of the random billiard indeed
correspond to the velocity process of the random flight in a cylindrical channel.
This is because, in principle, one would need to apply a random rotation to $\Psi_x(v)$
after each collision, to account for the fact that the moving frame ξ rotates from
a point on the channel to the next. To clarify this point, we first suppose that the
channel is an unspecified orientable differentiable surface embedded in $\mathbb{R}^3$, with
a given framing $\xi = (e_1, e_2, n)$. Denote by $p(\sigma^n(\omega))$, $\omega \in \Omega$, the point of n-th
collision with the surface for a random trajectory. Define $I_p : (v_1, v_2, v_3) \mapsto$
$v_1 e_1 + v_2 e_2 + v_3 n$ and write

$$R(\sigma^n(\omega)) := (I_{p(\sigma^{n+1}(\omega))})^{-1} \circ I_{p(\sigma^n(\omega))} \circ C.$$

The velocity after n collisions is given by

$$v \mapsto (I_{p(\sigma^n \omega)})^{-1} \circ \Psi_{\sigma^n(\omega)} \circ R(\sigma^{n-1}(\omega)) \circ \Psi_{\sigma^{n-1}(\omega)} \circ \cdots \circ \Psi_{\sigma(\omega)} \circ R(\omega) \circ \Psi_\omega \circ I_{p(\omega)} v.$$

If the surface is a cylinder and the moving frame is chosen as indicated in
Figure 8, then an elementary geometric argument shows that the orthogonal
transformations $R(\omega)$ are equal to the identity matrix. Therefore, the random
dynamical system defined by $\{\Psi_x : x \in \mathbb{T}^2\}$ actually describes the changing
velocity of a particle in a random flight inside the cylinder as claimed.

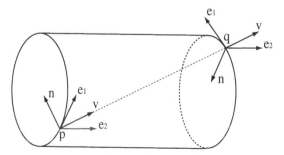

Figure 8. Standard frame over cylindrical surface.

2.5. The associated Markov chain. Define a Markov chain with state space V and transition probabilities

$$P(A|v) := (F_v)_* \lambda(A) := \lambda(F_v^{-1}(A)) = \lambda^{\mathbb{N}} \left(\{\omega \in \Omega : \Psi_\omega(v) \in A\} \right).$$

The conditional expectation of a function f, given v, is

$$\mathbb{E}[f|v] = \int_X f(F_v(x)) \, d\lambda(x).$$

The transition probability $P(A|v)$ should be interpreted as the probability that a particle will reflect with a velocity in A given that the pre-collision velocity is v. More generally, define $P_k(A|v) = \lambda^{\mathbb{N}}(\{\omega \in \Omega : \Psi_\omega^k(v) \in A\})$, for $k \in \mathbb{N}$. These k-step transition probabilities define a homogeneous Markov chain (see [Ar]). We have, in particular, the Chapman–Kolmogorov equation

$$P_{k+l}(A|v) = \int_V P_k(A|v') P_l(dv'|v),$$

for $k, l \in \mathbb{N}$.

If μ is a probability measure on V (not necessarily, for the moment, the velocity component of the Liouville measure), then there exists a unique probability measure, P_μ, on the space $\Omega_V = V^{\mathbb{N}}$ with the product σ-algebra such that the coordinate functions, $\pi_k : \Omega_V \to V$, $\pi_k(v_1, v_2, \ldots) = v_k$, constitute a Markov process with respect to the natural filtration $\mathcal{F}_k$ (generated by the coordinates $\pi_1, \ldots, \pi_k$) with transition probabilities P_k and initial distribution μ. That is, such that for any $l \le k$ and any measurable $f : V \to \mathbb{R}^+$,

$$\mathbb{E}_\mu[f \circ \pi_k | \mathcal{F}_l](\omega) = (P^{k-l} f)(v_l)$$

for P_μ almost all ω, where $(P^k f)(v) = \int_V f(w) P_k(dw|v)$ is the semigroup of positive linear operators corresponding to the kernel $(P_k)_{k \in \mathbb{N}}$ and P_μ is given on cylinders $U = \{\omega \in \Omega_V : \pi_i(\omega) \in A_i, i = 0, 1, \ldots, k\}$ by

$$P_\mu(U) = \int_{A_0} \int_{A_1} \cdots \int_{A_k} P(dv_k|v_{k-1}) \cdots P(dv_2|v_1) \mu(dv_1).$$

Notice that $(\pi_0)_* P_\mu = \mu$.

The measure P_μ is shift invariant if and only if μ is *stationary*, or P^k-invariant, for $k = 1, 2, \ldots$; that is, if

$$\mu = \int_V P_k(\cdot|v) \mu(dv).$$

A bounded measurable function f on V is said to be P^k-invariant if $P^k f = f$ μ-almost everywhere, for all k. A stationary measure μ is *ergodic* if invariant functions are μ-almost everywhere constant. It can be shown that μ is ergodic if and only if the shift map on Ω_V is ergodic for the invariant measure P_μ. This is also equivalent to invariance in the sense

$$\mu = \int_X (\Psi_x)_* \mu \, d\lambda(x).$$

(See below. Also see [Ki] or [Ar].)

Suppose, now, that μ is a measure on V such that $\nu := \lambda \otimes \mu$ is a T-invariant measure on E. The next remark is that μ is an invariant measure of the random billiard, hence also a stationary measure for the associated Markov chain.

PROPOSITION 2.1. *Suppose that μ is a measure on V such that $\nu = \lambda \otimes \mu$ is a T-invariant measure on E. Then μ is an invariant measure of the random billiard. In particular, μ is a stationary measure of the associated Markov chain.*

PROOF. Let f be any μ-integrable function on V. Then, from the equation $T_* \nu = \nu$ applied to $f \circ \pi_2$, we obtain

$$\mu(f) = \nu(f \circ \pi_2) = T_* \nu(f \circ \pi_2) = \nu(f \circ \pi_2 \circ T)$$

$$= \int_X \int_V f(\pi_2(T(x, v))) \, d\mu(v) d\lambda(x)$$

$$= \int_X \int_V f(\Psi_x(v)) \, d\mu(v) d\lambda(x) = \int_0^1 (\Psi_x)_* \mu(f) \, d\lambda(x).$$

Since f is arbitrary, we have $\mu = \int_X (\Psi_x)_* \mu \, d\lambda(x)$ as claimed. That μ is a stationary measure for the Markov chain is now a standard fact. We show it here for the sake of completeness. The claim is that

$$\mu(A) = \int_V P(A|v) \, d\mu(v),$$

for all measurable $A \subset V$. This is a consequence of the following calculation:

$$\int_V P(A|v) \, d\mu(v) = \int_V \int_X \chi_A(\Psi_x(v)) \, d\lambda(x) d\mu(v)$$

$$= \int_X \int_V \chi_A(\Psi_x(v)) \, d\mu(v) d\lambda(x) = \int_X (\Psi_x)_* \mu(A) \, d\lambda(x)$$

$$= \mu(A),$$

for all measurable $A \in V$. $\qquad \square$

From now on, we resume the earlier convention that μ denotes the velocity component of the Liouville measure on E.

3. Examples

We illustrate the concepts introduced above with a few simple examples. The simplest, shown in Figure 2, was discussed in the introduction.

3.1. Three-dimensional boxes. The billiard cell is defined by the surface of a parallelepiped without its top face, with sides h (height), b_1 and b_2. Using the vectors e_1, e_2, n as in Figure 8, the base of the parallelepiped is oriented as in Figure 9.

If $v = v'_1 u_1 + v'_2 u_2 - v_3 n$ is an incoming velocity expressed in the orthonormal frame aligned with the box, then the velocity after reflection is

$$v = \varepsilon_1 v'_1 u_1 + \varepsilon_2 v'_2 u_2 + v_3 n,$$

where ε_1 and ε_2 are independent random variables taking values in $\{1, -1\}$ with probabilities to be specified. Write

$$A = \begin{pmatrix} \cos\alpha & -\sin\alpha \\ \sin\alpha & \cos\alpha \end{pmatrix}$$

and let $v = v_1 e_1 + v_2 e_2 - v_3 n$ be an incoming velocity, now expressed in the frame of the channel. The velocity after reflection can now be written as a random variable $V(\varepsilon_1, \varepsilon_2) = w_1 e_1 + w_2 e_2 + v_3 n$ such that

$$\begin{pmatrix} w_1 \\ w_2 \end{pmatrix} = A \begin{pmatrix} \varepsilon_1 & 0 \\ 0 & \varepsilon_2 \end{pmatrix} A^t \begin{pmatrix} v_1 \\ v_2 \end{pmatrix}.$$

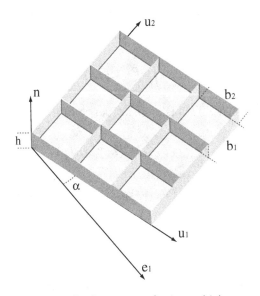

Figure 9. Geometry of microcubicles.

This can also be written as follows. Let u be the orthogonal projection of v to the linear span of $\{e_1, e_2\}$. Then $v = \|u\|(\cos \beta e_1 + \sin \beta e_2) - v_3 n$ (β is defined by this expression) and $V(\varepsilon_1, \varepsilon_2) = \|u\|U(\varepsilon_1, \varepsilon_2) + v_3 n$, where

$$U(\varepsilon_1, \varepsilon_2) = \frac{\varepsilon_1 + \varepsilon_2}{2}(\cos \beta e_1 + \sin \beta e_2) + \frac{\varepsilon_1 - \varepsilon_2}{2}(\cos(2\alpha - \beta)e_1 + \sin(2\alpha - \beta)e_2).$$

For a given initial v, the velocity after any later collision of the random flight in a cylindrical channel is one of the four vectors $\pm u + v_3 n, \pm u' + v_3 n$, where u' is defined by

$$u' = \|u\|(\cos(2\alpha - \beta)e_1 + \sin(2\alpha - \beta)e_2)$$

and the frame vectors are based at the collision point.

We assume that v is not parallel to u_1 or u_2 (the easier case of v parallel to either vector can be treated separately). Let

$$\frac{2h\langle v, n\rangle}{b_i \langle v, u_i\rangle} =: k_i(v) + s_i(v),$$

where k_i is the integer part and s_i is the fractional part of the number on the left-hand side of the equation. Define $p_i(v) = s_i(v)$ if $k_i(v)$ is odd and $p_i(v) = 1 - s_i(v)$ if $k_i(v)$ is even. The probability that $\varepsilon_i = 1$, given that v is the pre-collision velocity, is $p_i(v)$.

For simplicity of notation, we write below $p_i(u)$ instead of $p_i(u \pm v_3 n)$. We make explicit the dependence of ε_i on $w \in \{\pm u + v_3 n, \pm u' + v_3 n\}$ (the pre-collision velocity) by writing $\varepsilon_i(w)$. Define k_i, s_i as the integral and fractional

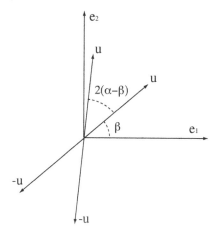

Figure 10. Possible values of the orthogonal projection of the reflected velocity.

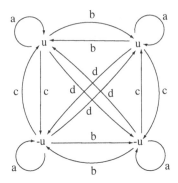

Figure 11. Transition probabilities for the velocity process.

parts of the numbers

$$\frac{2hv_3/\sqrt{1-v_3^2}}{b_1\cos(\beta-\alpha)} = k_1 + s_1, \qquad \frac{2hv_3/\sqrt{1-v_3^2}}{b_2\sin(\beta-\alpha)} = k_2 + s_2.$$

It can now be calculated that $p_i(u) = p_i(u') = p_i(-u) = p_i(-u') =: p_i$, where

$$p_i = \begin{cases} s_i & \text{if } k_i \text{ is even,} \\ 1 - s_i & \text{if } k_i \text{ is odd.} \end{cases}$$

Setting $a = p_1 p_2, b = p_1(1-p_2), c = (1-p_1)p_2, d = (1-p_1)(1-p_2)$, the transition probabilities can now be seen to be as shown in Figure 11.

The random walk on the real line, obtained by projecting the random flight inside the cylinder on a line parallel to e_2, can be described as follows. We suppose that the initial velocity is $v = u - v_3 n$. From any particular time and position, a particle moves a distance δ, which can be one of the four values: $\delta_1, \delta_2, \delta_3 = -\delta_2, \delta_4 = -\delta_1$, where

$$\delta_1 = \frac{2rv_3\|u\|\sin\beta}{v_3^2 + \|u\|^2\cos^2\beta},$$

$$\delta_2 = \frac{2rv_3\|u\|\sin(2\alpha-\beta)}{v_3^2 + \|u\|^2\cos^2(2\alpha-\beta)}.$$

A jump by $\pm\delta_i$ takes time τ_i, where

$$\tau_1 = \frac{2rv_3}{v_3^2 + \|u\|^2\cos^2\beta},$$

$$\tau_2 = \frac{2rv_3}{v_3^2 + \|u\|^2\cos^2(2\alpha-\beta)}.$$

Jumps of different lengths occur with probabilities specified by the transition matrix

$$P = (p(\delta_i|\delta_j)) = \begin{pmatrix} a & b & c & d \\ b & a & d & c \\ c & d & a & b \\ d & c & b & a \end{pmatrix},$$

where a, b, c, d are as above. The spectral decomposition of the transition matrix is easily obtained. If R is the orthogonal matrix

$$R = \frac{1}{2} \begin{pmatrix} 1 & 1 & 1 & 1 \\ 1 & 1 & -1 & -1 \\ 1 & -1 & 1 & -1 \\ 1 & -1 & -1 & 1 \end{pmatrix},$$

then $P = RDR^t$, where $D = \mathrm{diag}(\lambda_1, \lambda_2, \lambda_3, \lambda_4)$, with

$$\lambda_1 = a+b+c+d = 1,$$
$$\lambda_2 = a+b-c-d = 2p_1 - 1,$$
$$\lambda_3 = a-b+c-d = 2p_2 - 1,$$
$$\lambda_3 = a-b-c+d = (2p_1 - 1)(2p_2 - 1).$$

In particular, the stationary distribution assigns equal probability to each of the δ_i. By allowing different values of the initial velocity we can recover the geometric parameters, $b_1/h, b_2/h, \alpha$, from the spectrum.

3.2. Cavities and effusion. The example we give now is in a sense more representative than those seen so far in that the measure μ is ergodic, as explained later. Furthermore, its Markov operator admits a simple approximation that will allow for some explicit calculations in Section 7. The approximate operator, which by itself does not correspond to any random billiard system, is defined by

$$P_{\text{Maxwell}} = (1-\alpha)P_\mu + \alpha I,$$

where α is a constant in $(0, 1)$ and $P_\mu(dw|v) = d\mu(w)$. This is a model of gas-solid interaction proposed by Maxwell well-known in the gas kinetics and Boltzmann equation literature (except for the fact that we do not include an exponential term involving temperature) [C1]. The interpretation is that a particle will reflect in mirror-like fashion with probability α, and with probability $1 - \alpha$ it forgets the pre-collision velocity and rebounds along a random direction specified by the probability μ.

Notice that P_μ can be described as the orthogonal projection on the one-dimensional subspace spanned by μ in the Hilbert space, H, of signed measures

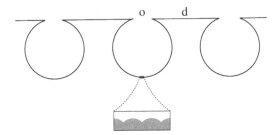

Figure 12. Randomizing cavity; o is the length of the open side and d is the length of the mirror-reflecting, flat side. The inside of the circular cavity is lined with some ergodic microgeometry.

with square integrable densities with respect to μ. Then

$$P_{\text{Maxwell}} = P_\mu + \alpha P_{\mu^\perp},$$

where $\mu^\perp$ denotes the subspace of H orthogonal to μ.

We now show in what sense P_{Maxwell} can be regarded as approximating the Markov operator of an actual billiard microgeometry. (Proposition 3.1 will not be used later in the paper other than for its heuristic value, so we only give a sketch of the proof.) Consider the contour shown in Figure 12. It represents an array of billiard cells consisting of circular cavities with a small opening on top. The inner surface of each cavity is itself lined with some other kind of microgeometry at a much smaller scale. The bumps lining the inner surface of the cavity are assumed to be so small that collisions with it are well approximated by a "second order" random billiard map, which we denote by Ψ. We assume that the associated Markov operator, P_{bumps}, is ergodic and aperiodic (see Section 5). In particular, P_{bumps}^k converges to μ as $k \to \infty$. This is the case for the particular example of circular bumps shown in the figure, according to the ergodicity criterion given in Section 5.

The random dynamical system, Φ, is defined as follows. Let v be an initial velocity and $x \in [0, 1]$ a random point on the open side of the cavity chosen with probability λ. Set $a = o/l$, where l is the perimeter of the circular cavity and o is the length of its open side. With probability 1, a particle that falls into the cavity leaves it after a finite number, $\#_a$, of collisions. This number is a random variable depending on v. It can be shown that the limit, as $a \to 0$, of the probability of the event $\{\#_a < n\}$ is 0 for each finite n. Let $\Theta_1, \Theta_2, \ldots, \Theta_{\#_a}$ be the post-collision random angles inside the cavity, measured with respect to the tangent line at each (random) collision point. Let W denote the random vector in $\mathbb{R}^2$ corresponding to the angle $\Theta_{\#_a}$. Then $\Phi(v)$ is the angle of W measured relative to the direction parallel to open side. For small a, $\Phi(v)$ is very nearly $\Theta_{\#_a}$.

Notice that Θ_i is both the post-collision velocity of the i-th collision and the pre-collision velocity of the $i + 1$-st collision. (This might not hold were the cavities not circular.) Therefore, P_{bumps}^k actually describes the probability distribution of post-collision velocity after k collisions inside the cavity.

Let, now, P_a be the Markov operator of the random billiard conditioned on the particle actually entering the cavity. This is an event of probability $o/(o + d)$. Let $\alpha = d/(o + d)$ be the probability of an incident particle being reflected specularly. Then, the Markov operator associated to Φ can be written as $P = (1 - \alpha) P_a + \alpha I$, where I is the identity operator. By the above argument, $P_a = P_{\text{bumps}}^{\#a}$. Thus we conclude:

PROPOSITION 3.1. *Denote by $P_a(\cdot|v) = P_{\text{cav}}(\cdot|v)$ the conditional measures for a cavity with ratio $o/l = a$. Suppose that Ψ has a unique stationary measure, μ, and that the conditional measures are absolutely continuous with respect to μ. (See Section 5; the "microbumps" of Figure 12 are one example for which these assumptions hold.) Then, for μ-a.e. $v \in V$, $\lim_{a\to 0} P_a(\cdot|v) = \mu$.*

Thus, it makes sense to think of $P_{\text{Maxwell}} = (1-\alpha) P_\mu + \alpha I$ as an approximation to the Markov operator of the random billiard of Figure 12, for a small opening at the top.

Another interpretation of Proposition 3.1 has to do with the phenomenon of *effusion* from cavities. If we imagine the circular cavity filled with gas at low pressure, then gas will escape through the small opening with a distribution of directions given by the probability density $d\mu/d\theta = \frac{1}{2} \sin \theta$. Here θ is the angle measured counterclockwise from the direction parallel to the open side. In particular, the probability of leaving at shallow angles approaches 0.

4. Reversibility and self-adjointness

The deterministic billiard system has the property of being reversible. In dimension 2, this means:

$$T(x, \theta) = (y, \phi) \iff T(y, \pi - \phi) = (x, \pi - \theta),$$

and in dimension 3, replace v for θ, $-w$ for $\pi - \phi$, etc, in the above expression. This can also be written as $T^{-1} \circ J = J \circ T$, where $J(x, v) = (x, -v)$ is the *flip map*. If Φ_t is a flow describing the time evolution of, say, a conservative mechanical system on its phase space or a geodesic flow, then reversibility means that $\Phi_t(x, v) = (y, w) \iff \Phi_t(y, -w) = (x, -v)$ for all (x, v) and t. In this case, $J \circ \Phi_t = \Phi_{-t} \circ J$.

Reversibility of the deterministic billiard system implies reversibility of the random billiard as well, in the following sense. First recall that a Markov process with state space V and transition probability measure P is said to be *reversible* with respect to a probability measure μ on X if

$$P(dw|v)d\mu(v) = P(dv|w)d\mu(w),$$

as measures on $V \times V$. More precisely,

$$\int_V \int_V f(v, w) P(dw|v)\, d\mu(v) = \int_V \int_V f(v, w) P(dv|w)\, d\mu(w)$$

for any integrable function $f(v, w)$ on V^2.

PROPOSITION 4.1. *The Markov process of a random billiard is reversible.*

PROOF. We actually show that $P(dw|v)d\mu(v) = P(dv|w)d\mu(w)$ for any measure μ on V such that $\nu := \lambda \otimes \mu$ is invariant under the billiard first return map $T : E \to E$ and the flip map J. First observe that

$$\int_V \int_V f(v, w) P(dw|v)\, d\mu(v) = \int_V \int_X f(v, \Psi_x(v))\, d\lambda(x) d\mu(v)$$

$$= \int_E F(\xi, T(\xi))\, d\nu(\xi),$$

where $F(\xi_1, \xi_2) = f(\pi_2(\xi_1), \pi_2(\xi_2))$ for all $\xi_1, \xi_2 \in E$. Therefore, the claim amounts to the property

$$\int_E F(\xi, T(\xi))\, d\nu(\xi) = \int_E F(T(\xi), \xi)\, d\nu(\xi).$$

But this is a consequence of the invariance of ν under T^{-1} and the flip map. $\square$

Notice the following well-known consequence of reversibility of the Markov process with respect to the stationary measure μ.

PROPOSITION 4.2. *Let $\langle f, g \rangle = \int_0^\pi f\bar{g}\, d\mu$ denote the inner product on the Hilbert space $H = L^2(V, \mu)$. Then, the linear operator P on H defined by*

$$(Pf)(v) := \int_X f(\Psi_x(v))\, d\lambda(x)$$

is self-adjoint.

We regard H as the space of densities of signed measure on V absolutely continuous with respect to μ. With this in mind, we write $P\eta$ for the action of the *Markov operator* P on a probability measure η on V. If η describes the probability distribution of incoming directions of a particle, then $P\eta$ is interpreted as the probability distribution of reflected directions.

5. Ergodicity

Given a Markov process with state space V and transition probability measure P, let $P_n(A|v)$ denote the transition probability measure after n steps. We are interested in the convergence of $P_n(A|v)$ to $\mu(A)$, where μ is the velocity component of the Liouville measure ν on E. Recall that μ is a stationary measure for P.

LEMMA 5.1. *Suppose that a measurable $A \subset V$ satisfies $P(A|v) = 1$ for μ-a.e. $v \in A$. Let $\bar{A} = X \times A$. Then, after possibly changing A on a set of measure 0, we have that $\bar{A}$ is T-invariant.*

PROOF. For almost all $v \in A$, $\lambda(\{x \in X : \Psi_x(v) \in A\}) = P(A|v) = 1$. Consequently, for almost all $(x, v) \in \bar{A}$, we have $T(x, v) \in \bar{A}$. By changing $\bar{A}$ on a set of measure zero it can be insured that $\bar{A}$ is T-invariant. $\square$

We say that the process is *indecomposable* if any measurable set $A \subset V$ such that $P(A|v) = 1$ for almost all $v \in A$ either has measure zero or its complement has measure zero.

COROLLARY 5.2. *Suppose that $T : E \to E$ is ergodic with respect to the Liouville measure. Then the Markov process of the random billiard is indecomposable.*

PROOF. Apply the previous proposition to $\bar{A} = X \times A$, where $P(A|v) = 1$ for μ-a.e. v to conclude that $\bar{A} = X \times A$ has measure either zero or one. Therefore, A has measure either zero or one. $\square$

COROLLARY 5.3. *If $T : E \to E$ is ergodic (in particular, if the billiard dynamical system on the closed table, with its top side included, is ergodic), then μ is the unique stationary distribution of the billiard Markov operator, and the Markov process with initial distribution μ is ergodic.*

PROOF. See [Br], Theorem 7.16. $\square$

Applying the previous result to powers of T, gives the next proposition.

PROPOSITION 5.4. *Suppose that all the iterates T^n of T are ergodic. Then Markov process is* aperiodic, *that is, V is indecomposable for all the iterates $P^n(\cdot|v)$, $n = 1, 2, \ldots$.*

COROLLARY 5.5. *Suppose that T^n is ergodic for $n = 1, 2, 3, \ldots$, and that $P(\cdot|v)$ is absolutely continuous with respect to μ, for almost all v. Then*

$$\lim_{n \to \infty} P^n(A|v) = \mu(A)$$

for all measurable A and almost all $v \in V$.

PROOF. This is a consequence of [Br], Theorem 7.18.   □

As an example, if the doubling of the billiard table at its open side gives a Sinai (dispersing) billiard, then the Markov process has a unique stationary measure given by μ.

It is to be expected that ergodicity and aperiodicity should hold under very general conditions. It is certainly clear that it is a much more general (and easier to prove) property for the random billiard than it is for the corresponding deterministic billiard. We give next a sufficient condition that is easy to verify in many examples.

We say that p is a *point of maximal height* of the (open) billiard cell if the entire cell is contained in one of the two closed half-spaces bounded by the plane through p parallel to the distinguished, or top, side. We also say that the billiard is *nondegenerate* at p if the height function (relative to the normal direction to the open side) is nondegenerate there. This means that at the point p of maximal height the billiard surface, or curve, the curvature is defined and nonzero.

PROPOSITION 5.6. *Suppose that the billiard table contains a point of maximal height at which the billiard is nondegenerate. Then the associated random billiard is ergodic and aperiodic, and it admits a unique stationary probability measure, which is μ (the velocity factor of the Liouville measure).*

PROOF. Given $A \subset V$ measurable, define $\mathcal{I}(A) = \{\Psi_x(v) : x \in X, v \in A\} \subset V$. If the billiard table is differentiable at a point p of maximal height, then $A \subset \mathcal{I}(A)$, since $v = \Psi_x(v)$ for an x such that the ray with initial condition (x, v) reflects at p. We claim that, for all $v \in V$,

$$V = \bigcup_{n=1}^{\infty} \mathcal{I}^n(v).$$

Granted the claim, it follows that P^n is indecomposable for $n = 1, 2, \ldots$. In fact, let $A \subset V$ satisfy $P(A|v) = 1$ for μ-a.e. $v \in A$. By Lemma 5.1, $\bar{A} = X \times A$ is T-invariant. Consequently, for almost all $v \in A$, $\mathcal{I}^n(v) \subset A$, up to sets of measure 0. Therefore, $A = V$ up to a set of measure 0.

We now turn to the proof of the claim. The description below applies to dimension 3, although it should be clear what modifications are need for dimension 2. The surface boundary of the billiard table will be denoted by M. This is a smooth surface near p. Recall that V is the open hemisphere of unit radius. Let $\Pi : V \to D$ denote the orthogonal projection onto the open unit disc. The essential point is that, for any compact subset, $K \subset V$, there is $\varepsilon > 0$ such that for any $v \in K$, the set $\Pi(\mathcal{I}(v))$ contains a disc of radius ε centered at $\Pi(v)$. To see this, it suffices to show that the Jacobian of $x \in \mathbb{T}^2 \mapsto \Pi(\Psi_x(v))$ is uniformly bounded away from 0 for v in any compact subset of V.

Let $\gamma(s)$ be a differentiable curve in $\mathbb{T}^2$ such that the ray with initial condition $(\gamma(0), v)$ hits M at p. By varying s, for s small enough, a differentiable path, η, through p is traced on M. The reflected angle is given by

$$\Psi_{\gamma(s)}(v) = v - 2\langle v, n(\eta(s))\rangle n(\eta(s)).$$

Let $G : M \to S^2$ denote the Gauss map of M near p. Note that, by definition, the differential of G at p is

$$dG_p \eta'(0) = \frac{d(n \circ \eta)}{ds}|_{s=0}.$$

We can now write

$$\frac{d}{ds}\Pi \circ \Psi_{\gamma(s)}(v)|_{s=0} = \Pi(-2\langle v, dG_p\eta'(0)\rangle n(p) - 2\langle v, n(p)\rangle dG_p\eta'(0))$$

$$= -2\langle v, n(p)\rangle \Pi dG_p\eta'(0).$$

The norm of Π at n is 1 and $\gamma'(0) = \eta'(0)$. Consequently,

$$\left\|\frac{d}{ds}\Pi \circ \Psi_{\gamma(s)}(v)|_{s=0}\right\| \geq 2|\langle v, n(p)\rangle| \min\{k_1, k_2\}\|\gamma'(0)\|,$$

where k_1 and k_2 are the principal curvatures of M at p, that is, the eigenvalues of dG_p, which are necessarily positive. Since the term $|\langle v, n(p)\rangle|$ is bounded away from 0 for all v on a compact $K \subset V$, the claim holds. □

We have so far assumed that the particle undergoing a random flight is point-like. There is obviously no loss of generality in making this assumption since a spherical billiard ball can be regarded as point-like by following the motion of its center. If we wish to vary the size of the probing particle, then the problem we are studying is the same as having a one parameter family of microgeometries parametrized by the particle radius, and billiard balls that are point-like. Figure 13 illustrates this. The thick contour represents the actual surface relief, and the thin line represents this geometry as viewed by the center of a disc-like billiard ball.

It is interesting to remark that if p is a point of maximal height at which the billiard is not-differentiable, say, a cone point, then viewed by the center of a spherical particle of positive radius, there is a point of maximal height (just above p) which is nondegenerate. Thus, for example, the comb geometry of Figure 13 does not correspond to an ergodic random billiard, but when probed

Figure 13. Microgeometry for positive particle radius.

by particles of arbitrary nonzero radius, the thickened table (represented by the thin line in the figure) does.

6. Spectrum

A fundamental issue to address is the relationship between the geometry of the billiard cell and the spectrum of the (self-adjoint) Markov operator P on $L^2(V, \mu)$. We take here a first step in this direction.

As before, μ refers to the (velocity part of the) Liouville measure on V. Let $\pi_2 : E \to V$ denote the projection map. By duality, P acts on signed measures on V with square integrable densities with respect to μ. If η is such a measure, we write ηP. (The notation $P\eta$ was also used earlier.)

6.1. Generalities.

LEMMA 6.1. *Let* $\varphi, \psi \in L^2(V, \mu)$, *and* η *the signed measure such that* $d\eta = \varphi \, d\mu$. *Then*

$$\int_V \psi \, d(\eta P) = \int_E (\psi \circ \pi_2)(\varphi \circ \pi_2 \circ T^{-1}) \, dv.$$

PROOF. Let ψ be any element of $L^2(V, \mu)$. Then

$$\int_V \psi \frac{d(\eta P)}{d\mu} d\mu = \int_V \psi \, d[(\pi_2 \circ T)_* \lambda \otimes \eta] = \int_E \psi \circ \pi_2 \circ T \, d(\lambda \otimes \eta)$$

$$= \int_E (\psi \circ \pi_2 \circ T)(\varphi \circ \pi_2) \, dv$$

$$= \int_E (\psi \circ \pi_2)(\varphi \circ \pi_2 \circ T^{-1}) \, dv. \qquad \square$$

PROPOSITION 6.2. *The spectrum of the Markov operator* P *is contained in the interval* $[-1, 1]$, *and* 1 *is an eigenvalue. If the associated Markov chain is ergodic, then* 1 *is a simple eigenvalue of* P.

PROOF. Clearly, 1 is the eigenvalue associated to the stationary measure μ, and is a simple eigenvalue if the Markov chain is ergodic. That $\|P\|_2 \leq 1$ is shown by the following standard calculation:

$$\|\eta P\|_2 = \sup_{\psi \in L^2(V, \mu), \|\psi\|_2 = 1} \int_V \psi \frac{d(\eta P)}{d\mu} d\mu$$

$$= \sup_{\substack{\psi \in L^2(V,\mu) \\ \|\psi\|_2 = 1}} \int_E (\psi \circ \pi_2)(\varphi \circ \pi_2 \circ T^{-1}) \, dv$$

$$\leq \sup_{\substack{\psi \in L^2(V,\mu) \\ \|\psi\|_2 = 1}} \sqrt{\int_E (\psi \circ \pi_2)^2 \, dv} \sqrt{\int_E (\varphi \circ \pi_2 \circ T^{-1})^2 \, dv}$$

$$= \sqrt{\int_E (\varphi \circ \pi_2)^2 \, dv} = \sqrt{\int_V \varphi^2 \, d\mu} = \|\eta\|_2.$$

Therefore, $\|\eta P\|_2 \leq \|\eta\|_2$, for all η. As P is self-adjoint, its spectrum is real, contained in $[-1, 1]$. $\qquad \square$

PROPOSITION 6.3. *Suppose that the conditional measure $P(\cdot|v)$ is absolutely continuous with respect to μ, for μ-almost every $v \in V$, and that the function $\kappa(w|v) := P(dw|v)/d\mu(v)$ (the Radon–Nikodým derivative of $P(\cdot|v)$ with respect to μ) is in $L^2(V^2, \mu \otimes \mu)$. Then, $P : H \rightarrow H$ is a Hilbert–Schmidt operator.*

PROOF. Under these assumptions,

$$(Pf)(v) = \int_V f(w)\kappa(w|v) \, d\mu(w).$$

Since κ is in $L^2(V^2, \mu \otimes \mu)$, P is a Hilbert–Schmidt operator. $\qquad \square$

6.2. Integral kernel and spectrum. We make at this point more specific and stronger assumptions about the deterministic billiard than we have so far. We restrict ourselves to the 2-dimensional case. One simple example for which all the assumptions to be made here hold is the billiard cell shown in Figure 14. It consists of two arcs of circle of positive curvature meeting at an angle $\gamma > \pi/2$, and forming angles α and β with the line segment representing the open side, where $\alpha, \beta \in [0, \pi/2)$. Notice that every trajectory collides with the curved sides at most twice before returning to the open side. Also notice that for each initial condition (x_0, θ) at which the return map is differentiable, the function of x defined by $F_\theta(x) := \Psi_x(\theta)$ is differentiable at x_0 and $|F'_\theta(x_0)|$ is bounded away from 0 by a constant that depends on θ and on the curvatures of the two sides.

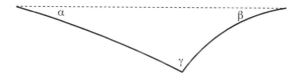

Figure 14. A simple example satisfying the requirements of this section.

The results to be proved later under the following assumptions are by no means optimal. They should be expected to hold for much more general billiard geometries.

ASSUMPTION 1. *For each $\theta \in (0, \pi)$ there exists a countable set, $X_\theta = \{x_i^\theta\}$, where $0 = x_0^\theta < x_1^\theta < \cdots < x_l^\theta = 1$ and $l = l_\theta \leq \infty$, such that*

(1) *F_θ is differentiable and $F_\theta' \neq 0$ over each interval $(x_i^\theta, x_{i+1}^\theta)$; and*
(2) *if $\mathscr{E}_\theta = \{\psi_0^+, \psi_1^-, \psi_1^+, \psi_2^-, \psi_2^+, \ldots\}$ denotes the set of right and left limits of F_θ at the x_i^θ, then*

$$\sum_{x \in F_\theta^{-1}(\psi)} |F_\theta'(x)|^{-1} < \infty$$

for each $\psi \in (0, \pi) \setminus \mathscr{E}_\theta$ for which $F_\theta^{-1}(\psi)$ is nonempty.

ASSUMPTION 2. *There exists a constant $c > 0$ such that all sides have curvature bounded below by c. In particular, all sides are concave.*

ASSUMPTION 3. *Above a certain height, which is strictly below the height of the open side, the billiard table contour consists of two smooth, concave curves with nonzero curvature at the endpoints p and q. (See Figure 15.) The angles α and β lie in the interval $[0, \pi/2)$.*

Figure 15. Assumption 3.

These assumptions are satisfied by the example of Figure 14. The number of x_i^θ, not counting 0 and 1, is at most 2 for each θ, and $|F_\theta'(x)|$ is bounded away from zero by a constant that depends on theta and on the curvatures of the arcs of circles. It should be noted that the sum increases with $1/\sin\theta$ as θ approaches 0 and π.

LEMMA 6.4. *Fix $\theta \in (0, \pi)$. Under Assumption 1, the push-forward measure $(F_\theta)_*\lambda$ is absolutely continuous with respect to μ. The Radon–Nikodým derivative, $\kappa(\psi|\theta) = [d(F_\theta)_*\lambda/d\mu](\psi)$, is given by*

$$\kappa(\psi|\theta) = \begin{cases} \dfrac{2}{\sin\psi} \sum_{x \in F_\theta^{-1}(\psi)} |F_\theta'(x)|^{-1} & \text{if } F_\theta^{-1}(\psi) \text{ is nonempty,} \\ 0 & \text{if } F_\theta^{-1}(\psi) \text{ is empty,} \end{cases}$$

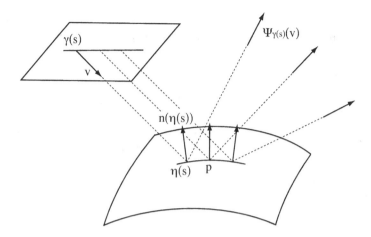

Figure 16. Reflection near a point of maximum height.

for all $\psi \in (0, \pi) \setminus \mathcal{E}_\theta$. *If, moreover, f is a bounded measurable function on V,*

$$\int_0^1 f(F_\theta(x)) \, d\lambda(x) = \int_0^\pi f(\psi) \kappa(\psi | \theta) \, d\mu(\psi).$$

PROOF. This is an immediate consequence of the definitions. □

LEMMA 6.5. *Under Assumption 2, for all $\theta \in (0, \pi)$ and $\psi \in (0, \pi) \setminus \mathcal{E}_\theta$, the function $\kappa(\psi | \theta)$ satisfies*

$$\kappa(\psi | \theta) \leq \frac{N/c}{\sin \psi \sin \theta}.$$

PROOF. Denote by $c(\theta)$ the infimum of $\{|F_\theta'(x)| : x \in (0, 1)\}$. From the expression of κ given in Lemma 6.4, it follows that $\kappa(\psi | \theta) \leq 2N/(c(\theta) \sin \psi)$. The lemma will be proved if we show that $c(\theta) \geq 2c \sin \theta$.

Since all sides of the billiard table are strictly concave, we can estimate $|F_\theta'(x)|$ by estimating the change of direction of the billiard trajectory due to the first collision only; the subsequent collisions can only magnify the angle variation to first order. Let $F(x) = F_{\theta,1}(x)$ denote the angle immediately after the first collision, measured with respect to some direction, say, a tangent vector to the side at the point of first collision. (This choice is immaterial since we are interested in the derivative of $F(x)$.)

The situation can be pictured by imagining the two-dimensional version of Figure 16. In the present case, p denotes the point of first collision and $\Psi_{\gamma(s)}(v)$ should be replaced with $F(\gamma(s))$. Also, the normal vector at p is not necessarily normal to the open side, as is the case in that figure. We continue to denote by $\gamma(s)$ a differentiable curve such that $p_0 = \gamma(0)$ is a point on the open side where F is differentiable, and by $\eta(s)$ the curve traced on the side of first collision by the first segment of the billiard trajectory. If v is, as in the figure, the unit vector representing the initial direction of the billiard trajectory, η can be written as $\eta(s) = \gamma(s) + t(s)v$, for a positive differentiable function $t(s)$. The trajectory's direction immediately after the first collision now reads

$$w(s) = v - 2\langle v, n(\eta(s))\rangle n(\eta(s)).$$

Notice that $|F'(x)| = \|w'(0)\|$. Taking the square norm of $w'(0)$, and using that $dn/ds = k(p)\eta'(0)$, where $k(p)$ denotes the curvature of the billiard contour at p, gives

$$\|w'(0)\|^2 = 4k(p)^2 \left[\left\langle v, \frac{\eta'(0)}{\|\eta'(0)\|} \right\rangle^2 + \langle v, n(p)\rangle^2 \right] \|\eta'(0)\|^2.$$

The quantity between square brackets is the norm of v, since $\eta'(0)$ and $n(p)$ are orthogonal, and $\|\eta'(0)\|$ assumes its minimum value when $n(p)$ and v are parallel, in which case $\|\eta'(0)\| = \|\gamma'(0)\| \sin\theta$. Therefore, $|F'(x)| = \|w'(0)\| \geq 2|k(p)| \sin\theta$, as needed to finish to proof.   □

We note the following simple geometric fact derived in the proof of Lemma 6.5.

LEMMA 6.6. *Using the notations defined in the proof of Lemma 6.5, it holds that* $\|w'(0)\| = 2|k(p)|\|\eta'(0)\|$.

LEMMA 6.7. *Under Assumptions 1, 2 and 3, κ belongs to* $L^2(V \times V, \mu^2)$.

PROOF. Due to Lemma 6.5, we only need to study the behavior of κ for incident and reflected rays having angles close to 0 or π. It is sufficient to carry the analysis for angles close to 0; angles close to π are similarly treated. There are four cases to study, depending on whether α and β are positive or 0. We consider only two: (i) $\alpha > 0$ and $\beta > 0$, and (ii) $\alpha = \beta = 0$. The two remaining cases follow by applying the same arguments used for these.

Case (i). The exit angle ψ can only be close to 0 when the incidence angle θ is close to 2α and x is close to the left end point of the open side. We refer to this point as the *left corner*, and its opposite as the *right corner*. Therefore, we only need to study $\kappa(\psi|\theta)$ for θ near 0 and 2α.

If θ_0 is sufficiently small, then for all $0 < \theta \leq \theta_0$ the function $F_\theta(x)$ has the following qualitative properties: it is smooth (assuming that the geometry of the billiard table is smooth near the ends), and it decreases monotonically from a maximum value, ψ_{max}^θ, to be estimated in a moment, to a minimum value, $\psi_{min}^\theta = \theta + 2\beta$. Therefore, using Lemma 6.5 again, we arrive at the upper bound for κ:

$$\kappa(\psi|\theta) \leq \frac{1}{c \sin\psi \sin\theta} \chi_{[\psi_{min}, \psi_{max}]}(\psi),$$

for $\theta \in (0, \theta_0)$ and a sufficiently small θ_0. Here $\chi_{[\psi_{\min}, \psi_{\max}]}$ denotes the characteristic function of the interval $[\psi_{\min}, \psi_{\max}]$.

We claim that $\psi_{\max} - \psi_{\min} \leq C \sin \theta$, for some constant C. Suppose for the moment that this is the case. It easily follows that

$$\int_0^{\theta_0} \int_0^{\pi} \kappa(\psi | \theta)^2 \, d\mu(\psi) d\mu(\theta) \leq \frac{Cc^2\theta_0}{2\sin(2\beta)}.$$

Therefore, the proposition will be established once we justify the claim.

Using the expression of Lemma 6.6 gives

$$|F_\theta'(x)| \leq 2K\|\eta'(x)\|,$$

where K is the maximum value of the curvature of the side of the table contour making an angle β with the open side and $\eta(x)$ is the curve traced along that side by the ray with initial condition (x, θ), $x \in [0, 1]$. Let $L = L(\theta)$ denote the length of this curve. Observe that

$$\psi_{\max} - \psi_{\min} = \int_0^1 |F_\theta'(x)| \, dx.$$

By elementary geometry, we have

$$L = \int_0^1 \|\eta'(x)\| \, dx \leq 2 \sin \theta / \sin \beta.$$

Therefore, $\psi_{\max} - \psi_{\min} \leq C \sin \theta$, where $C = 4K/\sin \beta$, proving the claim.

Still in case (i), it remains to consider θ near 2α. When θ is close to 2α from below, and x is near the left corner, ψ is close to $\pi - 2\beta > 0$. Therefore, only the interval $(2\alpha, 2\alpha + a)$, for an arbitrarily small $a > 0$, needs to be considered. We take a small enough that the exit angle ψ, after reflection on the concave side near the left corner, satisfies $0 \leq \psi \leq \pi/4$ (in particular, $\psi \leq \sqrt{2} \sin \psi$), and also small enough that $(\sin \theta)^{-1} < S$, for a constant $S < \infty$, for all $\theta \in (2\alpha, 2\alpha + a)$.

The same argument used in the first part of this proof gives

$$\kappa(\psi | \theta) \leq \frac{S}{c \sin \psi} \chi_{[\psi_{\min}, \psi_{\max}]}(\psi) \leq \frac{\sqrt{2}S}{c\psi} \chi_{[\psi_{\min}, \psi_{\max}]}(\psi),$$

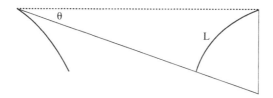

Figure 17. Definition of L.

where it is understood that the maximum and minimum values of ψ are functions of θ. Therefore,

$$\int_{2\alpha}^{2\alpha+a} \int_0^{\pi} \kappa(\psi|\theta)^2 \, d\mu(\psi) d\mu(\theta) \le \frac{2S^2}{c^2} \int_{2\alpha}^{2\alpha+a} \ln \frac{\psi_{\max}}{\psi_{\min}} \, d\theta.$$

The proof of case (i) will be concluded once we establish the following claim: if $\bar{k}$ denotes the maximum curvature of the side of the table the particle is reflecting from, then

$$\frac{\psi_{\max}}{\psi_{\min}} \le 1 + \frac{8\bar{k}}{\sin \alpha},$$

for all $\theta \in (2\alpha, 2\alpha + a)$. We refer to Figure 18.

In the figure, L is the length of the segment of curve in the side of first reflection going from the left corner to the point of collision. Notice that the angle $\alpha(L)$ between the tangent at that point and the horizontal direction (parallel to the open side) satisfies $\alpha(L) \le \alpha + 2\bar{k}L$. It also holds that

$$L \le 2 \sin \psi_{\min} / \sin \alpha \le \frac{2}{\sin \alpha} \psi_{\min}.$$

The maximum value of ψ (the angle of reflection when the collision takes place at the left corner) is $\psi_{\max} = \theta - 2\alpha$. The minimum angle is $\psi_{\max} = \theta - 2\alpha(L)$. Therefore

$$\psi_{\max} = \psi_{\min} + 2(\alpha(L) - \alpha) \le \psi_{\min} + \frac{8\bar{k}}{\sin \alpha} \psi_{\min},$$

and the claim about the ratio of these two angles holds.

Case (ii). We suppose now that $\alpha = \beta = 0$. In this case, the exit angle ψ can only approach 0 as θ itself approaches 0. With Lemma 6.5 in mind, to prove that κ is square integrable, it suffices to study what happens when θ is close to 0. (The same argument treats the case of θ close to π.)

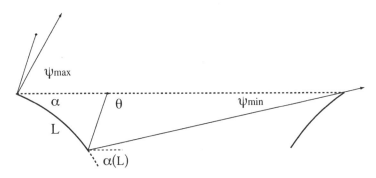

Figure 18. Exit at small angles.

In case (i) the main part of the analysis concerned values of θ for which only one collision between particle and table boundary took place. This is no longer the case here. For a sufficiently small θ_0, and $\theta \in (0, \theta_0)$, the function $\psi = F_\theta(x)$ has the qualitative properties shown in Figure 19. This is a piecewise smooth, continuous function, with two points of discontinuity of $F_\theta'(x)$, denoted $x_1(\theta), x_2(\theta)$. For simplicity, we omit reference to θ in x_1, x_2. Over the intervals $(0, x_1)$ and $(x_2, 1)$ the function is monotone decreasing, and it is monotone increasing over (x_1, x_2).

The discontinuities observed in this graph are due to the different types of trajectories, as follows. For a small θ, if $x \in (x_2, 1)$, the particle will collide with the concave segment of wall adjacent to the right corner and exit the cave with an angle ψ determined by this unique collision. At $x = x_2$, the trajectory first grazes the concave segment of wall adjacent to the left corner before hitting the concave right end wall and exiting. For $x \in (x_1, x_2)$, the trajectory first bounces off this left wall, then bounces off again at the right wall, then exits. For $x \in (0, x_1)$, there is only one collision, in this case with the left wall, before exiting.

We claim that the following estimates hold, for positive constants A, B and C, and all $\theta \in (0, \theta_0)$ for a sufficiently small θ_0:

(1) $\psi_{\max} \leq A\theta^{1/2}$;
(2) $A^{-1}\theta^2 \leq \psi_{\min}$;
(3) $|F_\theta'(x)|^{-1} \leq B\theta^{-1/2}$, for all $x \in (x_2, 1)$;
(4) $|F_\theta'(x)|^{-1} \leq C$, for all $x \in (0, x_1) \cup (x_1, x_2)$.

These estimates will be proven after it is shown how they give the desired result. An upper bound for κ can now be given by

$$\kappa(\psi|\theta) \leq \frac{1}{c \sin \psi}(2C + B\theta^{-1/2})\chi_{[\psi_{\min}, \psi_{\max}]}(\psi).$$

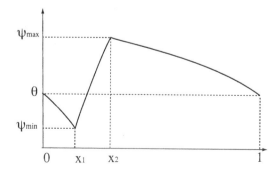

Figure 19. Qualitative features of $\psi = F_\theta(x)$ for small values of θ.

For small enough θ_0 we have $2C + B\theta^{-1/2} \leq 2B\theta^{-1/2}$ and $\sin\psi \geq \sqrt{2}\psi$. This yields

$$
\begin{aligned}
\int_0^{\theta_0} \int_0^{\pi} \kappa(\psi|\theta)^2 \, d\mu(\psi) d\mu(\theta) &\leq \frac{4\sqrt{2}B^2}{c^2} \int_0^{\theta_0} \int_0^{\pi} \frac{1}{\psi} \chi_{[\psi_{\min},\psi_{\max}]}(\psi) d\psi \, d\theta \\
&= \frac{4\sqrt{2}B^2}{c^2} \int_0^{\theta_0} \ln\frac{\psi_{\max}}{\psi_{\min}} d\theta \\
&\leq \frac{4\sqrt{2}B^2}{c^2} \int_0^{\theta_0} \ln(A^2\theta^{-3/2}) d\theta \\
&= \frac{6\sqrt{2}B^2\theta_0}{c^2}\left(1 - \ln(A^{-4/3}\theta_0)\right) < \infty.
\end{aligned}
$$

It remains to verify the estimates enumerated above. Let $\xi(x,\theta)$ denote the angle between the normal to the boundary surface of the billiard and the vertical direction (i.e., the direction perpendicular to the open side, oriented toward the outside of the billiard cell). Then, for small values of θ and for $x \in (x_2, 1)$, we have $F_\theta(x) = \theta + 2|\xi(x,\theta)|$. Ignoring the left wall and using concavity of the right wall, we obtain that $\psi_{\max} \leq \theta + 2|\xi(0,\theta)|$. By an elementary geometric argument, it can be shown that $\lim_{\theta \to 0} \xi(0,\theta)/\sqrt{\theta} = \sqrt{2k}$, where k is the absolute value of the curvature of the right wall at the point of tangency with the open side. Therefore, we can find $D > 0$ independent of θ such $D^{-1}\theta^{1/2} \leq \xi(0,\theta) \leq D\theta^{1/2}$. (We note that if the right wall is replaced with a circumference of curvature k, then it can be shown by elementary geometry that $1 - \cos\xi + \sin\xi \tan\theta = k\sin\theta$, where $\xi = \xi(0,\theta)$. It follows from this that $\theta = \xi^2/2k + o(\xi^2)$.) Therefore, there exists $A > 0$ independent of θ such that $\psi_{\max} \leq A\theta^{1/2}$ as claimed.

The inequality $A^{-1}\theta^2 \leq \psi_{\min}$ is proved in a similar way and we omit the details. Notice that the trajectory with exit angle $\psi_{\min}$ is the one that first reflects off the left wall at a point very close to the left corner and then grazes the right wall before exiting through the open side. In proving the second inequality, the roles of ψ and θ are essentially reversed, giving θ^2 instead of the $\theta^{1/2}$ of the first inequality.

The third inequality is obtained as follows. Fix θ and let $L(x)$, $x \in (x_2, 1)$, denote the length of the segment of wall starting at the point $p(x,\theta)$ at which the trajectory with initial condition (x,θ) bounces off and ending at the right corner. Then it can be shown by elementary geometry that

$$
|L'(x)| = \frac{\sin\theta}{\sin(\theta + \xi(x,\theta))}.
$$

We know from Lemma 6.6 that $|F'_\theta(x)| = 2|k(p(x,\theta))||L'(x)|$. Therefore, we can find a $D' > 0$ such that

$$|F'_\theta(x)|^{-1} = \frac{1}{2|k(p(x,\theta))|}\frac{\sin(\theta+\xi(x,\theta))}{\sin\theta} \leq D'\left(1+\frac{\xi(x,\theta)}{\theta}\right)$$

$$\leq D'\left(1+\frac{\xi(0,\theta)}{\theta}\right) \leq D'\left(1+\frac{D\theta^{1/2}}{\theta}\right) \leq B\theta^{-1/2},$$

for some $B > 0$, as claimed.

For $x \in (0, x_2)$, $|F'_\theta(x)|$ can be bounded from below by estimating the variation on the angle of reflection after collision with the left wall. That variation can still be written as $|F'_\theta(x)| = 2|k(p(x,\theta))||L'(x)|$, where $L(x)$ is now the length of the segment of left wall from the left corner to the point of first collision. Due to the downward bending of the wall, it holds that $|L'(x)| \geq |L'(0)| = 1$. Therefore $|F'_\theta(x)|^{-1} \leq C$ for some $C > 0$ which depends only on the curvature of the wall. $\qquad\square$

THEOREM 6.8. *Under Assumptions 1, 2 and 3, the Markov operator of a random billiard is a Hilbert–Schmidt operator. In particular, its spectrum consists of eigenvalues $\lambda_i \in [-1, 1]$, each of finite multiplicity, with 0 as the only accumulation point. Further, 1 is a simple eigenvalue.*

PROOF. This is now a corollary of the previous lemma. Observe that the random billiard system is ergodic under these assumptions, so that 1 has multiplicity one. $\qquad\square$

It would be interesting to try to extract from this proof an estimate of the *spectral gap* of the Markov operator, something which we do not do here. In any event, it is worth making the following remarks. We recall that P is said to have spectral gap γ if there is $\rho = 1-\gamma < 1$ such that, for $\eta \in H$ satisfying $\eta(V) = 0$ (that is, such that $\langle \eta, \mu \rangle = 0$), the inequality

$$\|P\eta\|_2 \leq \rho\|\eta\|_2$$

holds. In other words, $(\rho, 1)$ is the largest interval in $[-1, 1]$, having 1 as a boundary point, that does not intersect the spectrum of P.

A large spectral gap implies fast convergence to the stationary measure μ, in the following sense: for each $\eta \in H$ (the Hilbert space of signed measures on V with square integrable densities with respect to μ) such that $\eta(V) = 0$, there is C_η such that

$$\|P^k\eta - \mu\|_2 \leq C_\eta\rho^k,$$

for all $n \in \mathbb{N}$.

It is possible to express γ in terms of the random billiard map as follows. Define the *Dirichlet form*

$$\mathcal{E}(\eta, \eta)_\mu := \langle (I - P)\eta, \eta \rangle.$$

Note that, for any constant c,

$$\mathcal{E}(\eta - c\mu, \eta - c\mu) = \mathcal{E}(\eta, \eta),$$

since μ is in the kernel of the self-adjoint operator $I - P$. A simple calculation shows that if $d\eta = \phi d\mu$ and $\Phi = \phi \circ \pi_2$, then

$$\mathcal{E}(\eta, \eta)_\mu = \int_E (\Phi - \Phi \circ T)\Phi \, dv = \frac{1}{2} \int_E (\Phi - \Phi \circ T)^2 \, dv$$

$$= \frac{1}{2} \int_X \int_V [\phi(v) - \phi(\Psi_x(v))]^2 \, d\mu(v)d\lambda(x).$$

The spectral gap of the random billiard is now given by

$$\gamma = \inf \left\{ \frac{\mathcal{E}(\eta, \eta)_\mu}{\|\eta\|_2^2} : \eta(V) = 0 \right\}$$

$$= \inf \left\{ \frac{\int_E [\phi(v) - \phi(\Psi_x(v))]^2 dv}{2 \int_V \phi^2 d\mu} : \phi \in L^2(V, \mu), \int_V \phi \, d\mu = 0 \right\}.$$

7. Diffusion limit

From a random billiard system we obtain a random flight in a channel, as explained in Section 2.4, and to this random flight is associated a time-of-escape function $f(x)$, defined in the introduction. We expect $f(x)$ to contain information about the channel surface microgeometry. We have already seen in the introduction how such geometric information can be extracted from $f(x)$ in a simple example. In that case, the mean square displacement of a billiard particle along the axis of the channel is finite and the standard central limit theorem for Markov chains can be used to obtain the dependence of the diffusion constant

$$D = \lim_{x \to \infty} \frac{x^2}{f(x)}$$

on the ratio b/h, which is the sole geometric parameter of the family of Figure 2. Before showing further details of this and indicating how something similar can be accomplished for more general geometries, it is perhaps useful to try to describe intuitively how the microgeometry can affect the diffusion constant D by means of the following random walk model.

7.1. A coin model of random flight. Fix a number α between 0 and 1, which will be referred to as the surface *slippage*, and consider the random walk on a one-dimensional lattice with spacing δ between neighbor cells defined as follows. A point particle moves by jumping one step of length δ either backward or forward, the time between jumps being given by a constant $\bar{t}$. The direction of jump is decided by the rule that at each step we flip two coins simultaneously, one biased, with probability for heads equal to α, and the other unbiased. If the biased coin comes up heads the particle repeats the behavior of the previous jump regardless of the outcome of the unbiased coin. We say in this case that it *slips* with probability α. If the biased coin comes tails the particle *forgets* what it did in the previous jump and moves forward or backward depending on the outcome of the unbiased coin (say, heads = forward, tails = backward.)

There are only two directions of motion, represented by ± 1. The scattering operator P for this example is the transition probabilities matrix

$$P = \begin{pmatrix} \dfrac{1+\alpha}{2} & \dfrac{1-\alpha}{2} \\ \dfrac{1-\alpha}{2} & \dfrac{1+\alpha}{2} \end{pmatrix}.$$

The top left entry, $(1+\alpha)/2$, is the probability that the particle direction in the next move is $+1$ given that it was $+1$ in the last move; the lower-left entry, $(1-\alpha)/2$, is the probability that the next direction is $+1$ given that in the last move it was -1, etc. Zero slippage corresponds to ordinary random walk on a one-dimensional lattice, in which no memory of past moves is kept. The case $\alpha = 1$ corresponds to purely deterministic uniform motion forward or backward.

The Markov chain specified by P has the same unique stationary distribution of directions for all $\alpha \neq 1$, namely, probability $1/2$ for both $+1$ and -1. The limit diffusion process, however, has different diffusion constants depending on α. More precisely, let D_α denote the constant for the process given as a limit of this random walk, by making δ and $\bar{t}$ go to 0 so that the ratio $D_0 = \delta^2/\bar{t}$ remains constant. Then it can be shown, using the central limit theorem for Markov chains (stated in the next section), that

$$D_\alpha = \frac{1+\alpha}{1-\alpha} D_0.$$

The reference diffusivity, D_0, in that of standard random walk.

It is worth emphasizing the following property of this example: the stationary distribution of directions is the same for all α, but the speed of convergence to the stationary distribution varies greatly with slippage. In fact, the exponential rate of convergence is proportional to $1 - \alpha$. Therefore, as α is close to 1 (high slippage), it will take longer to reach stationarity. The diffusivity constant will be

Figure 20. The top microgeometry produces relatively slow diffusion compared to the one at the bottom.

proportionally larger as particles effectively move a greater distance on average during a small interval of time.

We can draw a parallel between this example and the random flight for a random billiard system with microgeometries such as those of Figure 20. Although we do not yet have precise estimates, it can be seen numerically that the rate of convergence to the stationary distribution is faster for the top microgeometry, so that it can be assigned an effective value of α smaller than for the lower microgeometry. Numerical simulation of the random flight indicates, in fact, that the diffusion constant for the "flatter" microgeometry has larger diffusion constant. The example of Section 7.3 suggests that an effective measure of slippage α can be defined for random billiards as $1 - \gamma$, where γ is the spectral gap of P (which is $1 - \lambda$, where λ is the second largest eigenvalue of P after 1.)

7.2. Random flight with finite mean free path. Brownian motion on the real line has the following property: let B_t denote the position of a particle undergoing standard Brownian motion starting at the origin at time $t = 0$ and denote by $f(x)$ the mean time it takes B_t to reach a distance x from 0. Then it is well-known that $D = x^2/2f(x)$, where $D = \sigma^2/2$ and σ^2 is the variance of B_1. If a discrete time process has finite mean square jumps and satisfies the central limit theorem, then a limit diffusion is obtained with constant given by the limit of $x^2/2f(x)$ as $x \to \infty$. Thus the key issue we face is to find a central limit theorem that applies to the random flight of a random billiard system. Ideally, such a theorem will allow to relate D to the spectrum of P.

The random displacement X_i between two consecutive collisions with the channel wall has infinite mean square with respect to the stationary measure μ whenever μ is ergodic. Nevertheless, there are examples where this mean square displacement is finite, so that the considerations of the previous paragraph are valid. In such cases Theorem 7.1 can be used. The theorem applies to examples such as the 2 and 3-dimensional rectangular cells of the introduction and Section 3.1, as well as other rational polygonal billiard cells. This is an adaptation of a result due to Kipnis and Varadhan, [KiVa]. A similar theorem that applies to the more general μ-ergodic case will be studied in a future paper.

We consider a particle undergoing random flight in a two or three-dimensional channel of infinite length and radius r. Let V be, as before, the set of directions of the particle at the open side of a billiard cell and write $\Omega = V^{\mathbb{N}}$, with projections $\pi_k : \Omega \to V$. The particle has constant scalar velocity, v. We fix reference values r_0 and v_0 and write $r = r_0/\zeta$, $v = g(\zeta)v_0$, where $g(\zeta)$, until further specified, is an unbounded monotone increasing function. (It is mostly the case here that $g(\zeta) = \zeta$, but it will be convenient for later to define it more generally at this point.) The position of the particle along the axis of the channel at time t is represented by $X^*_{\zeta,t}(\omega) \in \mathbb{R}$, for any $\omega \in \Omega$. We wish to study the limit of $X^*_{\zeta,t}$ as $\zeta \to \infty$.

Let P be a bounded self-adjoint operator on $L^2(V, \mu')$, for some probability measure μ', with spectrum in $[-1, 1]$. Let Π denote the spectral measure of P and define, for a given $h \in L^2(V, \mu')$, the measure on $[-1, 1]$ given by

$$\varepsilon_h(d\lambda) := \langle h, \Pi(d\lambda)h \rangle.$$

THEOREM 7.1. *Give Ω the probability measure associated to (P, μ'), where P is the Markov operator for a choice of random billiard and μ' is an ergodic stationary measure. Let $h(u) = 2r_0 u \cdot e/u \cdot n$, $|u| = 1$. Suppose that h is square integrable with respect to μ', has zero mean, and that*

$$\delta_0^2 := \int_{-1}^{1} \frac{1+\lambda}{1-\lambda} \varepsilon_h(d\lambda) < \infty.$$

Also suppose that $g(\zeta) = \zeta$. Then, for any sequence $\zeta_n \to \infty$, the process $X^*_{\zeta_n,t}$ converges to Brownian motion with variance $(\delta_0^2/\tau_0)t$, where τ_0 is the mean value of $2r_0/(v_0 u \cdot n)$ with respect to μ'. If (P, μ') is replaced with (P, δ_θ), where the initial probability distribution is a delta-measure concentrated at an angle θ in the support of μ', then the distribution of $X^*_{\zeta_n,t}$ will again converge to Brownian motion (with same variance), the convergence now being in measure as functions of θ, relative to μ'.

As already noted, up to minor modifications having to do with our need to consider random times, Theorem 7.1 is due to Kipnis and Varadhan, [KiVa]. Clearly, requiring h to be square integrable with respect to μ' is very restrictive since it excludes all systems for which $\mu' = \mu$. It is an open question whether there are any examples of billiard cells with square integrable h that are not rational polygons. We apply below Theorem 7.1 to a couple of simple examples.

2-D boxes. This is the example of the introduction (Figure 2). For an initial $\theta_0 \in (0, \pi)$, the possible states are θ_0 and $\pi - \theta_0$, with transition probabilities shown on the Markov diagram of Figure 2. The transition probabilities matrix,

$$P = \begin{pmatrix} p & 1-p \\ 1-p & p \end{pmatrix},$$

has eigenvalues 1 and $\lambda = 2p - 1$, where p depends on the geometric parameters as described in the introduction. The stationary probability distribution is $(1/2, 1/2)$.

The function $h(\theta)$ can take values $\pm 2r_0 \cot \theta_0$, and ε_h is the measure supported on λ, so that $\delta_0^2 = (2r_0 \cot \theta_0)^2 (1 + \lambda)/(1 - \lambda)$. The mean time between collisions is $\tau_0 = 2r_0/(v_0 \sin \theta_0)$. Then Theorem 7.1 gives that the diffusion limit is σB_t, where B_t is normalized Brownian motion (variance 1) and $\sigma^2 = 2r_0 v_0 p (1 - p)^{-1} \cos^2 \theta_0 / \sin \theta_0$.

As a special case, suppose that $\theta_0 = \pi/4$, and that $b > 2h$. Then $k = 0$, $s = 2h/b$, and $p = 1 - s$, so that

$$\sigma^2 = \sqrt{2} r_0 v_0 \left(\frac{b}{2h} - 1 \right).$$

The interpretation of this result in terms of gas kinetics is that, if a pulse of noninteracting particles are released inside the channel with the same angle θ_0, the transport along the channel can be approximated by the diffusion equation

$$\frac{\partial u}{\partial t} = \frac{1}{2} \sigma^2 \frac{\partial^2 u}{\partial x^2},$$

where $u(x, t)$ is the particle linear density, t is time and x is the coordinate along the axis of the channel.

3-D boxes. We consider now the example of Section 3.1. We refer to notation defined there. For any given initial direction, the function $h(v)$ can assume only four values: $\pm \delta_1, \pm \delta_2$. The spectral decomposition of the function h is

$$\begin{pmatrix} \delta_1 \\ \delta_2 \\ -\delta_2 \\ -\delta_1 \end{pmatrix} = \frac{\delta_1 + \delta_2}{2} \begin{pmatrix} 1 \\ 1 \\ -1 \\ -1 \end{pmatrix} + \frac{\delta_1 - \delta_2}{2} \begin{pmatrix} 1 \\ -1 \\ 1 \\ -1 \end{pmatrix},$$

where $(1/2)(1, 1, -1, -1)^t$ and $(1/2)(1, -1, 1, -1)^t$ are associated to eigenvalues λ_2 and λ_3, respectively. The stationary measure is given by the probability distribution $(1/4, 1/4, 1/4, 1/4)^t$ and the mean time between collisions is $\tau_0 = (\tau_1 + \tau_2)/2$. We have $(1 + \lambda_2)/(1 - \lambda_2) = p_1/(1 - p_1)$ and $(1 + \lambda_3)/(1 - \lambda_3) = p_2/(1 - p_2)$. The diffusion limit in this case is Brownian motion with variance σ^2 given by

$$\sigma^2 = \left[\frac{p_1}{1 - p_1} (\delta_1 + \delta_2)^2 + \frac{p_2}{1 - p_2} (\delta_1 - \delta_2)^2 \right] \frac{1}{\tau_0}.$$

The geometric parameters of the microgeometry are contained in the expressions for p_i, δ_i and τ_i given in Section 3.1.

7.3. The ideal cavity model. The passage from the random flight to the diffusion process is in most cases not entirely standard if under the stationary distribution μ the mean square displacement is infinite. Nevertheless a central limit theorem still holds under appropriate normalization. More precisely, one needs here a central limit theorem that gives limit in distribution for sums of the form $(X_1 + \cdots + X_n)/\sqrt{n \ln n}$ and expresses the variance of the limit random variable as a function of the spectrum of the Markov operator P. This will be discussed in a future paper. Here we only consider a single but representative example.

Nonstandard central limit theorems of this kind are used in a number of other studies. Most closely related to our concerns is [BGT]. The example discussed later, which is the Markov process defined by P_{Maxwell} (see Section 3.2) is essentially the system studied in that paper. A central limit theorem similar to what we need is used in [Bl] and [SV], where the Lorentz gas model with infinite horizon is investigated See also [AD] and [Go].

We consider now the example of Section 3.2, given by $P_\mu + \alpha P_{\mu\perp}$, where P_μ is the orthogonal projection to the 1-dimensional subspace spanned by the stationary measure μ and $P_{\mu\perp}$ is orthogonal projection to the orthogonal complement. Recall that α is interpreted as the probability that a particle bounces off the flat side in a specular way, and $1 - \alpha$ the probability that it falls into the cavity and eventually exits with a probability distribution of angles given by μ.

Theorem 7.1 does not apply here since the function $h(\theta) = 2r_0 \cot \theta$ has infinite variance with respect to μ. It turns out, however, that a diffusion limit can be obtained after appropriate scaling of the particle velocity. More precisely, the function $g(\zeta)$ that is required for the limit of $X^*_{\zeta,t}$ to be normally distributed is $g(\zeta) = \zeta / \ln \zeta$.

PROPOSITION 7.2. *Let $X^*_{\zeta,t}$ be as defined previously, now corresponding to the Markov operator $P = P_\mu + \alpha P_{\mu\perp}$. Then for any $\zeta_n \to \infty$ and any fixed $t > 0$, $X^*_{\zeta_n,t}$ converges in distribution to normal distribution with variance $t\sigma^2$, where*

$$\sigma^2 = C \frac{1+\alpha}{1-\alpha},$$

where the constant C only depends on r_0 and v_0.

See [BGT] for details. Although the context of that paper is different from ours, their proof carries on to this case, with only minor modifications.

The ideal cavity model does not correspond, strictly speaking, to any microgeometry, but it is useful for intuitively understanding actual random billiard systems. Observe, for example, the two geometries of Figure 20. They can be compared with the ideal cavity by assigning a larger α for the one with smaller curvature.

8. Remark about residence time in cells

The random walk on the line, driven by a random billiard map, only depends on the Markov operator P. For the strict random billiard model and its diffusion limit studied so far, the time a particle spends inside a billiard cell is considered negligible. But if we wish to go from an infinitesimal billiard geometry, in the strict sense defined above, to a small, but finite, deterministic billiard system at a scale comparable to that of the channel radius, it becomes important to estimate how much time, on average, a particle spends in the billiard cell. The mean time spent inside billiard cells, which we refer to as the *residence time* of the random billiard, would then be a numerical factor to be taken into account when describing the diffusion limit. This mean time can be calculated exactly, for ergodic billiards, by standard ergodic theory arguments. We show below how this average depend on the shape of the billiard cell.

Consider an ordinary billiard system with piecewise smooth boundary of total length l, a distinguished boundary component consisting of a line segment of length o, and enclosed area A. The distinguished flat side is parametrized by $[0, 1]$. Let $E = [0, 1] \times [0, \pi]$, as before, denote the part of the phase space for that flat side and suppose that the billiard system is ergodic. For each $(x, \theta) \in E$, let $S(x, \theta)$ and $N(x, \theta)$ be, respectively, the time of first return and the number of collisions before returning to the distinguished side, counting the arrival at the side as one collision. Denote by $\bar{E}$ the whole unit phase space (i.e., with scalar velocity equal to 1) and define for any $(p, v) \in \bar{E}$ the function $\tau(p, v)$ that gives the time duration of the free flight with velocity v from p to the point of next collision. The average values of N, S are defined by

$$\langle N \rangle_E := \frac{1}{2} \int_0^1 \int_0^\pi N(x, \theta) \sin\theta \, d\theta dx,$$

$$\langle S \rangle_E := \frac{1}{2} \int_0^1 \int_0^\pi S(x, \theta) \sin\theta \, d\theta dx.$$

Similarly, define the average of τ (over $\bar{E}$), by

$$\langle \tau \rangle_{\bar{E}} := \int_{\bar{E}} \tau(\xi) d\bar{v}(\xi),$$

where $\bar{v}$ is the normalized Liouville measure on $\bar{E}$. Let $B : \bar{E} \to \bar{E}$ denote the billiard map for the billiard table that includes the top flat side. The first return map to E will be written $T : E \to E$. The billiard flow will be denoted φ_t. Elements of $\bar{E}$ will be written $\xi = (x, v)$. Note that

$$S(\xi) = \tau(\xi) + \tau(B(\xi) + \cdots + \tau(B^{N(\xi)-1}(\xi)).$$

A standard application of Birkhoff's ergodic theorem gives the following answer to the question posed at the top of the section.

THEOREM 8.1. *Suppose that the billiard cell has finite diameter and that the system is ergodic. Let u be the scalar velocity of the billiard particle. Then*

(1) $\langle N \rangle_E = l/o_;,$
(2) $\langle S \rangle_E = A\pi/ou,$
(3) $\langle \tau \rangle_{\bar{E}} = A\pi/lu,$ *and*
(4) $\langle S \rangle_E = \langle N \rangle_E \langle \tau \rangle_{\bar{E}}.$

PROOF. We give a brief sketch of the proof, which for the most part is a standard argument. For each $\xi \in E$ and positive integer l, define

$$N^l(\xi) := N(\xi) + N(T(\xi)) + \cdots + N(T^l(\xi));$$
$$S^l(\xi) := S(\xi) + S(T(\xi)) + \cdots + S(T^l(\xi)).$$

Then $N^l(\xi)$ is the total number of collisions with the table boundary during the period of l returns to the distinguished (top) side, and $S^l(\xi)$ is the total time during the same period. It is immediate that $\lim_{l \to \infty} N^l(\xi)/l = \langle N \rangle_E$ and $\lim_{l \to \infty} S^l(\xi)/l = \langle S \rangle_E$. Therefore, for v-a.e. $\xi \in E$,

$$\langle N \rangle_E^{-1} = \lim_{l \to \infty} \frac{l}{N^l(\xi)} = \lim_{l \to \infty} \frac{1}{N^l(\xi)} \sum_{i=0}^{N^l(\xi)} \chi_E(B^i(\xi))$$

$$= \bar{v}(E) = \frac{\text{length of distinguished side}}{\text{total perimeter of table boundary}}.$$

This shows (1). To obtain (4), start with

$$\sum_{k=0}^{N^l(\xi)} \tau(B^k(\xi)) = S(\xi) + S(T(\xi)) + \cdots + S(T^{l-1}(\xi))$$

and average both sides over E, using T-invariance of the measure v. This gives

$$\left\langle \sum_{k=0}^{N^l(\xi)} \tau(B^k(\xi)) \right\rangle_E = l \langle S \rangle_E.$$

Consequently,

$$\langle S \rangle_E = \lim_{l \to \infty} \left\langle \left(\frac{N^l(\xi)}{l} \right) \left(\frac{1}{N^l(\xi)} \sum_{k=0}^{N^l(\xi)} \tau(B^k(\xi)) \right) \right\rangle_E = \langle N \rangle_E \langle \tau \rangle_{\bar{E}}.$$

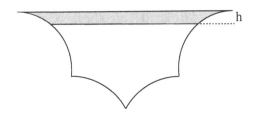

Figure 21. Define U_h as the strip of width h with the distinguished side as one of the boundary lines.

To show (2), it is convenient to first introduce the collar region U_h shown in Figure 21, where h is a small positive number.

Except for a set of small measure (which goes to zero with h), the time it takes for the ray with initial condition $\xi = (x, \theta) \in E$ to traverse U_h is $\eta(\xi) = h/u \sin \theta$. An explicit integral calculation gives

$$\lim_{h \to 0} \frac{1}{h} \langle \eta \rangle_E = \pi/2u.$$

We can now conclude that, for ν-a.e. $\xi \in E$,

$$\langle S \rangle_E \frac{\text{length of top side}}{\text{area of billiard cell}}$$

$$= \lim_{h \to 0} \lim_{m \to \infty} \left(\frac{S^m(\xi)}{m} \right) \left(\frac{1}{h S^m(\xi)} \int_0^{S^m(\xi)} \chi_{U_h}(\varphi_t(\xi)) \right) dt$$

$$= \lim_{h \to 0} \lim_{l \to \infty} \frac{1}{hm} \int_0^{S^m(\xi)} \chi_{U_h}(\varphi_t(\xi)) dt$$

$$= \lim_{h \to 0} \lim_{l \to \infty} \frac{1}{hm} \sum_{i=0}^{m-1} 2\eta(T^i(\xi)) = \lim_{h \to 0} \frac{2}{h} \langle \eta \rangle_E = \frac{\pi}{u}.$$

This gives the average value of S claimed in (2). $\square$

References

[AD] J. Aaronson and M. Denker. Local limit theorems for partial sums of stationary sequences generated by Gibbs–Markov maps, *Stoch. Dyn.* **1** (2001), 193–237.

[Ar] L. Arnold. *Random dynamical systems*, Springer, Berlin, 1998.

[ACM] G. Arya, H.-C. Chang, and E. J. Maginn. Knudsen diffusivity of a hard sphere in a rough slit pore, *Phys. Rev. Lett.* **91**:2 (July 2003), 026102(4).

[Ba] H. Babovsky. On Knudsen flows within thin tubes, *J. Stat. Phys.* **44**:5–6 (1986), 865–878.

[BGT] C. Börgers, C. Greengard, and E. Thomann. The diffusion limit of free molecular flow in thin plane channels, *SIAM J. Appl. Math.* **52**:4 (1992), 1057–1075.

[Bi] P. Billingsley. *Convergence of probability measures*, Wiley, New York, 1968.

[Bl] P. M. Bleher. Statistical properties of two-dimensional periodic Lorentz gas with infinite horizon, *J. of Stat. Physics* **66**:1 (1992), 315–373.

[Br] L. Breiman. *Probability*, Classics in Applied Mathematics 7, SIAM, Philadelphia, 1992.

[C1] C. Cercignani. *The Boltzmann equation and its applications*, Springer, New York, 1988.

[C2] C. Cercignani and D. H. Sattinger. *Scaling limits and models in physical processes*, DMV Seminar **28**, Birkhäuser, Basel, 1998.

[CFS] I. P. Cornfeld, S. V. Fomin and Y. G. Sinai. *Ergodic theory*, Grundlehren der Mathematischen Wissenschaften (Fundamental Principles of Mathematical Sciences) **245**, Springer, New York, 1982.

[FY1] R. Feres and G. Yablonsky. Knudsen's cosine law and random billiards, *Chem. Eng. Sci.* **59** (2004), 1541–1556.

[FY2] R. Feres and G. Yablonsky. Probing surface structure via time-of-escape analysis of gas in Knudsen regime, *Chem. Eng. Sci.* **61** (2006), 7864-7883.

[Go] S. Gouëzel. Central limit theorem and stable laws for intermittent maps, preprint, September, 2003.

[HK] B. Hasselblatt and A. Katok. *Introduction to the modern theory of dynamical systems*, Encyclopedia of Mathematics and its Applications **54**, Cambridge University Press, Cambridge, 1995.

[KS] A. Katok and J.-M. Strelcyn. *Invariant manifolds, entropy and billiards: Smooth maps with singularities*, Lecture Notes in Mathematics **1222**, Springer, Berlin, 1986.

[Ki] Y. Kifer. *Ergodic theory of random transformations*, Birkhäuser, Boston, 1986.

[KiVa] C. Kipnis and S. R. S. Varadhan. Central limit theorem for additive functionals of reversible Markov processes and applications to simple exclusions, *Commun. Math. Phys.* **104** (1986), 1–19.

[Kn] M. Knudsen. *Kinetic theory of gases: Some modern aspects*, Methuen's Monographs on Physical Subjects, Methuen, London, 1952.

[Re] D. Revuz. *Markov chains*, North-Holland, Amsterdam, 1975.

[SV] D. Szász and T. Varjú. Markov towers and stochastic properties of billiards, *Modern dynamical systems and applications*, M. Brin, B. Hasselblatt, Y. Pesin (eds.), 433–445, Cambridge University Press, Cambridge, 2004.

[Ta] S. Tabachnikov. *Billiards*, Panor. Synthèses **1**, Société Mathématique de France, Paris, 1995.

[Wa] P. Walters. *An introduction to ergodic theory*, Springer, New York, 1982.

RENATO FERES
WASHINGTON UNIVERSITY
DEPARTMENT OF MATHEMATICS
ST. LOUIS, MO 63130
UNITED STATES
  feres@math.wustl.edu

Recent Progress in Dynamics
MSRI Publications
Volume **54**, 2007

An aperiodic tiling using a dynamical system and Beatty sequences

STANLEY EIGEN, JORGE NAVARRO, AND VIDHU S. PRASAD

ABSTRACT. Wang tiles are square unit tiles with colored edges. A finite set of Wang tiles is a valid tile set if the collection tiles the plane (using an unlimited number of copies of each tile), the only requirements being that adjacent tiles must have common edges with matching colors and each tile can be put in place only by translation. In 1995 Kari and Culik gave examples of tile sets with 14 and 13 Wang tiles respectively, which only tiled the plane aperiodically. Their tile sets were constructed using a piecewise multiplicative function of an interval. The fact the sets tile only aperiodically is derived from properties of the function.

1. Introduction

There is a vast literature connecting dynamical systems and tilings of the plane. In this paper, we give an exposition of the work of Kari [7] and Culik [3] to show how by starting with a piecewise multiplicative function f, with rational multiplicands defined on a finite interval, we can produce a finite set of Wang tiles which tiles the plane. Further, a choice of multiplicands and interval, so that the dynamical system f has no periodic points, results in a set of Wang tiles that can only tile the plane aperiodically. In this manner, Kari and Culik produce a set of 13 Wang tiles. This is currently, the smallest known set of Wang tiles which only tiles the plane aperiodically.

The Kari–Culik construction is different from earlier constructions of aperiodic tilings — see Grunbaum and Shephard's book [5, Chapt 11] for a survey of these earlier results. Johnson and Madden [6], provide an accessible presentation

Mathematics Subject Classification: Primary 52C20, 05B45, 52C23; Secondary 37E05.

Keywords: Wang tiles, aperiodic tiling of the plane, piecewise multiplicative, dynamical system, Beatty difference sequence.

of Robinson's 1971 [11] example of 6 polygonal tiles which force aperiodicity (allowing rotation and reflection). These 6 tiles convert to a set of 56 Wang tiles which allow only aperiodic tilings of the plane. Kari and Culik's construction uses a dynamical system and Beatty sequences to label the sides of the Wang tiles. The properties of the dynamical system are used to conclude the collection tiles the plane and does so only aperiodically.

1.1. The Kari–Culik tile set.

Consider the dynamical system given by the function f defined on the interval $[\frac{1}{3}, 2)$,

$$f(x) = \begin{cases} 2x, & \frac{1}{3} \le x < 1 \\ \frac{1}{3}x, & 1 \le x < 2 \, . \end{cases}$$

This gives rise (Section 5 shows how) to a set $\mathcal{T}$ of thirteen Wang tiles, which we call the K-C tile set (see figure below). These thirteen tiles do tile the plane, but only aperiodically.

K-C Tile Set.

The proof that the tile set tiles the plane will follow from the existence of infinite orbits for f. The proof that the tile set tiles only aperiodically relies on the fact that f has no periodic points on the interval $[\frac{1}{3}, 2)$. We note that Kari and Culik [3; 7] use Mealy machines describe these tile sets. We give their description at the end of this paper. In the language of computer science a Mealy machine is a finite state machine where the output is associated with a transition; in symbolic dynamics a Mealy machine is referred to as a finite-state code [8].

2. Wang tiles: definitions and history

Wang tiles are square unit tiles with colored edges. All tiles in this paper are assumed to be Wang tiles. In Kari and Culik's tile set, numbers are used to color the edges: edges will have a color and a numerical value. Thus, the colored edges 0, $0'$ and $\frac{0}{3}$ are considered different colors, but these edges have a numerical value, which in this case is zero.

A *tiling set* $\mathcal{T}$ consists of a collection of finitely many Wang tiles $T \in \mathcal{T}$, each of which may be copied as much as needed. When used to tile the plane, the tiles must be placed edge-to-edge with common edges having matching colors. Rotations and flips (reflections) of the tiles are not permitted.

A tiling set $\mathcal{T}$ which can tile the plane is said to have a *valid tiling*, and $\mathcal{T}$ is called a *valid tile set*. A valid tiling is a map on the integer lattice, $\tau : \mathbb{Z} \times \mathbb{Z} \to \mathcal{T}$ such that, at each lattice point (i, j) we have a tile $\tau(i, j) = T_{i,j} \in \mathcal{T}$ whose neighboring tiles have matching colors along common edges.

If rotations were permitted, then any tile and its 180 degree rotation forms a valid tile set for the plane, as the following argument shows. Label the four colors of a tile a, b, c, d, (not necessarily distinct). Take two copies of the tile and two copies of its rotation through 180 degrees and construct the following two-by-two block.

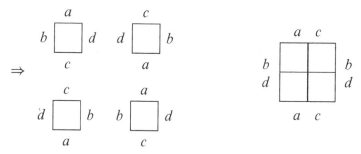

Rotation Example.

The two-by-two block has the same colors on the top as the bottom, and the same colors on the left as the right. The two-by-two block tiles the plane.

2.1. Periodicity. A valid tiling is *periodic* with period $(h, v) \in \mathbb{Z}^2 \setminus \{(0, 0)\}$ if the tile at position (i, j) is the same as the tile at position $(i + h, j + v)$ for all $(i, j) \in \mathbb{Z}^2$. That is, $\tau(i, j) = T_{i,j} = T_{i+h,j+v} = \tau(i + h, j + v)$.

Needless to say a tile set may have more than one valid tiling; some of which may be periodic and some of which may not. A tile set is called *aperiodic* if it has at least one valid tiling, but does not have a valid tiling which is periodic. The K-C tile set is aperiodic (Theorem 10).

Hao Wang [14] conjectured in 1961 that if a set of tiles has a valid tiling τ then it has a valid tiling τ' which is periodic. However, in 1966 R. Berger showed that there exists a tile set which only tiles aperiodically, and this aperiodic tile set contained 20,426 tiles. Since that time, the size of the smallest known set of aperiodic Wang tiles has been reduced considerably. By 1995, J. Kari [7] and K. Culik [3] constructed a set of 14 and 13 Wang tiles respectively that tiles only aperiodically.

An open problem is to determine W such that any set of Wang tiles $\mathcal{T}$ of size $w \le W$ which has a valid tiling must also have a periodic tiling. As far as the authors are aware, $4 \le W < 13$.

2.2. One-dimensional result.

In one dimension, Wang's conjecture that any valid tile set for the line must have a periodic tiling, is true. In one dimension the tiles are unit intervals colored on the left and right. A valid tiling is a map $\tau : \mathbb{Z} \to \mathcal{T}$ with adjoining left right edges having the same color. Periodicity of $\mathcal{T}$, in this case, means there exists a $p > 0$ so that $\tau(i) = \tau(i + p)$ for all i.

THEOREM 1 (WANG). *If a set of one-dimensional tiles $\mathcal{T}$ has a valid tiling of the line, then $\mathcal{T}$ has a periodic tiling of the line.*

Let τ be a valid tiling for $\mathcal{T}$, $\tau : \mathbb{Z} \to \mathcal{T}$. Since there are only a finite number of tiles in $\mathcal{T}$, there must be an $n > 0$ such that $\tau(0) = \tau(n)$. Hence the block of tiles $\tau(0)\tau(1) \cdots \tau(n-1)$, endlessly repeated, tiles the line.

A slight strengthening of the hypotheses yields one-dimensional tiling sets that tile only periodically — this shows how different the two-dimensional aperiodic tiling sets are.

PROPOSITION 2. *If $\mathcal{T}$ is valid tile set of one-dimensional tiles and no proper subset of $\mathcal{T}$ is a valid tile set of the line, then the tiles can tile only periodically.*

The proof follows the previous argument. Let m be the shortest length from any tile to its first repetition in a valid tiling τ of the line. Clearly the pigeonhole principle implies $m \le |\mathcal{T}| + 1$, where $|\mathcal{T}|$ is the cardinality of $\mathcal{T}$. The hypothesis that no proper subset is a valid tile set implies that $m = |\mathcal{T}| + 1$. Let $\tau(1)\tau(2) \cdots \tau(m) = \tau(1)$, be a shortest repeated block. Note that the right colors of all of these tiles in the block must be distinct. Indeed, suppose that two tiles $\tau(i)$ and $\tau(j)$ were the same, so that $\tau(1) \cdots \tau(i) \cdots \tau(j)\tau(k) \cdots \tau(1)$ could be replaced by $\tau(1) \cdots \tau(i)\tau(k) \cdots \tau(1)$, where the tile $\tau(j)$ does not appear. But then $\tau(j)$ is not needed for a valid tiling. Hence all right hand colors are distinct, and similarly we can show all left colors are distinct. Hence, there is exactly one way for these tiles to fit together, and that is with the block $\tau(1)\tau(2) \cdots \tau(m-1)$ endlessly repeated.

A *minimal* tiling set is one that is a valid tile set but no proper subset is a valid tile set. It is an open question whether the K-C tile set is minimal.

2.3. Rectangular tilings. In the Rotation Example given in Section 1.1, the constructed two-by-two block extends to a valid tiling of the plane which has the two linearly independent periods $(2, 0)$ and $(0, 2)$. A *rectangular* tiling of the plane is a valid tiling τ which has two periods $(n, 0)$, $(0, m)$, $n, m > 0$, that is, $\tau(i, j) = \tau(i + n, j)$ and $\tau(i, j) = \tau(i, j + m)$. In other words, it has a rectangular block of size $n \times m$ which tiles the plane.

It is well known that having a rectangular tiling is not stronger than having a periodic tiling [5].

PROPOSITION 3. *If a set of tiles admits a periodic tiling τ of the plane, then it also admits a rectangular tiling.*

We propose the following higher dimensional result (which may be already be known): *If a set of n-dimensional Wang cubes has a valid tiling τ of n-dimensional space and this tiling has $n - 1$ linearly independent periods, then*

(i) *there is another tiling with n linearly independent periods, and*
(ii) *there is another tiling which is rectangular, in the sense that there are n periods, $(p_1, 0, \ldots, 0)$, $(0, p_2, 0, \ldots, 0), \ldots, (0, \ldots, 0, p_n)$.*

3. Aperiodicity

The aperiodicity of the K-C tile set is easy to see and does not require understanding how the tiles are derived from the dynamical system. It follows the same reasoning as the following proof that f has no periodic points.

LEMMA 4. *The dynamical system f has no periodic points.*

Suppose $f^n(x) = x$ for $n > 0$. From the definition of f as a piecewise multiplicative function, it follows that $f^n(x) = q_n \cdot q_{n-1} \cdots q_1 \cdot x$ where $q_i \in \{\frac{1}{3}, 2\}$. Hence

$$f^n(x) = \frac{2^{n-k}}{3^k} \cdot x = x$$

for some $0 \le k \le n$. Dividing by $x \in [\frac{1}{3}, 2)$ gives $2^{n-k}/3^k = 1$, a contradiction.

To understand how this applies to the tiles, we consider the notion of a multiplier tile.

3.1. Multiplier tiles. A tile $b \begin{array}{c} a \\ \Box \\ c \end{array} d$ is a *multiplier tile* with *multiplier $q > 0$* if

$$q \cdot a + b - d = c \tag{3-1}$$

Note that this notion requires only the numerical value of the edges. The multiplier for a tile is unique if $a \neq 0$. If $a = 0$ then every real q is a multiplier for the tile when $b - d = c$.

A direct examination of the thirteen tiles in the K-C tile set reveals two facts:

LEMMA 5. *The first six tiles all have multiplier $\frac{1}{3}$. We call these Tile Set $\frac{1}{3}$.*

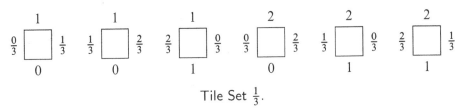

Tile Set $\frac{1}{3}$.

LEMMA 6. *The last seven tiles all have multiplier 2. We call these Tile Set $2'$.*

Tile Set $2'$.

Observe that the six tiles in Tile Set $\frac{1}{3}$ have side colors $\{\frac{0}{3}, \frac{1}{3}, \frac{2}{3}\}$ while the seven tiles in Tile Set $2'$ have side colors $\{0, -1\}$. Since these two sets of side colors are disjoint the next lemma is immediate (and is the reason why the two zeros $\{\frac{0}{3}, 0\}$ are defined to be different colors).

LEMMA 7. *If τ is a valid tiling for the tiles in the K-C tile set, then each horizontal row $\{\tau(i, j) : i \in \mathbb{Z}\}$, for j fixed, consists either exclusively of the tiles in Tile Set $\frac{1}{3}$ or exclusively of the tiles in Tile Set $2'$.*

Next, consider the row directly below a given row in a valid tiling. This requires the bottom colors of the higher row to match exactly the colors on the top of the lower row. There are restrictions on the tiles that can appear in the lower row.

LEMMA 8. *Let τ be a valid tiling for the tiles in the K-C tile set. If a horizontal row consists exclusively of tiles from Tile Set $\frac{1}{3}$ then the row immediately below it consists exclusively of tiles from Tile Set $2'$.*

The proof is simply a matter of inspecting the colors on the tiles. Suppose there are two consecutive rows of tiles from Tile Set $\frac{1}{3}$. We examine the colors along the common edge between the two rows. Since the colors along the top of tiles from Tile Set $\frac{1}{3}$ are $\{1, 2\}$ and the colors on the bottom of these tiles are $\{0, 1\}$, the only way the colors along the common edge can match is if they are all 1. However the tiles in Tile Set $\frac{1}{3}$ cannot produce a complete row with all 1's along the bottom, and so there cannot be two consecutive rows of tiles from Tile Set $\frac{1}{3}$. This lemma is related to the dynamics of f in the following manner: if $f(x) = y = \frac{1}{3}x$, then $f(y) = 2y$.

LEMMA 9. *Let τ be a valid tiling for the tiles in the K-C tile set. Then there must exist rows with tiles exclusively from Tile Set $\frac{1}{3}$.*

Lemma 9 is related to the dynamics of f in the following way: given x, $f(x)$, $f^2(x)$, at least one of these three terms must be in the interval $[1, 2)$. Any point in $[1, 2)$ will be mapped by multiplying by $1/3$. This can be used to prove the Lemma. However, we prove the lemma by directly analyzing the tiles.

Assume there are three consecutive rows of tiles from Tile Set $2'$. First consider the common edge between the highest row and the middle row. In particular, observe that the colors along the top of Tile Set $2'$ are $\{0, 0', 1\}$ while the numbers along the bottom of Tile Set $2'$ are $\{0', 1, 2\}$. To match, the common colors must be $\{0', 1\}$. The same argument shows that the colors along the common edge between the middle and lowest row must also be $\{0', 1\}$. This forces the middle row to be restricted to the two tiles

$$
0 \; \boxed{\begin{array}{c} 0' \\ \\ 1 \end{array}} \; {-1} \qquad {-1} \; \boxed{\begin{array}{c} 1 \\ \\ 1 \end{array}} \; 0
$$

from Tile Set $2'$, which means the middle row has only 1 as a bottom color and the pattern $(0', 1)$ repeated as the top colors.

The only way the third row can have a top row of all 1's is if it uses one of the two tiles

$$
{-1} \; \boxed{\begin{array}{c} 1 \\ \\ 2 \end{array}} \; {-1} \qquad 0 \; \boxed{\begin{array}{c} 1 \\ \\ 2 \end{array}} \; 0
$$

This forces the fourth row to be restricted to tiles in Tile Set $\frac{1}{3}$.

We are now able to show:

THEOREM 10. *The K-C tile set does not have a valid periodic tiling of the plane.*

The proof is by contradiction and follows the reasoning that shows the function f has no periodic points (Lemma 4).

Let τ be a periodic tiling. From Proposition 3 we can assume that τ has two periods $(n, 0)$ and $(0, m)$ with $n, m > 0$, and there is an $n \times m$ block with the same colors on both the top and bottom and the same colors on the left and right. For convenience we refer to this block as B.

Denote the top colors of Block B by $a_{i,1}$, $1 \leq i \leq n$ and the colors along the left side by $b_{1,j}$, $1 \leq j \leq m$. By the periodicity assumption, the colors along the bottom are also $\{a_{i,1}\}$ and the colors along the right side are $\{b_{1,j}\}$.

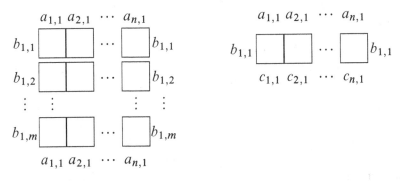

Block B.

First row of Block B.

Consider the first row of Block B. Each edge common to two tiles has the same color for the left tile and the right tile.

$$
\begin{array}{ccccccc}
& a_{1,1} & & a_{2,1} & & & a_{n,1} \\
b_{1,1} \Box & d_{1,1} \; d_{1,1} \Box & d_{2,1} & \cdots & d_{n-1,1} \Box & b_{1,1} \\
& c_{1,1} & & c_{2,1} & & & c_{n,1}
\end{array}
$$

First Row of Block B Expanded.

From Lemma 7, all the tiles in a row have the same multiplier q_1. Apply the multiplier rule (3–1) to each tile in the row.

$$
\begin{aligned}
q_1 a_{1,1} + b_{1,1} - d_{1,1} &= c_{1,1} \\
q_1 a_{2,1} + d_{1,1} - d_{2,1} &= c_{2,1} \\
q_1 a_{3,1} + d_{2,1} - d_{3,1} &= c_{3,1} \\
&\;\;\vdots \\
q_1 a_{n,1} + d_{n-1,1} - b_{1,1} &= c_{n,1}
\end{aligned}
$$

Summing results in

$$q_1 \sum_{i=1}^{n} a_{i,1} = \sum_{i=1}^{n} c_{i,1}.$$

First Two Rows of Block B Expanded.

Similarly, all the tiles in the second row of Block B have a common multiplier q_2 giving

$$q_2 \cdot \sum_{i}^{n} a_{i,2} = \sum_{i=1}^{n} c_{i,2}.$$

Combining these two equations and using $c_{i,1} = a_{i,2}$ yields

$$q_2 q_1 \sum_{i=1}^{n} a_{i,1} = \sum_{i=1}^{n} c_{i,2}.$$

Repeating for the rest of the rows in Block B results in

$$q_m \cdots q_2 q_1 \sum_{i=1}^{n} a_{i,1} = \sum_{i=1}^{n} a_{i,1}.$$

By Lemma 9 and the periodicity of the tiling, we can assume the very top row of the block B consists of tiles exclusively from Tile Set $\frac{1}{3}$. Since the top colors of the tiles in Tile Set $\frac{1}{3}$ are $\{1, 2\}$, we can divide by $\sum_{i=1}^{n} a_{i,1}$ getting $\prod_{j=1}^{m} q_j = 1$. As the $q_j \in \{2, \frac{1}{3}\}$ we have a contradiction and conclude that no periodicity can occur.

4. Existence of a valid tiling

In this section we show how to construct the tile set $\mathcal{T}$, and prove that the tile set thus constructed has valid tilings.

The K-C Tile Set is derived from the Basic Tile Construction (given in the next section) resulting in a tile set $\mathcal{T}_f$. The tile colors in $\mathcal{T}_f$ are "tweaked", to give the K-C tile set. This refers to the fact that the zeros $0, 0', \frac{0}{3}$ are considered different colors. We have already seen the reason $\frac{0}{3}$ is not the same color as 0, namely Lemma 7, which ensures that each row of tiles consists of tiles with the same multiplier. The second "tweaking" concerns $0'$ and will be explained in Section 5.2. We will see the property that $\mathcal{T}_f$ is a valid tile set, is preserved even after the colors are "tweaked".

4.1. The basic tile. All the tiles in the example are constructed as follows. We refer to this as the *Basic Tile Construction*, and it gives the values of the edges of a Basic Tile which we call $B(x, q, n)$.

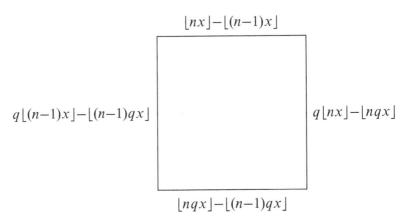

$$\lfloor nx \rfloor - \lfloor (n-1)x \rfloor$$

$$q\lfloor (n-1)x \rfloor - \lfloor (n-1)qx \rfloor \qquad\qquad q\lfloor nx \rfloor - \lfloor nqx \rfloor$$

$$\lfloor nqx \rfloor - \lfloor (n-1)qx \rfloor$$

Basic Tile $B(x, q, n)$.

Here, $x > 0$ is a real number, $q > 0$ is a rational, n is an integer and $\lfloor x \rfloor$ denotes the greatest integer less than or equal to x.

A straightforward calculation gives:

LEMMA 11. *The Basic Tile $B(x, q, n)$, is a multiplier tile with multiplier q.*

Recall a tile $b \begin{array}{c} a \\ \square \\ c \end{array} d$ has multiplier q if $q \cdot a + b - d = c$. For the Basic Tile we have

$$q\left(\lfloor nx \rfloor - \lfloor (n-1)x \rfloor\right) + \left(q\lfloor (n-1)x \rfloor - \lfloor (n-1)qx \rfloor\right) - \left(q\lfloor nx \rfloor - \lfloor nqx \rfloor\right)$$
$$= \lfloor nqx \rfloor - \lfloor (n-1)qx \rfloor.$$

4.2. A finite number of tiles. Clearly when x, q, n are fixed, one gets a single tile. Surprisingly for q rational and x in a bounded interval one gets only a finite number of tiles.

For example, if we set $q = \frac{1}{3}$ and bound $x \in [1, 2)$ then there are only six tiles resulting from the above Basic Tile construction (see Tile Set $\frac{1}{3}$)—despite the fact that x is ranging over all reals in the interval $[1, 2)$ and n is ranging over all integers.

THEOREM 12. *Let q be a rational number and $k > 0$ an integer. If we restrict $x \in [k, k+1)$ then there are only a finite set of tiles derived from the Basic Tile construction*

To prove this, we simply show that the four sides of the Basic Tile can assume only a finite number of values. We use this simple fact:

LEMMA 13. *For all $n \in \mathbb{Z}$ and for all $x \in [k, k+1)$,*

$$\lfloor nx \rfloor - \lfloor (n-1)x \rfloor \in \{k, k+1\}$$

Lemma 13 applies to both the bottom and top of the Basic Tile. The bottom uses the real qx which is bounded by $[qk, q(k+1))$. For example with $q = \frac{1}{3}$ and $x \in [1, 2) \Rightarrow qx \in [\frac{1}{3}, \frac{2}{3}) \subset [0, 1)$. Hence the top of the tiles take values in $\{1, 2\}$ while the bottom of the tiles have values in $\{0, 1\}$.

LEMMA 14. *For $q > 0$ rational, $q\lfloor nx \rfloor - \lfloor nqx \rfloor$ takes on only a finite number of values. To be more precise,*

- *if q is an integer then $q\lfloor nx \rfloor - \lfloor nqx \rfloor \in \{1-q, 2-q, \ldots, 0\}$;*
- *if $q = \frac{r}{s}$, in reduced form, then $q\lfloor nx \rfloor - \lfloor nqx \rfloor \in \{-\frac{1-r}{s}, -\frac{2-r}{s}, \ldots, \frac{s-1}{s}\}$.*

First observe that if q is an integer then clearly $q\lfloor nx \rfloor - \lfloor nqx \rfloor$ is an integer and if $q = \frac{r}{s}$ is rational then $q\lfloor nx \rfloor - \lfloor nqx \rfloor$ is limited to rational numbers of the form $\frac{i}{s}$.

It remains to show that $q\lfloor nx \rfloor - \lfloor nqx \rfloor$ is bounded above and below. From the definition of the greatest integer function $\lfloor \cdot \rfloor$,

$$q\lfloor nx \rfloor \leq qnx < \lfloor qnx \rfloor + 1$$

Subtracting $\lfloor qnx \rfloor$ gives the upper bound

$$q\lfloor nx \rfloor - \lfloor qnx \rfloor < 1$$

Again, from the definition of $\lfloor \cdot \rfloor$,

$$\lfloor qnx \rfloor \leq qnx = q(nx) < q(\lfloor nx \rfloor + 1).$$

Multiplying by -1 and adding $q\lfloor nx \rfloor$ gives the lower bound

$$q\lfloor nx \rfloor - \lfloor qnx \rfloor > -q$$

These bounds clearly place the value of $q\lfloor nx \rfloor - \lfloor qnx \rfloor$ in the ranges listed in the lemma.

4.3. Applying the basic tile construction using f. For x in the domain of f, set

$$q(x) = \begin{cases} 2, & \frac{1}{3} \leq x < 1 \\ \frac{1}{3}, & 1 \leq x < 2 \end{cases}$$

Denote by $\mathcal{T}_f = \{B(x, q(x), n)\}$ the set of tiles constructed for $\{x, q(x), n\}$ with x in the domain of f. Note that this is not yet the K-C tile set $\mathcal{T}$ because there has been no color tweaking yet, *i.e.*, there is only one 0 at this stage.

By Lemma 12, this is a finite set of tiles.

It can be seen quite easily that the tiles for a specific $\{x, q(x)\}$ fit together, with a "natural order", to form a row denoted by $\mathcal{R}(x)$.

LEMMA 15. *Fix x in the domain of f and let n range through the integers to produce a row of valid tiles $\mathcal{R}(x)$.*

By "natural order" we mean that the tile constructed using $n + 1$ is to the immediate right of the tile constructed using n. The tile constructed for n, $B(x, q(x), n)$, has the right side color $\big(q(x) \cdot \lfloor nx \rfloor - \lfloor n \cdot q(x) \cdot x \rfloor\big)$ which is the same as the left side color of the tile constructed for $n + 1$, $B(x, q(x), n + 1)$, $\big(q(x) \cdot \lfloor (n + 1 - 1)x \rfloor - \lfloor (n + 1 - 1) \cdot q(x) \cdot x \rfloor\big)$.

4.4. Beatty difference sequences. To complete the proof of the existence of valid tilings we use the notion of a Beatty difference sequence. For any real number x, the *Beatty difference sequence* of x is the two-sided sequence $\{\lfloor nx \rfloor - \lfloor (n - 1)x \rfloor : n \in \mathbb{Z}\}$. Recalling Lemma 13, if $x \in [k, k + 1)$ then the Beatty difference sequence for x belongs to $\prod_{-\infty}^{\infty}\{k, k + 1\}$.

Beatty difference sequences and Beatty sequences $\{\lfloor nx \rfloor : n \in \mathbb{Z}\}$ (see [1]) are related to the continued fraction expansion of x. There is a vast literature on Beatty sequences and their applications; see [4] and references therein.

By using the Beatty difference sequence, we see how the rows fit together. That is, the n-th tile in row $\mathcal{R}(x)$ has top $\lfloor nx \rfloor - \lfloor (n - 1)x \rfloor$ which is the n-th term in the Beatty difference sequence of x.

The bottoms of the tiles in this row give the Beatty difference sequence for $q(x) \cdot x$, *i.e.*, $\{\lfloor n \cdot q(x) \cdot x \rfloor - \lfloor (n - 1) \cdot q(x) \cdot x \rfloor$. But this is also the top of the row of tiles $\mathcal{R}(f(x))$ and the two rows fit together.

THEOREM 16. *Every infinite orbit of the dynamical system f corresponds to a valid tiling of the plane using the tiles in the tile set $\mathcal{T}_f$.*

5. Tweaking the colors

Referring back to the K-C tile set there are two color changes for $\mathcal{T}_f$ that will be incorporated to get $\mathcal{T}$. That is, there are the three "zeros" $\{\frac{0}{3}, 0', 0\}$ — two of which are color changes from the original 0. The first, $\frac{0}{3}$, is concerned with side colors.

5.1. Side color changes. The purpose of changing the color 0 to color $\frac{0}{3}$ is to ensure that each row corresponds to a single multiplicand.

The function f is defined in two pieces:

$$f(x) = \begin{cases} f_1(x) = 2x & \text{if } \frac{1}{3} \le x < 1, \\ f_2(x) = \frac{1}{3}x & \text{if } 1 \le x < 2, \end{cases}$$

with two different multiplicands $\{\frac{1}{3}, 2\}$. When the side colors are calculated, for the two pieces in the Basic Tile Construction one gets

$$\frac{1}{3}\lfloor nx \rfloor - \lfloor \frac{1}{3}nx \rfloor \in \{0, \frac{1}{3}, \frac{2}{3}\} \text{ for all } n \quad \text{and} \quad 2\lfloor nx \rfloor - \lfloor 2nx \rfloor \in \{0, -1\} \text{ for all } n.$$

The problem is that 0 appears as a side color for both pieces. This would allow tiles with a multiplier of $\frac{1}{3}$ to appear on the same row as tiles with multiplier 2. The solution is to change one of the 0's to a different color which explains the new color $\frac{0}{3}$ (see also Section 6).

5.2. Top-bottom color changes. In this section, we will change some of the top and bottom 0's to 0' in the tile set $\mathcal{T}_f$: such changes are called top-bottom changes. This is necessary because the tile set $\mathcal{T}_f$ (without top-bottom changes) is not aperiodic. By introducing these top-bottom color changes (and possibly additional tiles) periodicity may be avoided.

Note that the top-bottom color changes will not affect the multiplier property of any tile (since the numerical value of an edge will not be changed) but will only be concerned with the "colors" of the tiles. Thus the existence of valid tilings will not be affected.

The final K-C tile set is obtained from $\mathcal{T}_f$ by incorporating both the side color changes and the top-bottom color changes.

Consider the piece f_1 of f. Recall that $f_1(x) = 2x$ has domain $[\frac{1}{3}, 1)$ and range $[\frac{2}{3}, 2)$. The Basic Tile Construction for $x \in [\frac{1}{3}, 1)$ yields the six tiles

Tile Set 2.

Unfortunately, this tile set is not aperiodic. The first tile (and the second tile) tiles the plane periodically.

The reason is that the tile set has lost the information that the domain of the piece $f(x) = 2x$ is restricted to $[\frac{1}{3}, 1)$. More specifically, the Basic Tile Construction for $x \in [0, 1)$ yields exactly the same 6 tiles (recall Lemmas 13 and 14) and enlarging the interval would add more tiles. Hence, Tile Set 2 is really the tile set for $f'(x) = 2x$ with domain $[0, 1)$ and range $[0, 2)$. The periodic tiling τ of the plane given by the single tile

$$\tau(i, j) = 0 \quad \boxed{\begin{array}{c} 0 \\ \\ 0 \end{array}} \quad 0 , \quad -\infty < i, j < \infty,$$

corresponds to the fixed point $f'(0) = 2 \cdot 0 = 0$.

More generally, any tile of the form $a \quad \boxed{\begin{array}{c} 0 \\ \\ 0 \end{array}} \quad a$ can tile the plane periodically.

Such tiles arise when there are points $x \in [0, 1)$ in the domain of f' such that $f'(x) \in [0, 1)$; Lemma 13 shows that these points may give rise to tiles having 0 on both the top and bottom. It is to avoid such tiles that the additional color changes are made (and additional tiles added to the set).

Examining a portion of a typical orbit for the function f, for example

$$1 \Rightarrow \frac{1}{3} \Rightarrow \frac{2}{3} \Rightarrow \frac{4}{3} \Rightarrow \cdots$$

reveals immediately that the function f has at most two consecutive images in the interval $[\frac{1}{3}, 1) \subset [0, 1)$.

Rewrite the function $f =: F$ in four pieces as

$$F_1(x) = 2x, \quad [\tfrac{1}{3}, \tfrac{1}{2}) \to [\tfrac{2}{3}, 1), \qquad F_2(x) = 2x, \quad [\tfrac{1}{2}, \tfrac{2}{3}) \to [1, \tfrac{4}{3}),$$

$$F_3(x) = 2x, \quad [\tfrac{2}{3}, 1) \to [\tfrac{4}{3}, 2), \qquad F_4(x) = \tfrac{1}{3}x, \quad [1, 2) \to [\tfrac{1}{3}, \tfrac{2}{3}).$$

Only piece F_1 has points with $x, F(x) \in [0, 1)$. Consequently it is only this piece that gives rise to tiles with 0 on both the top and bottom.

In this case, we will make only one color change and that is on the interval $[\frac{2}{3}, 1)$ which is the range of F_1. Specifically, any $x \in [\frac{2}{3}, 1)$ has a Beatty difference sequence using just $0, 1$. We change this 0 to $0'$. That is, for any point $x \in [\frac{2}{3}, 1)$, $\lfloor nx \rfloor - \lfloor (n-1)x \rfloor \in \{0', 1\}$.

This color change will also change the colors for tiles constructed from F_3 because the domain of F_3 is the interval $[\frac{2}{3}, 1)$ where the color change was performed.

Hence the tiles constructed for F_1 via the Basic Tile Construction with multiplier 2, will have top colors $\{0, 1\}$ and bottom colors $\{0', 1\}$. This results in the following four tiles.

The piece F_2 will have tiles with top colors $\{0, 1\}$ and bottom colors $\{1, 2\}$. This gives the following four tiles.

The third and fourth of these two tiles are already in the tile set for F_1, so the combined tile set for F_1 and F_2 is only six tiles.

The piece F_3 has tiles with top colors $\{0', 1\}$ and bottom colors $\{1, 2\}$. This gives the following four tiles.

The first three of these tiles are already in the set of tiles for F_1 and F_2. The combined set of tiles for F_1, F_2, F_3 consists of only seven tiles and these are the tiles given in Tile Set $2'$ (see figure for Lemma 6).

Finally we examine piece F_4. This will have tiles with top colors $\{1, 2\}$ and bottom colors $\{0, 1\}$, and will give the 6 tiles in Tile Set $\frac{1}{3}$ (the side color change has already been incorporated).

Together these result in the thirteen tiles for the K-C tile set.

6. Generalization

In this section we present generalizations of the previous work. Detailed proofs are omitted as the essential ideas have already been given.

Consider a function

$$
g(x) = \begin{cases} q_1 x & \text{if } x_0 \le x < x_1, \\ q_2 x & \text{if } x_1 \le x < x_2, \\ \quad \vdots \\ q_k x & \text{if } x_{k-1} \le x < x_k, \end{cases}
$$

defined on a finite interval $[x_0, x_k)$ where the $\{q_1, \ldots, q_k\}$ are positive, rational numbers chosen so that g is an invertible bijection of $[x_0, x_k)$ onto itself.

THEOREM 17. *For g as above, the Basic Tile Construction defines a finite set of tiles $\mathcal{T}_g$, and every infinite two-sided orbit of g gives a valid tiling of the plane.*

An obvious question, which we do not pursue at this time, is whether every valid tiling corresponds to a two-sided orbit or if the tile set can be modified to have this property.

However, we remark that one-to-oneness is not precisely necessary for the existence of valid tilings. If g were defined as above but was only required to be *onto* $[x_0, x_k)$, it would still have a tiling set which has valid tilings; however these valid tilings need not correspond to two-sided orbits of g. Under additional assumptions they will correspond to the two-sided orbit of the Rokhlin invertible extension of g.

Side-color tweaking is always possible.

LEMMA 18. *Given g with pieces g_i defined for $x_{i-1} \leq x < x_i$ it is always possible to change the side colors so that the tiles for each piece have disjoint side colors. These color changes will not affect the existence of valid tilings nor the number of tiles in the tile set $\mathcal{T}_g$.*

THEOREM 19. *Let g be a piecewise, rationally multiplicative, invertible function such that*

(i) $1 \leq x_0$,
(ii) $q_1^{n_1} q_2^{n_2} \cdots q_k^{n_k} = 1$ for $n_i \geq 0$ only if $n_i = 0$ for all $i = 1, \ldots, k$.

If $\mathcal{T}_g$ is the tile set constructed for g with side color changes incorporated then $\mathcal{T}_g$ is aperiodic.

PROOF. Same as that of Theorem 10 — that is, the arguments about the colors of the periodic block B are exactly the same. The assumption $1 \leq x_0$ means that there are no zeros in the Beatty sequence for any x in the domain of g (Lemma 13). Which in turn means the tops of all the tiles are nonzero, and this allows the division by $\sum a_{i,1} \neq 0$. $\qquad \square$

This theorem does not apply to f in the K-C tile set because $x_0 = \frac{1}{3} < 1$. This required the additional Top-Bottom color tweaking.

The function f has a maximum consecutive orbit of length 2 wholly contained within the interval $[0, 1)$. Because of this, we used two top-bottom colors $\{0, 0'\}$.

Suppose g has a maximum consecutive orbit of length $0 \leq M < \infty$ wholly contained within $[0, 1)$. We then use M different 0's for the top and bottom colors, $0, 0', 0'', \ldots, 0^{(M-1)}$.

Define

$$I_0 = \{x \in [0, 1): g^{-1}(x) \notin [0, 1)\},$$
$$I_1 = \{x \in [0, 1): g^{-1}(x) \in [0, 1), \ g^{-2}(x) \notin [0, 1)\},$$

$$\vdots$$

$$I_{M-1} = \{x \in [0, 1): g^{-i}(x) \in [0, 1), i = 1, \cdots M-1, \ g^{-(M)}(x) \notin [0, 1)\}.$$

Then, for $x \in I_j$, use the colors $\{0^{(j)}, 1\}$ when calculating the colors

$$\lfloor nx \rfloor - \lfloor (n-1)x \rfloor$$

in the Basic Tile Construction.

THEOREM 20. *Assume for g as above that*

(i) $q_1^{n_1} q_2^{n_2} \cdots q_k^{n_k} = 1$ *for $n_i \geq 0$ only if $n_i = 0$ for all $i = 1, \ldots, k$;*
(ii) *there is an $M \geq 0$ such that the longest consecutive orbit wholly contained in $[0, 1)$ is of length M.*

Then by incorporating both side and top-bottom color changes the resulting tile set $\mathcal{T}_g$, is aperiodic.

7. Mealy machine representation

Kari and Culik present their tile set using Mealy machines, a type of finite-state automaton where the output is associated with a transition. The K-C tile set can be represented by a pair of Mealy machines, the first of which describes Tile Set $\frac{1}{3}$:

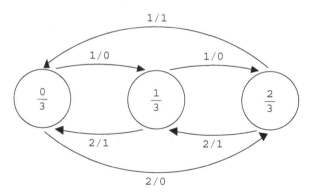

Each edge of the graph represents a tile. The label "i/j" gives the bottom and top numbers of the tile respectively. The tail state (vertex) of the transition arrow is the label of the left side of the tile. The head state (vertex) of the transition arrow is the label of the right side of the tile. This Mealy machine has six edges and these edges correspond to the first six tiles of the K-C tile set.

The following Mealy machine has seven edges which in turn correspond to the last seven tiles of the K-C tile set, namely Tile set $2'$.

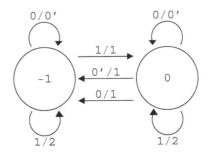

An infinite 2-sided path through either machine defines an infinite row of tiles. The labels along the tops of this infinite row give an admissible input sequence to the machine. The bottom labels of the row give an admissible output sequence.

Our analysis is essentially a discussion of when an infinite 2-sided output sequence of either machine can be admissible as an input sequence to either machine.

References

[1] Beatty, S. Problem 3173. *Amer. Math. Monthly* **33**:3 (1926), 159. Solutions by Ostrowski, A., Hyslop, J., and Aitken, A. C. in **34**:3 (1927), 159–160.

[2] Berger, R. *The undecidability of the domino problem.* Memoirs of the American Mathematical Society **66**, American Mathematical Society, Providence, RI, 1966.

[3] Culik, K. An aperiodic set of 13 Wang tiles. *Discrete Math.* **160**:1–3 (1996), 245–251.

[4] Fraenkel, A. S. Iterated floor functions, algebraic numbers, discrete chaos, Beatty subsequences, semigroups. *Trans. Amer. Math. Soc.* **341**:2 (1994), 639–664.

[5] Grunbaum, B. and Shephard, G. C. *Tilings and patterns.* Freeman, New York, 1987.

[6] Johnson, A. and Madden, K. Putting the pieces together: Understanding Robinson's nonperiodic tilings. *The College Mathematics Journal* **28**:3 (1997), 172–181.

[7] Kari, J. A small aperiodic set of Wang tiles. *Discrete Math.* **160**:1–3 (1996), 259–264.

[8] Lind, D. and Marcus, B. *Symbolic dynamics and coding.* Cambridge University Press, Cambridge, 1995.

[9] Radin, C. *Miles of tiles.* Student Mathematical Library **1**, American Mathematical Society, Providence, RI, 1999.

[10] Robinson, E. A. The dynamical properties of Penrose tilings. *Trans. Amer. Math. Soc.* **348** (1996), 4447–4464.

[11] Robinson, R. M. Undecidability and nonperiodicity for tilings of the plane. *Invent. Math.* **12** (1971), 177–209.

[12] Schmidt, K. Tilings, fundamental cocycles and fundamental groups of symbolic $\mathbb{Z}^d$-actions. *Ergodic Theory Dynam. Systems* **18** (1998), 1473–1525.

[13] Schmidt, K. Multi-dimensional symbolic dynamical systems. *Codes, systems, and graphical models* (Minneapolis, MN, 1999), 67–82, IMA Vol. Math. Appl. **123**, Springer, New York, 2001.

[14] Wang, H. Proving theorems by pattern recognition, II. *Bell System Technical Journal* **40**:1 (1961), 1–41.

STANLEY EIGEN
NORTHEASTERN UNIVERSITY
BOSTON, MA 02115
UNITED STATES
  eigen@neu.edu

JORGE NAVARRO
UNIVERSITY OF TEXAS
BROWNSVILLE, TX 78520
UNITED STATES
  jorge.navarro@utb.edu

VIDHU S. PRASAD
UNIVERSITY OF MASSACHUSETTS
LOWELL, MA 01854
UNITED STATES
  vidhu_prasad@uml.edu

Recent Progress in Dynamics
MSRI Publications
Volume **54**, 2007

A Halmos–von Neumann theorem for model sets, and almost automorphic dynamical systems

E. ARTHUR ROBINSON, JR.

Dedicated to Anatole Katok in celebration of his 60th birthday.

ABSTRACT. A subset $\Sigma \subseteq \mathbb{R}^d$ is the spectrum of a model set $\Lambda \subseteq \mathbb{R}^d$ if and only if it is a countable subgroup. The same result holds for $\Sigma \subseteq \widehat{G}$ for a large class of locally compact abelian groups G.

1. Introduction

A *model set* Λ is a special kind of uniformly discrete and relatively dense subset of a locally compact abelian group G. Model sets were first studied systematically in 1972 by Yves Meyer [15], who considered them in the context of Diophantine problems in harmonic analysis. More recently, model sets have played a prominent role in the theory of quasicrystals, beginning with N. G. de Bruijn's 1981 discovery [4] that the vertices of a Penrose tiling are a model set. Much of the interest in model sets is due to the fact that although they are aperiodic, model sets have enough "almost periodicity" to give them a discrete Fourier transform. This corresponds to spots, or Bragg peaks, in the X-ray diffraction pattern of a quasicrystal.

In this paper, we study model sets from the point of view of ergodic theory and topological dynamics. A translation invariant collection X of model sets in G has a natural topology, and with respect to this topology the translation action of G on X is continuous. We think of this action as a *model set dynamical system*.

Keywords: Aperiodic tilings, symbolic dynamics, quasicrystals.

Model set dynamical systems are closely related to tiling dynamical systems (see [30], [24]), and the first examples that were worked out, Penrose tilings [25], "generalized Penrose tilings" [25], [10], and chair tilings [26], [1], were tilings with model sets as vertices. In each of these examples, the corresponding dynamical system can be shown to be an almost 1:1 extension of a *Kronecker dynamical system* (a rotation action on a compact abelian group). A general proof of this fact for model set dynamical systems was given by Schlottman [29].

Almost automorphic functions were defined by Bochner around 1955 as a generalization Bohr almost periodic functions. In 1965, Veech [31] defined an *almost automorphic dynamical system* to be the orbit closure of an almost automorphic function. This generalizes the fact that the orbit closure of an almost periodic function is an equicontinuous dynamical system (i.e., a dynamical system topologically conjugate to a Kronecker system). Veech showed that almost automorphic dynamical systems are always minimal. He also proved a structure theorem [31], which says that a dynamical system is almost automorphic if and only if it is topologically conjugate to an almost 1:1 extension of a Kronecker system (which is its unique maximal equicontinuous factor).

It follows from Veech's structure theorem, combined with Schlottman's result above, that all model set dynamical systems are almost automorphic. Some older well known examples also turn out to be almost automorphic: the *Sturmian sequences*, defined in 1955 by Gotschalk and Hedlund [7], and the *Topeplitz sequences*, introduced in 1952 by Oxtoby [20] (see [32]). Perhaps not surprisingly, these sequences can be interpreted as model sets[1] in $G = \mathbb{Z}$.

Later work on almost automorphic dynamical systems (see e.g., [21], [14], [32], [2]) brought to light the existence of a dichotomy. In one case, which we will regard in this paper as somewhat pathological, a lack of unique ergodicity can lead to positive topological entropy (see [14], [32]). On the other hand, in the uniquely ergodic case one has metric isomorphism to the Kronecker factor. This implies topological entropy zero, and also pure point spectrum. The latter suggests the Halmos-von Neumann Theorem.

Published in 1942, the *Halmos-von Neuman theorem* [8] is one of the oldest results of ergodic theory. Extended to actions of locally compact metric abelian groups G, it says that every ergodic measure preserving dynamical system with pure point spectrum $\Sigma \subseteq \widehat{G}$ is metrically isomorphic to a unique Kronecker system. This Σ is always a countable dense subgroup, of $\widehat{G}$, and (this part is sometimes called the *Halmos-von Neumann representation theorem*) any $\Sigma \subseteq \widehat{G}$

---

[1] As de Bruijn [4] observed, Sturmian sequences are closely related to Penrose tilings. Also, Toeplitz sequences are closely related to chair tilings (see [1], [26]).

can arise as the spectrum of a Kronecker dynamical system. The main goal of this paper is to obtain a similar result for model set dynamical systems.

Spectral problems about model sets naturally divide into two parts[2], which correspond to the two parts of the model set construction. The first part is algebraic, and is called a *cut and project scheme*. By itself, a cut and project scheme determines the Kronecker factor of a model set dynamical system. In Theorem 3.8 we obtain a Halmos-von Neumann representation theorem for cut and project schemes, showing (modulo a few algebraic assumptions) that any countable dense subgroup $\Sigma \subseteq \widehat{G}$ can be the spectrum. Equivalently, every Kronecker dynamical system can be realized in terms of a cut and project scheme. However, unlike the classical Halmos-von Neumann theorem, our result is not unique, and different cut and project schemes can have the same spectrum. In particular, all the cut and project schemes constructed via Theorem 3.8 have a compact group H as their *internal space*. Yet in some cases there exist cut and project schemes with the same spectrum, but where H is not compact. Ultimately, these lead to completely different model sets.

The second part of the model set construction is geometric, and involves a choice of (topologically nice) subset W, called a *window*, of the internal space H. Expressed in terms of dynamical systems theory, the choice of a window is equivalent to the construction of a particular almost 1:1 extension. It is well known in topological dynamics (see [21], [14]) that if the boundary of W has Haar measure zero, then the almost automorphic dynamical system is uniquely ergodic, and hence metrically isomorphic to its Kronecker factor (the same result is also known in context of model sets, see [17]). We show in Theorem 6.6 that if H is compact, then such a window W always exists. Combining this with Theorem 3.8, we obtain our main result, Theorem 6.7, which is a Halmos-von Neumann Theorem for model sets. In particular, it says that for any countable dense subgroup $\Sigma \subseteq \widehat{G}$, there exists a model set $\Lambda \subseteq G$ with Σ as its spectrum.

A second goal of this paper is to provide an exposition of the theory of model sets and model set dynamical systems. While this material is mostly not new (we follow [15], [29] and [17]) our emphasis is different because we concentrate on the connections to dynamical systems theory, and in particular, to the theory of almost automorphic dynamical systems. However, our exposition is not exhaustive, and there are several interesting topics we do not cover. These include the question of which model sets have a local matching rule (see [12]), and questions about the algebraic topology of model set spaces (see [5]). There is also an interesting new topology on collections of model sets, called the autocorrelation topology, that gives the Kronecker factor [18].

---

[2]An important third part, pertaining to the diffraction spectrum, is briefly discussed in Section 7.

E. ARTHUR ROBINSON, JR.

2. Cut and project schemes

Locally compact abelian groups. Most of the groups G discussed in this paper will be a Hausdorff, locally compact abelian groups (abbreviated LCA groups). We will always assume groups are metrizable and separable, which is equivalent to satisfying the second countability axiom. Since every σ-compact metric space is separable, and every separable, locally compact metric space is σ-compact, our groups are the same as the metrizable σ-compact locally compact abelian groups. Let $\widehat{G}$ denote the dual of G (see [19], [23]). Since metrizability is dual to σ-compactness (see [23]), it follows that $\widehat{G}$ is also a separable metric group. Metrizable compact (Hausdorff) abelian groups G (abbreviated CA groups) satisfy all these hypotheses. A discrete abelian group G is an LCA group in our sense if and only if it is countable. We abbreviate these as DA groups.

Kronecker dynamical systems. A *Kronecker dynamical system* is an action R of a noncompact LCA group G on a CA group Y by *rotations* (i.e., translations). We can write this as $R^g y = y + \varphi(g)$, where $\varphi : G \to Y$ is a continuous homomorphism. We are going to want R to be a *free action* (i.e., $R^g = I$ implies $g = 0$), which is equivalent to φ being an injection. We are also going to want R to be *minimal*[3], which is equivalent to φ having a dense range. We call a topological group homomorphism with dense range *topologically surjective*. For Kronecker systems, minimality is equivalent to unique ergodicity, and the combination of these two properties is called *strict ergodicity*.

DEFINITION 2.1. A continuous LCA group homomorphism $\varphi : G \to Y$ is called a *compactification* of G if Y is CA, and φ is injective and topologically surjective.

Thus, R being a free, strictly ergodic Kronecker G-action on Y is equivalent to $\varphi : G \to Y$ being a compactification of G.

The dual $\widehat{Y}$ of a CA group Y is a DA group (see [23]). Since φ injective and topologically surjective, so is the dual homomorphism $\widehat{\varphi} : \widehat{Y} \to \widehat{G}$. This follows from the fact—which we will use repeatedly—that injectivity and topological surjectivity are dual properties (see [23]).

Define $\Sigma_R := \widehat{\varphi}(\widehat{Y})$, which is a countable dense subgroup of $\widehat{G}$. It is called the *point spectrum* of R. In particular, if $\chi := \widehat{\varphi}(f) \in \Sigma_R$ for some $f \in \widehat{Y}$, (i.e., $f : Y \to \mathbb{T} \subseteq \mathbb{C}$), then

$$f(R^g y) = f(y + \varphi(g)) = f(\varphi(g)) f(y)$$
$$= \widehat{\varphi}(f)(g) f(y) = \chi(g) f(y),$$

---

[3] A G action on Y compact metric is minimal if and only if every orbit is dense.

or more briefly

$$f \circ R = \chi f. \tag{2-1}$$

We call f as an *eigenfunction* for R, corresponding to the *"eigenvalue"* χ. A standard argument shows that this construction produces all L^2 functions (for Haar measure θ_Y on Y) that satisfy (2–1). Moreover, since span$(\widehat{Y})$ is dense in L^2, it follows that R (i.e., any Kronecker system) has *pure point spectrum*.

Conversely, suppose Σ is a countable dense subgroup of $\widehat{G}$ (such a subgroup exists since $\widehat{G}$ is separable). Let $Y := \widehat{\Sigma}$ (which is compact) and let $\varphi := \widehat{i}$: $G \to Y$, where $i : \Sigma \to \widehat{G}$ is inclusion. Then φ is a compactification of G, and thus defines a free, strictly ergodic Kronecker system R with $\Sigma_R = \Sigma$. This is the *Halmos-von Neumann representation theorem* (see [8]).

REMARK 2.2. A CA group Y that is a compactification of $\mathbb{Z}$ is called *monothetic*. A CA group Y that is a compactification of $\mathbb{R}$ is called *solenoidal*. We use the terms d-*monothetic*[4] and d-*solenoidal* for compactifications of $\mathbb{Z}^d$ and $\mathbb{R}^d$ respectively.

Cut and project schemes. A cut and project scheme is the algebraic part of the definition of a model set. A discrete closed subgroup of a LCA group G is called a *lattice* if it has a compact quotient.

DEFINITION 2.3. A *cut and project scheme* is a triple $\mathscr{S} = (G, H, \Gamma)$ of LCA groups that satisfies:

(1) $\Gamma \subseteq G \times H$ is a lattice,
(2) $p_1|_\Gamma$ is injective and,
(3) $p_2(\Gamma)$ is dense in H.

Here, $p_1 : G \times H \to G$ and $p_2 : G \times H \to H$ denote the coordinate projections. The groups G and H are called (respectively) *physical space* and *internal space*. The subgroup $L := p_1(\Gamma) \subseteq G$ is called the *structure group*, and the mapping

$$g \mapsto g^* := p_2(p_1^{-1}(g)) : L \to H$$

is called the *star homomorphism*.

DEFINITION 2.4. A cut and project scheme is called *aperiodic* if any of the following (equivalent) conditions hold:

(1) $p_2|_\Gamma$ is injective,
(2) the *period group*, $J := \ker(*)$, is trivial, or
(3) $(\{0\} \times H) \cap \Gamma = \{0\}$.

---

[4]The definition of d-monothetic in [2], as a d-fold product of monothetic groups, is a special case of our definition.

We will usually assume a cut and project scheme is aperiodic.

Let $\mathcal{S} = (G, H, \Gamma)$ be a cut and project scheme. Let $Y := (G \times H)/\Gamma$, which by Definition 2.3 (1), is a CA group. Define

$$\varphi = \pi \circ i_1 : G \to Y, \tag{2-2}$$

where $i_1 : G \to G \times H$ is the coordinate injection and π is the canonical homomorphism.

PROPOSITION 2.5. *The homomorphism φ in (2–2) is a compactification, so the G action R on Y, defined by*

$$R^g y := y + \varphi(g), \tag{2-3}$$

is a free, strictly ergodic Kronecker system. We call this the action the Kronecker system associated with cut and project scheme $\mathcal{S}$. *The spectrum of R is* $\Sigma = \widehat{\varphi}(\widehat{Y})$.

PROOF. If $\varphi(g) = 0$, then $(g, 0) \in \Gamma$. Thus $g \in L$ and $g^* = p_2(g, 0) = 0$, so that $g = 0$ by the aperiodicity of $\mathcal{S}$. It follows that φ is injective.

To show φ is topologically surjective, it suffices to show $i_1(G) + \Gamma = (G \times \{0\}) + \Gamma$ is dense in $G \times H$. But

$$(G \times \{0\}) + \Gamma = (G \times \{0\}) \oplus (\{0\} \times p_2(\Gamma)) = G \times p_2(\Gamma),$$

which is dense by Definition 2.3, (3). Everything else follows. □

The spectrum of a cut and project schemes

DEFINITION 2.6. By the *spectrum $\Sigma_{\mathcal{S}}$ of a cut and project scheme $\mathcal{S}$* we mean the spectrum $\Sigma_R = \widehat{\varphi}(\widehat{Y})$ of the corresponding Kronecker dynamical system R.

Now we will describe $\Sigma_{\mathcal{S}}$ in terms of $\mathcal{S}$ itself. To this end, we follow Y. Meyer [15]. Consider the diagram

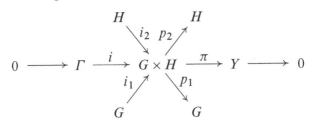

where i denotes inclusion map. Note that the horizontal row is exact. The dual of this diagram is given by

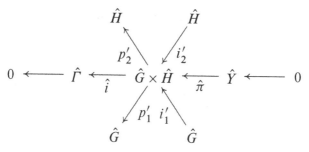

where the coordinate injections i'_k satisfy $i'_k = \widehat{p_k}$, and the coordinate projections satisfy $p'_k = \widehat{i_k}$. Since the dual of an exact sequence is exact (see [19]) the horizontal row of this sequence is also exact.

THEOREM 2.7 (MEYER, [15]). *Define* $\Delta := \widehat{\pi}(\widehat{Y}) \subseteq \widehat{G} \times \widehat{H}$. *Then* $\widehat{\mathscr{G}} = (\widehat{G}, \widehat{H}, \Delta)$ *is a cut and project scheme. We call it the* dual *of* $\mathscr{G}$.

PROOF. We need to verify Definition 2.3. By duality theory, Δ is a lattice, so (1) holds.

By Proposition 2.5, $\varphi = \pi \circ i_1$ is a compactification, so it is topologically surjective. Thus its dual, $\widehat{\varphi} = p'_1 \circ \widehat{\pi}$ is injective. This implies $p'_1|_\Delta$ is injective, proving (2).

We have $p'_2(\Delta) = p'_2(\widehat{\pi}(\widehat{Y}))$. To prove (3) we need to show this is dense, or equivalently that $p'_2 \circ \widehat{\pi}$ is topologically surjective. This is equivalent to showing that

$$(p'_2 \circ \widehat{\pi})\widehat{} = (\widehat{i_2} \circ \widehat{\pi})\widehat{} = \pi \circ i_2$$

is injective. If $(0, h) \in \ker(\pi)$ then $(0, h) \in \Gamma$, and also $p_1(0, h) = 0$. Since p_1 is injective, $h = 0$. □

COROLLARY 2.8. *The spectrum* $\Sigma_\mathscr{G}$ *of a cut and project scheme* $\mathscr{G}$ *is the structure group of the dual* $\widehat{\mathscr{G}}$ *of* $\mathscr{G}$.

PROOF. $\Sigma_\mathscr{G} = \Sigma_R = \widehat{\varphi}(\widehat{Y}) = p'_1(\widehat{\pi}(\widehat{Y})) = p'_1(\Delta)$. □

COROLLARY 2.9. (1) *A cut and project scheme* $\mathscr{G}$ *has a dense structure group* $L \subseteq G$ *if and only if its dual* $\widehat{\mathscr{G}}$ *is aperiodic.*

(2) *A cut and project scheme* $\mathscr{G}$ *is aperiodic if and only of its spectrum* Σ *is dense in* $\widehat{G}$.

REMARK 2.10. *The* flip $\widetilde{\mathscr{G}} := (H, G, \Gamma)$ *of a cut and project scheme* $\mathscr{G} = (G, H, \Gamma)$, *is a cut and project scheme if and only if* $\mathscr{G}$ *is aperiodic and has a dense structure group. In this case the flip of the dual* $\widehat{\mathscr{G}}$ *is also a cut and project scheme (the dual of the flip).*

3. Realization for cut and project schemes

The realization property

DEFINITION 3.1. Let G be a noncompact LCA group. We say G satisfies the *realization property* if, given a countable dense subgroup of $\Sigma \subseteq \widehat{G}$, there is an aperiodic cut and project scheme $\mathcal{S} = (G, H, \Gamma)$ with $\Sigma_{\mathcal{S}} = \Sigma$.

It will be convenient to phrase realization in terms of the dual cut and project scheme.

LEMMA 3.2. *Let G be a noncompact LCA group and let Σ be a countable dense subgroup of $\widehat{G}$. Suppose $\mathcal{S}' = (\widehat{G}, H', \Gamma')$ is a cut and project scheme with structure group Σ. Then the dual $\mathcal{S} := \widehat{\mathcal{S}'}$ is aperiodic and $\Sigma_{\mathcal{S}} = \Sigma$.*

PROOF. By Theorem 2.7, $\mathcal{S}$ is a cut and project with $\widehat{\mathcal{S}} = \mathcal{S}'$, and by Corollary 2.8, $\mathcal{S}$ has spectrum Σ. Since Σ is dense it follows from (2) of Corollary 2.9 that $\mathcal{S}$ is aperiodic. $\qquad\square$

The discrete case

THEOREM 3.3. *Every infinite DA group G satisfies the realization property.*

PROOF. Note that $\widehat{G}$ is CA. Given a countable dense subgroup $\Sigma \subseteq \widehat{G}$, we put $H' = \Sigma_d$ (the subscript d indicates Σ given the discrete topology). Using the fact that $H' \subseteq \widehat{G}$ we define

$$\Gamma' = \{(k, k) : k \in \Sigma\} \subseteq \widehat{G} \times H'.$$

By Lemma 3.2 it suffices to show $\mathcal{S}' = (\widehat{G}, H', \Gamma')$ is a cut and project scheme with structure group Σ. We then have $\mathcal{S} = \widehat{\mathcal{S}'}$.

First, we have that Γ' is a lattice since H' is discrete, and $(\widehat{G} \times H')/\Gamma' = \widehat{G}$ is compact. This shows Definition 2.3 (1). For (2), we have $p'_1(k, k) = k$ which implies $k = 0$ for $(k, k) \in \ker(p'_1)$. Clearly $p_1(\Gamma') = \Sigma$, so $\mathcal{S}'$ has structure group Σ. For (3), we note that p'_2 is surjective. $\qquad\square$

REMARK 3.4.

(1) The examples constructed in Theorem 3.3 all have $H = \widehat{\Sigma_d}$, which is compact. We call this kind of cut and project scheme *internally compact*.
(2) The cut and project scheme $\mathcal{S}'$ in Theorem 3.3 has a *compact* physical space $\widehat{G}$. While such model sets do not lead to useful quasicrystals (or to Kronecker systems), they do play an important role in the theory as duals to cut and project schemes with discrete physical spaces.

EXAMPLE 3.5. Let $G = \mathbb{Z}$ and for $\alpha \in \mathbb{R} \setminus \mathbb{Q}$ let

$$\Sigma = \{e^{2\pi i n \alpha} : n \in \mathbb{Z}\} \subseteq \mathbb{T} = \widehat{\mathbb{Z}}.$$

Identify $\Sigma_d = \mathbb{Z}$ and let $H = \widehat{\Sigma_d} = \widehat{\mathbb{Z}} = \mathbb{T}$, so that $G \times H = \mathbb{Z} \times \mathbb{T}$. Then

$$\Gamma = \{(n, e^{2\pi i n \alpha}) : n \in \mathbb{Z}\} \subseteq G \times H$$

is a lattice, and $\mathscr{S} = (\mathbb{Z}, \mathbb{T}, \Gamma)$ is a cut and project scheme with spectrum Σ. Here, $R = R_\alpha$ is the *irrational rotation* on the circle $\mathbb{T}$.

EXAMPLE 3.6. Again let $G = \mathbb{Z}$, but now let $\Sigma = e^{2\pi i \mathbb{Z}[\frac{1}{2}]}$, where $\mathbb{Z}[\frac{1}{2}] \subseteq \mathbb{Q}$ denotes the dyadic rationals. It is well known that $H := \widehat{\Sigma_d} = \mathbb{Z}_2$, the group of dyadic integers. Putting $\mathscr{S} = (\mathbb{Z}, \mathbb{Z}_2, \Gamma)$, where $\Gamma = \{(n, n) : n \in \mathbb{Z}\}$, we have that $\mathscr{S}$ is a cut and project scheme with spectrum Σ. Here R is the von Neumann adding machine, viewed as a rotation on $\mathbb{Z}_2$.

Realization in a large class of groups

DEFINITION 3.7. We say an LCA group G is *lattice dense* if every countable dense subgroup Σ of G has a subgroup $\Omega \subseteq \Sigma$ that is a lattice in G. A group *has lattice dense dual* if $\widehat{G}$ is lattice dense.

All CA groups and DA groups are both lattice dense and have lattice dense duals. Our main result in this section is the following theorem, which can be interpreted as a Halmos-von Neumann theorem for cut and project schemes.

THEOREM 3.8. *Suppose G is a noncompact LCA group with a lattice dense dual. Then G satisfies the realization property. In particular, for any countable dense subgroup $\Sigma \subseteq \widehat{G}$, there is an aperiodic cut and project scheme $\mathscr{S} = (G, H, \Gamma)$ with spectrum $\Sigma_{\mathscr{S}} = \Sigma$.*

Before proving the theorem, we discuss some examples.

PROPOSITION 3.9. *The group $\mathbb{R}^d$ is lattice dense and has a lattice dense dual.*

PROOF. Since $\widehat{\mathbb{R}^d} \cong \mathbb{R}^d$ it is enough to show $\mathbb{R}^d$ is lattice dense.

Any set of d vectors in $\mathbb{R}^d$ that are linearly independent over $\mathbb{R}$ generates a lattice. Let L' be a lattice in $\mathbb{R}^d$ generated by $\mathbf{v}'_1, \ldots, \mathbf{v}'_d$. For ε sufficiently small, if $\mathbf{v}_1, \ldots, \mathbf{v}_d$ satisfy $\|\mathbf{v}_j - \mathbf{v}'_j\| < \varepsilon$ for all j, then $\mathbf{v}_1, \ldots, \mathbf{v}_d$ are also linear independent over $\mathbb{R}$, and thus also generate a lattice L. $\square$

Recall that a LCA group G is called *compactly generated* if there is a compact subset $K \subseteq G$ that generates G. The *Structure Theorem* (see [9]) says that a (not necessarily metrizable) LCA group is compactly generated if and only if it has the form $G = \mathbb{R}^{d_1} \times \mathbb{Z}^{d_2} \times Y$, where Y is compact. Such a G is metrizable if and only if Y is. If $Y = \mathbb{T}^{d_3} \times F$, where F is finite, G is called "elementary".

COROLLARY 3.10. *Any metrizable compactly generated LCA group G, and in particular, any elementary group, is lattice dense and has a lattice dense dual. In particular, all such groups satisfy the realization property.*

Proposition 3.9 and Corollary 3.10 show that all the physical spaces G likely to be of interest in the theory of quasicrystals satisfy the realization property. There are, however, important LCA groups that are not lattice dense (and do not have lattice dense duals).

PROPOSITION 3.11. *The group (i.e., field) $\mathbb{Q}_p$ of p-adic numbers is not lattice dense and does not have a lattice dense dual.*

PROOF. For $\alpha \in \mathbb{Q}_p$, $\alpha \neq 0$, the subgroup $\alpha\mathbb{Z}$ is infinite (see [9] for a definition of $\mathbb{Q}_p$). Moreover, $\overline{\alpha\mathbb{Z}} = \alpha\overline{\mathbb{Z}} = \alpha\mathbb{Z}_p$, where $\mathbb{Z}_p \subseteq \mathbb{Q}_p$ denotes the set of p-adic integers, a compact subgroup. It follows that there can be no lattice in $\mathbb{Q}_p$ since the closed subgroup generated by any $\alpha \neq 0$ is compact. Since $\widehat{\mathbb{Q}_p} \cong \mathbb{Q}_p$, it follows that $\mathbb{Q}_p$ does not have a lattice dense dual. $\square$

The proof of Theorem 3.8 follows immediately from the next lemma.

LEMMA 3.12. *Let G be a noncompact LCA group and let Σ be a countable dense subgroup of $\widehat{G}$. Suppose there is a lattice Ω in $\widehat{G}$ with $\Omega \subseteq \Sigma$. Then there is a cut an project scheme $\mathcal{S} = (G, H, \Gamma)$ with spectrum Σ.*

PROOF. Let $\omega : \widehat{G} \to \widehat{G}/\Omega$ be the canonical homomorphism. Since Ω is a lattice, $\widehat{G}/\Omega$ is compact, and it contains a countable dense subgroup Σ/Ω. We put $H' := (\Sigma/\Omega)_d$ and define

$$\Gamma' := \{(g, h) \in \widehat{G} \times H' : \omega(g) \in H'\}. \tag{3-1}$$

Note that

$$p_1'(\Gamma') = \{g \in \widehat{G} : \omega(g) \in H'\} = \omega^{-1}(\Sigma/\Omega) = \Sigma. \tag{3-2}$$

By Lemma 3.2 it suffices to show that $\mathcal{S}' = (\widehat{G}, H', \Gamma')$ is a cut and project scheme, since (3–2) shows that $\mathcal{S}'$ has structure group Σ.

Clearly (1) Γ' is a lattice in $\widehat{G} \times H'$, (2) p_1' is injective (see (3–2)), and (3) p_2' is surjective (and thus topologically surjective) since $p_2'(\Gamma') = H_1$ by (3–1). $\square$

EXAMPLE 3.13. Let $G = \mathbb{R}$ and for $\alpha \in \mathbb{R} \setminus \mathbb{Q}$, let

$$\Sigma := \{n + m\alpha : (n, m) \in \mathbb{Z}^2\} \subseteq \widehat{G} = \widehat{\mathbb{R}} = \mathbb{R}.$$

For $\Omega = \mathbb{Z} \subseteq \Sigma \subseteq \mathbb{R}$ let

$$G_1 := \widehat{G}/\Omega = \mathbb{T} \cong \{z \in \mathbb{C} : |z| = 1\},$$

so that for $\Omega = \mathbb{Z} \subseteq \Sigma \subseteq \mathbb{R}$, we have

$$\Sigma_1 := \Sigma/\Omega \cong \left\{e^{2\pi i n\alpha} : n \in \mathbb{Z}\right\} \subseteq G_1,$$

is countable and dense. Letting $H' := (\Sigma_1)_d \cong \mathbb{Z}$ and $\Gamma_1 = (j \times j)(H')$, we note in passing that $\mathcal{S}_1 = (G_1, H', \Gamma_1)$ is the cut and project scheme dual to Example 3.5.

Following (3–1), we have

$$\Gamma' = \{(n + m\alpha, m) \in \mathbb{R} \times \mathbb{Z} : (n, m) \in \mathbb{Z}^2\},$$

so that $\mathcal{S}' = (\widehat{G}, H', \Gamma') = (\mathbb{R}, \mathbb{Z}, \Gamma')$ is a cut and project scheme with structure group Σ. It follows that $\mathcal{S} = \widehat{\mathcal{S}'} = (\mathbb{R}, \mathbb{T}, \Gamma)$, where $\Gamma = \{(n, e^{2\pi i n}) : n \in \mathbb{Z}\}$, is a cut and project scheme with spectrum Σ.

The structure group of $\mathcal{S}$ satisfies $L = \mathbb{Z}$. The associated Kronecker system R is the "irrational flow" on $\mathbb{T}^2$ with slope α. It is a suspension of the irrational rotation R_α from Example 3.5.

REMARK 3.14. Many features of Example 3.13 are characteristic of the cut and project schemes $\mathcal{S}$ produced by Theorem 3.3:

(1) The internal space H is always compact.
(2) If G is not discrete then the structure group L is a lattice in G. In particular it is never dense.
(3) The Kronecker system R is a suspension of the Kronecker system of another cut and project scheme, call it $\mathcal{S}_1$, which has physical space L, internal space H, and spectrum Σ/Ω.

EXAMPLE 3.15. Let $A = \left(\begin{smallmatrix} a & b \\ c & d \end{smallmatrix}\right) \in \mathrm{Gl}(2, \mathbb{R})$ and put $\Gamma = A\mathbb{Z}^2 \subseteq \mathbb{R}^2$. If $\frac{a}{c}$ and $\frac{b}{d}$ are irrational numbers, then $\mathcal{S}(A) = (\mathbb{R}, \mathbb{R}, \Gamma)$ is an aperiodic cut and project scheme with dense structure group $L = \{na + mc : (n, m) \in \mathbb{Z}^2\}$. Then $\widehat{\mathcal{S}(A)} = \mathcal{S}((A^{-1})^t)$ (the columns of $(A^{-1})^t$ are basis for the dual lattice for Γ).

Let us now specialize. Starting with the subgroup $\Sigma = \{n + m\alpha : (n, m) \in \mathbb{Z}^2\} \subseteq \widehat{\mathbb{R}}$ from Example 3.13, let $A = \left(\begin{smallmatrix} 1 & \alpha \\ -\alpha & 1 \end{smallmatrix}\right)$. Then we get an aperiodic cut and project scheme $\mathcal{S} = (\mathbb{R}, \mathbb{R}, \Gamma') = \mathcal{S}((A^{-1})^t)$, with dense structure group, and spectrum Σ.

Even though this example has the same spectrum as Example 3.13, it looks quite different. Notably, H is not compact. As we will see, although the model sets obtained from these two cut and project schemes have the same spectrum, geometrically they are considerably different.

Remarks on the general case. Although the structure group L of a cut and project scheme may not be dense in G, it will always at least be relatively dense. Suppose G is a LCA group and let $L \subseteq G$ be a dense or relatively dense, countable subgroup. Here we consider the following question: Are there groups H and Γ so that $\mathcal{S} = (G, H, \Gamma)$ is a cut and project scheme with structure group L? So far, we know that the answer is yes if L contains a lattice (this follows from Lemma 3.12).

Now assume that L is dense in G, but not equal to G. Let d denote the translation invariant metric for the topology on G, and pull it back to L. In this topology (which is not discrete) L is not locally compact. Every metrizable LCA group is, in fact, completely metrizable (see [3] where it is shown that every locally compact topological group is completely uniformizable). It follows that G is the unique completion of L with respect to d. Define the "norm" $|g| := d(g, 0)$.

To see what we are after, let $\mathcal{S} = (G, H, \Gamma)$ be an aperiodic cut and project scheme with structure group L. Note that $* = p_2 \circ p_1^{-1} : L \to H$ is injective. Pulling back the invariant metric d' on H, we get a second metric d' on L, and we denote the corresponding norm $|g|' := d'(g, 0)$. Then H is the completion of L in d'. The fact that Γ is a lattice in $G \times H$ can be expressed in terms these two norms:

LEMMA 3.16. *Let L be a countable abelian group, and suppose L has two norms $|\cdot|$ and $|\cdot|'$ (that come from two invariant topological group metrics). Let G and H (respectively) be the completions of L with respect to these norms (i.e., metrics). Define $\Gamma \subseteq G \times H$ by $\Gamma := (i \times i)(K)$. Then Γ is a lattice if and only if:*

(1) *There exists $r > 0$ so that if $\ell \in L$ satisfies $|\ell| < r$, then $|\ell|' > r$.*
(2) *There exists $R > 0$ so that for any $g, h \in L$ there exists $\ell \in L$ so that $|g - \ell| < R$ and $|h - \ell|' < R$.*

PROOF. It is easy to see that (2) is equivalent to the condition that for any $(g, h) \in G \times H$ there exists $\ell \in L$ so that $|g - \ell| < R$ and $|h - \ell|' < R$. Combined with (1), this shows that Γ is a *Delone set* in $G \times H$ (see Definition 4.5 below). The lemma follows, since for a subgroup, the Delone property is equivalent to Γ being a lattice (see Lemma 4.6).   □

In effect, (1) and (2) say that the identity map needs to be wildly discontinuous.

COROLLARY 3.17. *Let L be a countable dense subgroup of a LCA group G and let d denote the metric induced on L by G. Then there is an aperiodic cut and project scheme $\mathcal{S} = (G, H, \Gamma)$ with structure group L if and only if* (i) *H is the completion of L with respect to another metric d',* (ii) *conditions (1) and (2) of Lemma 3.16 are satisfied, and* (iii) *$\Gamma = (i \times i)(L)$.*

EXAMPLE 3.18. Consider $\Sigma = \mathbb{Z}[\frac{1}{p}] \subseteq \mathbb{Q}_p$. Then Σ inherits the p-adic valuation $|\cdot|_p$ from $\mathbb{Q}_p$. But $\mathbb{Z}[\frac{1}{p}]$ also has the usual absolute value $|\cdot|$, and with respect to that, its completion is $\mathbb{R}$. One can check that the conditions of Lemma 3.16 hold. Let $\Gamma' = (i \times i)(\mathbb{Z}[\frac{1}{2}]) \subseteq \mathbb{Q}_p \times \mathbb{R}$. Then $\mathcal{S}' = (\mathbb{Q}_p, \mathbb{R}, \Gamma')$ with is an aperiodic cut and project scheme with structure group Σ. From this it follows $\mathcal{S} = \widehat{\mathcal{S}}' = (\mathbb{Q}_p, \mathbb{R}, \Gamma)$ is a cut and project scheme with spectrum Σ.

4. Model sets

Let $\mathscr{S} = (G, H, \Gamma)$ be a cut and project scheme with G noncompact. A *precompact* subset $W \subseteq H$ is called the *window*. The *model set* corresponding to $\mathscr{S}$ and W is defined to be the subset of G (i.e., of L) given by the *selection* construction:

$$\Lambda(W) := p_1((G \times W) \cap \Gamma). \tag{4-1}$$

Selection is the geometric part of the definition of a model set.

More generally, any translation $\Lambda(W) - g$, $g \in G$ is also called a model set. We call

$$\mathcal{M}(W) := \{\Lambda(W + h) - g : (g, h) \in G \times H\} \tag{4-2}$$

the *model set family* associated with the window W.

EXAMPLE 4.1. Let $\mathscr{S}$ be the cut and project scheme in Example 3.5 and let $W := [e^{2\pi i a}, e^{2\pi i b}]$ be a proper closed arc in $H = \mathbb{T}$. Then

$$\Lambda(W) = \{n \in \mathbb{Z} : e^{2\pi i n \alpha} \in W\} \subseteq L = \mathbb{Z}.$$

The *Sturmian case* is $a = \theta$ and $b = \theta + \alpha$. Assume $\theta \neq e^{2\pi i n \alpha}$ for any n (we call this *nonsingular* below). Then the "characteristic sequence" (see [13]) $x = (x_k) \in \{0, 1\}^{\mathbb{Z}}$, defined by $x_k := \chi_{\Lambda(W)}(k)$, is the classical *Sturmian sequence*.

More generally, nonsingularity means

$$e^{2\pi i n \alpha} \notin \{e^{2\pi i a}, e^{2\pi i b}\}. \tag{4-3}$$

EXAMPLE 4.2. Take $\mathscr{S}$ as in Example 3.13, noting that $H = \mathbb{T}$ as in Example 3.5. Taking W from Example 4.1, we get the same $\Lambda(W)$, but this time as a subset of $\mathbb{R}$. Note that $L = \mathbb{Z} \subseteq \mathbb{R}$ is a lattice and is generated by $\Lambda(W)$.

EXAMPLE 4.3. For $\mathscr{S}(A)$ in Example 3.15 let $W = [p, q] \subseteq \mathbb{R}$. Then $\Lambda(W) = \{na + mc : (n, m) \in \mathbb{Z}^2, na + mc \in [p, q]\} \subseteq \mathbb{R}$. In this example, "nonsingularity" means p and q are chosen so that $na + mc \notin \{p, q\}$ for any $(n, m) \in \mathbb{Z}^2$.

Properties of model sets

LEMMA 4.4.

(1) $\Lambda(W) - g \subseteq L$ implies $g \in L$.
(2) $g \in L$ implies $\Lambda(W) - g = \Lambda(W - g^*)$.
(3) $g \in J$ implies $\Lambda(W) - g = \Lambda(W)$.
(4) $\Lambda(W) - \Lambda(W) = \Lambda(W - W)$.
(5) If $\Lambda' \subseteq \Lambda(W)$ then Λ' is a model set.

PROOF. For (1) $\Lambda(W) \subseteq L$ so $g = g_1 - g_2$ for $g_1, g_2 \in L$. For (2) we first note that $\Lambda(W) = \{g \in L : g^* \in W\}$. Thus $\Lambda(W - g^*) = \{g_1 \in G : (g_1 + g)^* \in W\} = \Lambda(W) - g$. For (3), we have $\Lambda(W) - g = \Lambda(W - g^*) = \Lambda(W)$ where the first equality is by (2), and the second is by $g^* = 0$ since $g \in J$. The equation in the proof of (2) also proves (4). Since any subset of a precompact set is precompact, (5) follows. $\square$

We are going to want model sets to have some additional properties. The most basic of these is the *Delone property*.

DEFINITION 4.5. A subset $\Lambda \subseteq G$ is called *uniformly discrete* if there exists an open neighborhood U of 0 in G such that $(U + g) \cap \Lambda = \{g\}$ for any $g \in \Lambda$.

   A subset $\Lambda \subseteq G$ is called *relatively dense* if there exists a compact $K \subseteq G$ such that $K + \Lambda = G$.

   A *Delone set* is any set that is both uniformly discrete and relatively dense.

It will often be useful to express uniformly discrete and relatively dense in terms of the metric on G. Uniformly discrete means that there exists $r > 0$ such that $B_r(g) \cap \Lambda = \{g\}$ for all $g \in \Lambda$. Similarly, relatively dense means that there exists $R > 0$ such that $B_R(g) \cap \Lambda \neq \varnothing$ for all $g \in G$.

   A set $\Lambda \subseteq G$ is called an (r, R)-*set* if it is a Delone set for the parameters $0 < r < R$. We denote the collection of all (r, R)-sets by $X_{r,R}$. We mention the following important fact without proof.

LEMMA 4.6. *A subgroup Λ of a LCA group G is a Delone set if and only if it is a lattice.*

The next result shows that all model sets are uniformly discrete. With an additional hypotheses, they are also relatively dense. Moreover, both of these properties occur in a uniform way in a model set family.

PROPOSITION 4.7. *Let $\mathcal{S} = (G, H, \Gamma)$ be a cut and project scheme and let $W \subseteq H$ be a window.*

(1) *There exists an open neighborhood V of 0 in G such that for any $\Lambda \in \mathcal{M}(W)$ and any $g \in \Lambda$, $(V + g) \cap \Lambda = \{g\}$.*
(2) *(Meyer [15]) If, in addition $W^\circ \neq \varnothing$, then there exists K compact so that for any $\Lambda \in \mathcal{M}(W)$, $K + \Lambda = G$.*

Before the proof we state a corollary and make some remarks.

COROLLARY 4.8. *If $W^\circ \neq \varnothing$ then there exist $R > r > 0$ so that $\mathcal{M}(W) \subseteq X_{r,R}$. In particular, every model set $\Lambda \in \mathcal{M}(W)$ is a Delone set.*

REMARK 4.9. (1) Nonempty interior is not necessary for a model set to be Delone. Consider a cut and project scheme $\mathcal{S} = (G, \{0\}, \Gamma)$, where $\Gamma \subseteq G$ is a lattice, and let $W = \{0\}$. Then one has $\Lambda(W) = \Gamma$.

(2) However, some additional hypotheses on a window W are necessary. In an aperiodic cut and project scheme $\mathcal{S}$, any finite window W leads to a finite (possibly empty) model set $\Lambda(W)$. This can not be relatively dense if G is not compact.

PROOF OF PROPOSITION 4.7. For Part (1), we follow Moody [17]. By taking the closure of W if necessary, we may assume without loss of generality that W is compact. By the local compactness of G, there is a neighborhood U of 0 in G with $\overline{U}$ compact. Thus $\overline{U} \times (W - W)$ is compact, and since Λ is a lattice, $(U \times (W - W)) \cap \Gamma$ is finite. Since p_1 is injective, we can choose U so small that $\left(\overline{U} \times (W - W) \right) \cap \Gamma = \{(0, 0)\}$. Let V be a neighborhood of 0 in G with compact closure such that $V - V \subseteq U$.

Suppose for some $(g_0, h_0) \in G \times H$ there are $g_1, g_2 \in \Lambda(W + h_0) - g_0$ such that $(V + g_1) \cap (V + g_2) \neq \varnothing$. Then

$$p_1^{-1}(g_1 - g_2) \in ((V - V) \times (W - W)) \cap \Gamma \subseteq \{(0, 0)\},$$

so $p_1^{-1}(g_1 - g_2) = (0, 0)$. Since p_1 is injective, $g_1 = g_2$. $\square$

Part (2) of Proposition 4.7 follows from the next lemma.

LEMMA 4.10 (MEYER [15], LEMMA 10, P 50). *Let $\mathcal{S} = (G, H, \Gamma)$ be a cut and project scheme. The for each $U \subseteq H$ open, with $\overline{H}$ compact, there exists a compact $K \subseteq G$ such that $G \times H = \Gamma + (K \times U)$.*

PROOF. By Lemma 4.6, Γ is a lattice, so by Lemma 4.6 there exist $K_1 \subseteq G$ and $K_2 \subseteq H$ compact, such that $\Gamma + (K_1 \times K_2) = G \times H$. Since $p_2(\Gamma)$ is dense in H, there is a finite subset $F \subseteq \Gamma$ so that $K_2 \subseteq p_2(F) + U$. Let $K = K_1 - p_1(F)$. We have $G \times H = \Gamma + K_1 \times (p_2(F) + U) = \Gamma + (K_1 - p_1(F)) \times U = \Gamma + K \times U$. $\square$

Usually, we will require a little more from a window W.

DEFINITION 4.11. A window $W \subseteq H$ is *topologically regular* if it is the closure of its interior: $W = \overline{W^\circ}$. A model set $\Lambda(W)$ is called topologically regular if W is topologically regular.

In particular, topological regularity is more than enough to ensure $\Lambda(W)$ is Delone.

LEMMA 4.12. *If W is topologically regular then $\overline{\Lambda(W)^*} = W$.*

PROOF. $\Lambda(W)^* := \{g^* \in H : g \in \Lambda(W)\} = p_2(\Gamma) \cap W$. Since W is topologically regular $W^\circ \neq \varnothing$, and $\Lambda(W)^* \cap W^\circ = p_2(\Gamma) \cap W^\circ$, which is dense in W by Definition 2.3, (3). $\square$

REMARK 4.13. The windows in Examples 4.1 and 4.3 are topologically regular. Thus the model sets in the examples are Delone sets.

5. Model set dynamical systems

The local topology. Let G^∞ denote the one point compactification of G. For a Delone set $\Lambda \subseteq G$, let $\Lambda^\infty := \Lambda \cup \{\infty\} \subseteq G^\infty$, noting that Λ^∞ is closed in G^∞. We give $X_{r,R}$ the topology obtained by pulling back the Hausdorff topology (see [3]) on the closed subsets of G^∞. We have that $X_{r,R}$ is compact since it is closed.

REMARK 5.1. This compactness result is essentially due to D. Rudolph [27], in the different looking but essentially equivalent context of tilings of $\mathbb{R}^d$. The terms *tiling topology* and *local topology* are often used in the literature for topologies this type. The definition given here, in terms of G^∞, is attributed in [5] to Johansen.

Define the translation action T of G on $X_{r,R}$ by

$$T^g \Lambda := \Lambda - g \qquad (5\text{--}1)$$

It is easy to see that the G-action T is a continuous action in the local topology. We call T acting on $X_{r,R}$ a *pattern dynamical system*. More generally, any closed T-invariant subspace $X \subseteq X_{r,R}$ is called a pattern dynamical system.

Given a Delone set Λ, we define its *orbit closure* (or *dynamical hull*) to be the set

$$X(\Lambda) := \overline{\{T^g \Lambda : g \in G\}}. \qquad (5\text{--}2)$$

Since $X(\Lambda) \subseteq X_{r,R}$ is clearly closed and T-invariant, $X(\Lambda)$ is a pattern dynamical system. If Λ is a model set, then $X(\Lambda)$ is called a *model set dynamical system*.

Finiteness conditions. A subset $\Lambda \subseteq G$ is called *locally finite* if $\Lambda \cap K$ is finite for any $K \subseteq G$ compact. Note that any uniformly discrete set (and thus any Delone set) is locally finite.

We say two subsets $A, B \subseteq G$ are *equivalent* if $B = A - g$.

DEFINITION 5.2. A subset $\Lambda \subseteq G$ is said to have *finite local complexity* if for any $Q > 0$, the number of equivalence classes of subsets of the form $B_Q(g) \cap \Lambda$ for all $g \in G$, is finite.

Note that if $\Lambda \in X_{r,R}$ then it suffices to check only $Q = R$ in Definition 5.2.

Fix $0 < r < R$ and let $Q \geq R$. Given a finite collection $\mathcal{F}$ of finite subsets of $B_Q(0) \subseteq G$, we say $Z \subseteq X_{r,R}$ has $\mathcal{F}$-local complexity if every $F \in \{B_Q(s) \cap \Lambda : g \in G, \Lambda \in X_{r,R}\}$ is equivalent to some $F' \in \mathcal{F}$. It is easy to see that $X(\mathcal{F})$, defined to be the collection of all $\Lambda \in X_{r,R}$ that have $\mathcal{F}$-local complexity, is closed. We say $X \subseteq X_{r,R}$ has *finite local complexity* if $X \subseteq X(\mathcal{F})$ for some $\mathcal{F}$.

Lagarias [11] showed that a set $\Lambda \subseteq G$ has finite local complexity if and only if $\Lambda - \Lambda$ is locally finite. A Delone set $\Lambda \subseteq G$ is called a *Meyer set* (see [16]) if $\Lambda - \Lambda$ is Delone. In particular, Meyer sets have finite local complexity.

PROPOSITION 5.3. *If Λ is a topologically regular Model set, then $\Lambda - \Lambda$ is a topologically regular model set. In particular, every topologically regular model set Λ is a Meyer set. Thus a topologically regular model set has finite local complexity.*

This follows from (4) of Lemma 4.4.

COROLLARY 5.4. *If Λ is a topologically regular model set then the model set dynamical system $X(\Lambda)$ has finite local complexity.*

REMARK 5.5. Lagarias used the term "finite type" for those $\Lambda \subseteq G$ that have the property we call "finite local complexity". However, calling $X(\Lambda)$ a "finite type" dynamical system conflicts with the way "finite type" is used in symbolic dynamics, where one would probably call the $X(\mathscr{F})$ a finite type dynamical system. It is only in a few cases that $X(\Lambda)$ is finite type in this more restricted sense (e.g., for Λ the vertices of a Penrose tiling, more generally see [12]).

Repetitivity and minimality. Suppose $\Lambda \subseteq X_{r,R}$ and $F \subseteq \Lambda$ is finite and nonempty. We denote the set of *occurrences* of F in Λ by

$$\Lambda[F] := \{g \in \Lambda : F + g \subseteq \Lambda\} = \bigcap_{f \in F} \Lambda - f. \tag{5-3}$$

Note that $0 \in \Lambda[F]$ and that $\Lambda[F]$ is uniformly discrete (since $\Lambda[F] \subseteq \Lambda$).

DEFINITION 5.6. We say Λ is *repetitive* if every nonempty $\Lambda[F]$ is relatively dense; equivalently, if every $\Lambda[F]$ is a Delone set.

Now suppose Λ has finite local complexity. A standard argument using *Gottschalk's Theorem* [6] (see [24], for instance) shows that the dynamical system $X(\Lambda)$ is minimal if and only if Λ is repetitive. A topologically regular model set is not necessarily repetitive without a further assumption.

DEFINITION 5.7. A topologically regular window $W \subseteq H$ (or the corresponding model set $\Lambda(W)$) is called *nonsingular* if $\Lambda(W) = \Lambda(W^\circ)$.

LEMMA 5.8. *If W is nonsingular and $g \in L$, then $W - g^*$ is nonsingular.*

PROOF. If $W - g^*$ is singular then $\Lambda(\partial(W - g^*)) \neq \varnothing$. But then $\Lambda(\partial(W) - g^*) = \Lambda(\partial(W)) - g \neq \varnothing$, so $\Lambda(\partial(W)) \neq \varnothing$ and W is singular. $\square$

PROPOSITION 5.9. *Let $\mathscr{S} = (G, H, \Gamma)$ be a cut and project scheme with $W \subseteq H$ a nonsingular topologically regular window. Suppose $F \subseteq \Lambda(W)$ is nonempty and finite. Then there exists a nonsingular topologically regular window $V \subseteq W^\circ$ so that $\Lambda(W)[F] = \Lambda(V)$.*

PROOF. Suppose $F \subseteq \Lambda(W) \subseteq L$. Then $\Lambda(W)[F] \subseteq L$ is nonempty and

$$\Lambda(W)[F] = \Lambda(W^\circ)[F] = \bigcap_{f \in F} \Lambda(W^\circ) - f = \bigcap_{f \in F} \Lambda(W^\circ - f^*)$$

$$= \Lambda\left(\bigcap_{f \in F} W^\circ - f^*\right) = \Lambda(V_0),$$

where $V_0 = \bigcap_{f \in F} W^\circ - f^*$, which is open since F is finite. Put $V := \overline{V_0}$, which is topologically regular. Since $V = \cap_{f \in F}(W - f^*)$, it is nonsingular, since $W - f*$ is nonsingular by Lemma 5.8. It follows that V is nonsingular, and $\Lambda(V) = \Lambda(V^\circ) = \Lambda(W)[F]$.   □

COROLLARY 5.10. *Every nonsingular topologically regular model set Λ is repetitive and has finite local complexity. Thus the corresponding model set dynamical system $X(\Lambda)$ is minimal.*

REMARK 5.11. In Examples 4.1 and 4.3, conditions for nonsingularity were already given. These guarantee the minimality of $X(\Lambda)$.

Aperiodicity

DEFINITION 5.12. Let A be a subset of a LCA group G. We call $g \in G$ a *period* for A if $A - g = A$. A subset $A \subseteq G$ is called *aperiodic* if its only period is $g = 0$.

Roughly speaking, a "crystal" is a Delone set whose periods form a lattice in G. Here we really want our model sets to be "quasicrystals," so we insist that they be aperiodic. Aperiodicity has the following dynamical interpretation.

LEMMA 5.13. *The orbit closure $X(\Lambda)$ of a Delone set Λ is free if and only if Λ is aperiodic.*

Here is the way that we ensure a model set is aperiodic.

LEMMA 5.14. *If $\mathcal{S} = (G, H, \Gamma)$ is an aperiodic cut and project scheme and the window $W \subseteq H$ is topologically regular and aperiodic, then the model set $\Lambda(W) \subseteq G$ is aperiodic. Conversely, suppose that either $\mathcal{S}$ is not aperiodic or that W has a nontrivial period in $*(L)$. Then $\Lambda(W)$ is not aperiodic.*

PROOF. Suppose $\Lambda(W) = \Lambda(W) - g$. Then $\Lambda(W) = \Lambda(W - g^*)$, and since W is regular, $W = W - g^*$ by Lemma 4.12. Since W is aperiodic $g^* = 0$, which implies $g = 0$ since $\mathcal{S}$ is aperiodic.

Conversely, if $g \in J \setminus \{0\} \subseteq L$ then $\Lambda(W) - g = \Lambda(W - g^*) = \Lambda(W)$ since $g^* = 0$. Similarly, if $W - g^* = W$ for some nonzero $g \in L$, then $\Lambda(W) - g = \Lambda(W - g^*) = \Lambda(W)$.   □

REMARK 5.15. In Example 4.1, $W = [e^{2\pi i a}, e^{2\pi i b}]$ is aperiodic if $b - a \in \mathbb{R} \setminus \mathbb{Q}$, a condition that always holds in the "Sturmian" case $a = \theta$, $b = \theta + \alpha$. (In fact, no $h \in *(L)$ can be a period of any arc $W \subseteq H$, so all $\Lambda(W)$ are aperiodic). In Example 4.3, aperiodicity is automatic.

Parameterization. Although the model set family $\mathcal{M}(W)$, defined in (4–2), is parameterized by $G \times H$, these parameters are redundant because $\Lambda(W + h) - g = \Lambda(W + h') - g'$ whenever $(g, h) - (g', h') \in \Gamma$. This suggests $Y = (G \times H)/\Gamma$ as a more natural parameter space.

Define $\alpha : Y \to \mathcal{M}(W)$ by

$$\alpha(y) := \Lambda(W + h) - g$$

for any $(g, h) \in G \times H$ that satisfies $y = \pi(g, h)$. This mapping is clearly well defined and satisfies $\alpha(R^g y) = T^g \alpha(y)$ for $g \in G$, $y \in Y$, where R is the Kronecker dynamical system (2–3), and T is the translation action (5–1).

LEMMA 5.16. *If $W \subseteq H$ is an aperiodic, topologically regular window, then α is injective (and thus a bijection). Conversely, whenever α is injective, W is aperiodic.*

PROOF. Suppose $\Lambda(W + h') - g' = \Lambda(W + h) - g$. We may assume without loss of generality (by translating) that $(g', h') = 0$. Thus we have $\Lambda(W) = \Lambda(W + h) - g$. It follows that $g \in L$ and, by regularity that $W = W + h - g^*$. Since W is aperiodic, $h - g^* = 0$ or $h = g^*$. Written differently, $(g, h) = (g, g^*)$, which means $(g, h) \in \Gamma$.

We prove the converse by contradiction. Suppose $W + h = W$ for $h \neq 0$. Let $h' = \pi(0, h) \in Y$ and note that $h' \neq 0$. This is because $(0, h') \notin \Gamma$, which is because π_1 is injective. But $\alpha(h') = \alpha(0)$, so α is not injective. □

Unfortunately, $\mathcal{M}(W)$ turns out to have some undesirable properties in the local topology. These are associated with the singular model sets. However, nonsingularity is common.

LEMMA 5.17. *Let $\mathcal{S} = (G, H, \Gamma)$ be a cut and project scheme. If $W \subseteq H$ is a topologically regular window then there exists a dense G_δ subset $H_0 \subseteq H$ so that for all $h \in H_0$, $W + h$ is nonsingular.*

Basically this follows from the countability of Γ.

We define the *nonsingular parameters* $Y_0 := \pi(G \times H_0)$ and the *nonsingular model sets* $\mathcal{M}_0(W) = \alpha(Y_0)$. Since π is open, $Y_0 \subseteq Y$ is dense G_δ.

PROPOSITION 5.18. *Let W be a topologically regular window. Then for any $\Lambda \in \mathcal{M}_0(W)$ the model set dynamical system $X(\Lambda) = \overline{\mathcal{M}_0(W)}$. In particular, all the nonsingular model sets in the same family generate the same model set dynamical system.*

PROOF. We will show that if $\Lambda, \Lambda' \in \mathcal{M}_0(W)$ then $\Lambda' \in X(\Lambda)$. We may assume without loss of generality that $\Lambda = \Lambda(W)$, where W is nonsingular and topologically regular, and that $\Lambda' = \Lambda(W + h)$ for some $h \in H_0$. By the definition of the local topology, it suffices to show that for any finite $F \subseteq \Lambda$, there exists $g \in G$ so that $F \subseteq \Lambda' - g$.

By Proposition 5.9, there exists a nonsingular, topologically regular window V so that $\Lambda(V) = \Lambda(W)[F] = \Lambda[F]$. We have

$$\Lambda(V + h) = \Lambda(W + h)[F] = \Lambda'[F].$$

But $V + h$ is topologically regular, so by Proposition 4.7, $\Lambda'[F]$ is relatively dense. By (5–3) we have that for any $g \in \Lambda'[F]$ that $F + g \subseteq \Lambda'$, which is the same as the condition above. $\qquad\square$

The inverse of α

THEOREM 5.19. *Let $\mathcal{S}$ be a cut and project scheme and let R be the corresponding Kronecker system. Let $W \subseteq H$ be an aperiodic, topologically regular window, and let $X(\Lambda)$ be the corresponding model set dynamical system. Then there is a continuous mapping $\beta : X(\Lambda) \to Y$ such that $\beta = \alpha^{-1}$ on $\mathcal{M}_0(W)$, and such that $\beta \circ T^g = R^g \circ \beta$ for all $g \in G$.*

PROOF. For $\Lambda = \Lambda(W + h) - g \in \mathcal{M}(W)$ we define $\beta = \alpha^{-1}$ by $\beta(\Lambda) := \pi(g, h)$. Clearly $\beta = \alpha^{-1}$, and thus $\beta \circ T^g = R^g \circ \beta$. To prove the theorem it suffices to show β is uniformly continuous on $\mathcal{M}_0(W)$.

First assume that $\Lambda = \Lambda(W + h) \in \mathcal{M}_0(W)$. Then $\Lambda \subseteq L$, and since $W + h$ is topologically regular, $\overline{\Lambda^*} = W + h$ by Lemma 4.12. Note that $q \in \Lambda$ if and only if $q^* \in W + h$, or equivalently $-h \in W - q^*$. Thus $-h \in \bigcap_{q \in \Lambda} W - q^*$.

Suppose now that $k \in \bigcap_{q \in \Lambda} W - q^*$. Then $\overline{\Lambda^*} = W + k$, so that $W - h = W + k$. Since W is aperiodic, $k = -h$. It follows that $\bigcap_{q \in \Lambda} W - q^* = \{-h\}$. Putting

$$I_s = - \bigcap_{q \in B_s(0) \cap \Lambda} W - q^*, \qquad (5\text{–}4)$$

we have that $\operatorname{diam}(I_s)$ is nonincreasing, and that $\{h\} = \bigcap_{s=0}^\infty I_s$. Fix $\varepsilon > 0$. Choose s so large that $\operatorname{diam}(I_s) < \varepsilon/2$. Let

$$\delta < \min\left(\frac{1}{s+R}, \frac{\varepsilon}{2}\right) \qquad (5\text{–}5)$$

where $\mathcal{M}(W) \subseteq X_{r,R}$.

Suppose $\Lambda_1, \Lambda_2 \in \mathcal{M}_0(w)$ with $d(\Lambda_1, \Lambda_2) < \delta$. Let $\Lambda_1 := \Lambda(W + h_1) - g_1$ and $\Lambda_2 := \Lambda(W + h_2) - g_2$. By the definition of d, there exist g_1', g_2' with $|g_1'|, |g_2'| < \delta$ so that

$$(\Lambda_1 - g_1') \cap B_{\frac{1}{\delta}}(0) = (\Lambda_2 - g_2') \cap B_{\frac{1}{\delta}}(0). \qquad (5\text{–}6)$$

Since $1/\delta > R$, there exists

$$g \in (\Lambda_1 - g_1') \cap (\Lambda_2 - g_2') \cap B_R(0) \neq \varnothing. \tag{5-7}$$

Thus

$$\begin{aligned} \Lambda_i - g_i' - g &= \Lambda(W + h_i) - (g_i + g_i' - g) \\ &= \Lambda\left(W + h_i - (g_i + g_i' + g)^*\right). \end{aligned} \tag{5-8}$$

Here we have used the fact that $g_i + g_i' + g \in L$, since by (5–7), $\Lambda_i - g_i' - g \subseteq L$. Now by (5–6)

$$(\Lambda_1 - g_1' - g) \cap B_s(0) = (\Lambda_2 - g_2' - g) \cap B_s(0),$$

so that by (5–4), $-\left(h_i - (g_i + g_i' + g)^*\right) \in I_s$, for $i = 1, 2$, which implies

$$|h_1 - (g_1 + g_1' + g)^* - (h_2 - (g_2 + g_2' + g)^*)| \leq \operatorname{diam}(I_s) < \frac{\varepsilon}{2}.$$

Note that $\pi(g, g^*) = 0$ for any $g \in L$, since $(g, g^*) \in \Gamma$. Because $g_i + g_i' + g \in L$ and π is a homomorphism, we have

$$\begin{aligned} \beta(\Lambda_i) &= \pi(g_i, h_i) = \pi(g_i, h_i) - \pi\left((g_i + g_i' + g), (g_i + g_i' + g)^*\right) \\ &= \pi\left(-(g_i' + g), h_i - (g_i + g_i' + g)^*\right). \end{aligned}$$

It follows that

$$\begin{aligned} |\beta(\Lambda_1) - \beta(\Lambda_2)| &= |\pi(g_1' - g_2', h_1 - (g_1 + g_1' + g)^* - (h_2 - (g_2 + g_2' + g)^*))| \\ &\leq |g_1' - g_2', h_1 - (g_1 + g_1' + g)^* - (h_2 - (g_2 + g_2' + g)^*)| \\ &\leq |g_1' - g_2'| + |h_1 - (g_1 + g_1' + g)^* - (h_2 - (g_2 + g_2' + g)^*)| \\ &< \tfrac{1}{2}\varepsilon + \tfrac{1}{2}\varepsilon = \varepsilon. \qquad \square \end{aligned}$$

COROLLARY 5.20. *For any $\Lambda \in \mathcal{M}_0(W)$ the model set dynamical system $X(\Lambda)$ is an almost 1:1 extension of the Kronecker system R (from the cut and project scheme $\mathcal{S}$). In particular, $X(\Lambda)$ is almost automorphic.*

We now describe, without proof, the structure of the sets $\Lambda' \in X(\Lambda) \setminus \mathcal{M}_0(W)$. These are the sets Λ' that are limits of nonsingular model sets.

THEOREM 5.21. *Suppose that $\Lambda_n \to \Lambda'$, where $\Lambda_n \in \mathcal{M}_0(W)$ and $\Lambda' \in X(\Lambda) \setminus \mathcal{M}_0(W)$. Let $y_n = \beta(\Lambda_n) \in Y_0$ and let $y = \lim y_n \in Y \setminus Y_0$. Let $(g, h) \in G \times (H \setminus H_0)$ with $\pi(g, h) = y$ and define $\Lambda = \Lambda(W - h) + g \in X(\Lambda)$.*

(1) *$\Lambda' \subseteq \Lambda$.*
(2) *There exists $W' \subseteq H$, with $W^\circ \subsetneq W' \subsetneq W$ (with each inclusion proper) such that $\Lambda' = \Lambda(W' - h) + g$.*
(3) *Λ' is a model set that is Delone but (generally) not topologically regular.*

Measurable properties of model sets. We call a model set *measurably regular* if $\theta_H(\partial W) = 0$, where θ_H denotes Haar measure on H. It is easy to see that if W is measurably regular, then $\theta_Y(Y \setminus Y_0) = 0$. As a consequence we have the following.

THEOREM 5.22. *Suppose $\mathcal{S} = (G, H, \Gamma)$ is an aperiodic cut and project scheme and W is a topologically and measurably regular, aperiodic window. Let $\Lambda \in \mathcal{M}_0(W)$. Then the model set dynamical system $X(\Lambda)$ is uniquely ergodic (and thus strictly ergodic).*

COROLLARY 5.23. *Under the conditions of* Theorem 5.22, *$X(\Lambda)$ and R (the Kronecker system for $\mathcal{S}$) are measurably isomorphic. In particular, they have the same point spectrum $\Sigma_{\mathcal{S}}$.*

Now we will discuss the geometric significance of unique ergodicity. A sequence $\mathcal{A} = \{A_n\}$ of compact subsets of G, with $\theta_G(A_n) > 0$, is called a *van Hove sequence* if for any $K \subseteq G$ compact,

$$\lim_{n \to \infty} \frac{\theta_G(\partial^K A_n)}{\theta_G(A_n)} = 0,$$

where

$$\partial^K A := (K + A) \setminus A^\circ \cup \left((-K + \overline{G \setminus A} \cap A\right),$$

(see [17]). We call a van Hove sequence *good* if for some $c > 0$,

$$\theta_G(A_n - A_n) \leq c\, \theta_G(A_n). \tag{5--9}$$

The existence of a good van Hove sequence follows, for example, from the assumption that G is σ-compact (see [28], [29], [17]).

Suppose $\Lambda \subseteq G$ is *locally finite* (e.g., $\Lambda \in X_{r,R}$). Given a good van Hove sequence $\mathcal{A} = \{A_n\}$, we define the *metric density* of Λ by

$$\mathrm{dens}(\Lambda) = \lim_{n \to \infty} \frac{1}{\theta_G(A_n)} \mathrm{card}\,(\Lambda \cap A_n).$$

If this limit exists, it is independent of the choice of van Hove sequence (see [17]).

PROPOSITION 5.24 (MOODY, [17]). *Let $\mathcal{S} = (G, H, \Gamma)$ be a cut and project scheme, let $\mathcal{A} = \{A_n\}$ be a good van Hove sequence for G, and let $W \subseteq H$ be a measurably regular window. Then*

$$\mathrm{dens}(\Lambda(W + h)) = \theta_H(W). \tag{5--10}$$

for θ_H a.e. $h \in H$. If, in addition, $\theta_H(\partial W) = 0$, then (5--10) holds for all $h \in H$.

The second part of this lemma says that if W is measurably regular, then the metric density of every model set $\Lambda \in \mathcal{M}(W)$ is $\theta_H(W)$. This comes close to proving the unique ergodicity of $X(\Lambda)$ (Theorem 5.22), but it is not quite equivalent. Unique ergodicity requires each nonempty $\Lambda'[F]$ for $\Lambda' \in X(\Lambda)$ to have positive density (see [28]). And, this will be the case if W is both topologically and measurably regular (Theorem 5.22).

6. Realization for model sets

The existence of regular aperiodic windows. In this section and the next, we study the question of when regular aperiodic windows exist. The next result shows that regularity comes almost for free.[5]

LEMMA 6.1. *Let H be a LCA group that is not discrete. Then there exists $r > 0$ so that $W = \overline{B_r(0)}$ is topologically and metrically regular.*

PROOF. Let $S_r(0) := \{g : d(g,0) = r\}$. Then $B_r(0) = \bigcup_{0 \le s < r} S_r(0)$, and since H is not discrete, $B_r(0) \setminus \{0\} \ne \varnothing$ for all $r > 0$. Thus $W = \overline{B_r(0)} = \{g : d(g,0) \le r\}$ and $W^\circ = B_r(0)$. It follows that W is topologically regular.

By local compactness, we can choose R so that $\overline{B_R(0)}$ is compact. It follows that $\theta_R(\overline{B_R(0)}) < \infty$. If $\theta_H(S_r(0)) > 0$ for all $r \in (0, R)$, then there exists $N \in \mathbb{N}$ so that $I_N := \{r : \theta_H(S_r(0)) > 1/N\}$ is infinite. For otherwise, $(0, R) = \bigcup_N I_N$ would be countable. But the existence of such an N implies that $\theta_H(B_R(0)) \ge \theta_H\left(\bigcup_{r \in I_N} S_r(0)\right) = \infty$. This contradicts the compactness of $\overline{B_R(0)}$. It follows that $\theta_H(S_r(0)) = 0$ for some (i.e., all but finitely many) $r \in (0, R)$. This implies that $\overline{B_r(0)}$ is topologically and metrically regular. $\qquad\square$

Aperiodicity is a bit more challenging. In this section we consider the following case.

DEFINITION 6.2. We say a LCA group H has *no compact subgroups* if the only compact subgroup of H is $\{0\}$.

PROPOSITION 6.3. *Suppose H has no compact subgroups. Then every topologically regular window is aperiodic.*

PROOF. $H_W := \{h : W + h = W\}$ is a closed subgroup of H. If $h \in W_H$ then $g_1, g_2 \in W$ so that $g_1 + h = g_2$. Equivalently, $h = g_2 - g_1$ so that $h \in W - W$, and $H_W \subseteq W - W$. Since $W - W$ is compact $H_W = \{0\}$. $\qquad\square$

COROLLARY 6.4. *If H has no compact subgroups then there is a topologically and measurably regular aperiodic window $W \subseteq H$.*

---

[5] My thanks to Nelson Markley for telling me this proof.

The following result from duality theory shows that the property no compact subgroups is, in fact, a relatively common.

LEMMA 6.5. *A LCA group H has no compact subgroups if and only if $\widehat{H}$ is connected.*

For example, $H = \mathbb{R}^d$ has no compact subgroups (and neither does $\mathbb{Z}^d$). However, $\mathbb{Q}_p$ has $\mathbb{Z}_p$ as a compact subgroup.

The compact case. One large class of LCA groups which fail to have no compact subgroups are the CA groups. Nevertheless, the next theorem shows that CA groups always have good windows. Our approach follows the 1973 dissertation of Michael Paul [22], (see [21]), which concerns the symbolic dynamics of rotation actions of $\mathbb{Z}$ on monothtic groups (i.e., Kronecker $\mathbb{Z}$-actions).

THEOREM 6.6. *If H is an infinite CA group then there exists an aperiodic, topologically and measurably regular window $W \subseteq H$.*

By combining Theorem 6.6 with Theorem 3.8 and Corollary 5.23, we obtain the main result of this paper. It is a Halmos–von Neumann theorem for model sets.

THEOREM 6.7. *Let G be a LCA group with a lattice dense dual. Let $\Sigma \subseteq \widehat{G}$ be a countable dense subgroup. Then there exists a cut and project scheme $\mathscr{S} = (G, H, \Gamma)$ and a topologically and metrically regular window $W \subseteq H$, so that for any $\Lambda \in \mathcal{M}_0(W)$, the model set dynamical system $X(\Lambda)$ has spectrum Σ.*

REMARK 6.8.

(1) $G = \mathbb{Z}^d$ and $G = \mathbb{R}^d$ satisfy the hypotheses of Theorem 6.7.
(2) The same result holds for any G satisfying the realization property.
(3) Since it is a corollary of Theorem 6.6, the internal spaces H constructed in Theorem 6.7 are always compact.

PROOF OF THEOREM 6.6. The proof divides into two cases.

In **Case 1** we assume H is totally disconnected. Since H is infinite and metrizable, it is homeomorphic to the standard (middle thirds) Cantor set $C \subseteq [0, 1]$. From now on we identify C and H.

Let $b \in H$ be an element of C that is not an endpoint of any complementary interval and define $W := [0, b] \cap H$. Note that we have $\partial W = \{b\}$ so that $\theta_H(\partial W) = 0$. Thus W is topologically and measurably regular.

There exists a sequence $\{g_n\}$ in H with the following properties:

(1) $g_n \to b$,
(2) $g_n \in W$ for infinitely many n, and
(3) $g_n \in W^c$ for infinitely many n.
(4) $g_n \neq b$,

Note that, in fact, *any* convergent sequence satisfying (2) and (3) must also satisfy (1).

Now suppose $W + h = W$. Then on the one hand, $g_n + h \to b + h$. On the other hand, $g_n + h \in W + h = W$ if and only if $g_n \in W$. It follows that $g_n + h$ is a convergent sequence satisfying (2) and (3), so $g_n + h \to b$. Then $b + h = b$, so $h = 0$, and W is aperiodic.

Case 2. For any $\chi \in \widehat{H}$, $\chi(H)$ is a compact subgroup of $\mathbb{T}$. There are just two possibilities:

(1) $\chi(H) = \mathbb{T}$, or
(2) $\chi(H) = \mathbb{Z}/N = \{e^{2\pi i k/N} : k = 0, \ldots, N - 1\}$ for some $N \in \mathbb{N}$.

By our assumption, H is not totally disconnected. It follows from the Structure Theorem for CA groups (see [23]) that H has a connected subgroup. Some character, call it $\chi_1 \in \widehat{H}$, must be nontrivial on this subgroup. Since the continuous image of a connected set is connected, χ_1 is of the first type. Let $\chi_0 = 1$ denote the identity in $\widehat{H}$, and note that this is of the second type.

Because H is an infinite CA group, $\widehat{H}$ is countably infinite, and we will write $\widehat{H} = \{\chi_0, \chi_1, \chi_2, \ldots\}$. Let $I_1 = \{1, i_2, i_3, \ldots\}$ be a list of all characters of the first type, and let $I_2 = \{j_2, j_3, \ldots\}$ be a list of all characters of the second type, excepting χ_0. Then $I_1 \cup I_2 = \mathbb{N}$, and at least one is infinite.

The *infinite dimensional torus* $\mathbb{T}^{\mathbb{N}}$ is a (nonmetrizable) compact abelian group with dual $\mathbb{Z}^{\mathbb{N}}$. Define the mapping $\varepsilon : H \to \mathbb{T}^{\mathbb{N}}$ by

$$\varepsilon(h) = (\chi_1(h), \chi_2(h), \ldots).$$

As a set of functions, $\widehat{H}$ separates points, so the mapping ε is an algebraic and topological isomorphism of H onto its image $\varepsilon(H)$ in $\mathbb{T}^{\mathbb{N}}$. We will identify H with its image. Let θ_H denote the Haar measure on H, viewed as a measure on $\mathbb{T}^{\mathbb{N}}$.

Identify $\mathbb{T}$ with $[0, 2\pi)$. For $i = 1, 2, \ldots$ let $U_i = (a_i, b_i)$ be a collection of intervals in $(0, \pi)$ such that $a_{i+1} > b_i$. Let $U_i' = (c_i, d_i)$ be such that $d_i - c_i < \pi$, and such that

$$\chi_i(\chi_1^{-1}(U_i)) \cap U_i' \neq \varnothing, \tag{6–1}$$

Then (6–1) guarantees that

$$U_i' \cap \chi_i(H) \neq \varnothing. \tag{6–2}$$

Without loss of generality we may assume that if $i \in I_2$, then U_i' is small enough that $U_i' \cap \chi_i(H) = \{e_i\}$ is a single point.

Define $\mathbb{N}_i = \{i + 1, i + 2, \ldots\}$ and put

$$V_1 = (U_1 \times \mathbb{T}^{\mathbb{N}_1}) \cap \varepsilon(H),$$

$$V_2 = (U_2 \times U'_2 \times \mathbb{T}^{\mathbb{N}_2}) \cap \varepsilon(H),$$
$$V_3 = (U_3 \times \mathbb{T} \times U'_3 \times \mathbb{T}^{\mathbb{N}_3}) \cap \varepsilon(H),$$
$$\vdots$$
$$V_i = (U_i \times \mathbb{T}^{i-2} \times U'_i \times \mathbb{T}^{\mathbb{N}_i}) \cap \varepsilon(H),$$

where (6–2) guarantees $V_i \neq \varnothing$ and the choice of U_i implies the sets V_i have disjoint closures. Define $W_0 := \bigcup_{i \in \mathbb{N}} V_i$, which is open, and define $W := \overline{W_0}$. It is clear that W is topologically regular.

By the construction we have $\partial(W) = \bigcup_i \partial(V_i)$ is a disjoint union. Thus, to show W is metrically regular, it suffices to show $\theta_H(\partial V_i) = 0$.

For $i = 1$, we have $\partial(V_i) = (\{a_1, b_1\} \times \mathbb{T}^{\mathbb{N}_1}) \cap \varepsilon(H)$. Let θ_1 be projection of θ_H to the first $\mathbb{T}$ factor of $\mathbb{T}^{\mathbb{N}}$, namely, $\theta_1(E) = \theta_H(E \times \mathbb{T}^{\mathbb{N}_1})$. It is easy to see that θ_1 is translation invariant and $\theta_1(\mathbb{T}) = 1$, so $\theta_1 = \lambda$, the normalized Lebesgue measure on $\mathbb{T}$. We can decompose

$$\theta_H = \int_{\mathbb{T}} \eta_t \, d\lambda(t),$$

where η_t is a finite measure on the t-th fiber $F_t := \{t\} \times \mathbb{T}^{\mathbb{N}_1}$. Then

$$\theta_H(\partial(V_1)) = \int_{\{a_1, b_1\}} \eta_t(F_t \cap \varepsilon(H)) \, d\lambda(t) = 0,$$

since λ has no atoms.

Now suppose $i > 1$, and without loss of generality, put $i = 2$. Let $\varepsilon_2(h) = (\chi_1(h), \chi_2(h))$ be the projection to the first $\mathbb{T}^2$ factor of $\mathbb{T}^{\mathbb{N}}$. Then $\varepsilon_2(H)$ is a closed subgroup of $\mathbb{T}^2$, and since $\chi_1(H) = \mathbb{T}$, $\varepsilon_2(H)$ is not discrete. As above, the projection θ_2 of θ_H to $\varepsilon_2(H)$, namely, $\theta_1(E) = \theta_H(E \times \mathbb{T}^{\mathbb{N}_2})$ for $E \subseteq \mathbb{T}^2$, is normalized Haar measure on $\varepsilon_2(H)$. Since $\varepsilon_2(H)$ is not discrete, θ_2 is nonatomic. Again we can decompose

$$\theta_H = \int_{\mathbb{T}^2} \eta_t \, d\theta_2(\mathbf{t}),$$

where $\eta_{\mathbf{t}}$, $\mathbf{t} \in \mathbb{T}^2$, is the $\mathbf{t}$-th finite measure on the fiber $F_{\mathbf{t}} := \{\mathbf{t}\} \times \mathbb{T}^{\mathbb{N}_2}$.

If $2 \in I_2$ then

$$\partial(V_2) = (\{(a_2, e_2), (b_2, e_2)\} \times \mathbb{T}^{\mathbb{N}_2}) \cap \varepsilon(H),$$

and we have

$$\theta_H(\partial(V_2)) = \int_{\{(a_2, e_2), (b_2, e_2)\}} \eta_t(F_{\mathbf{t}} \cap \varepsilon(H)) \, d\theta_2(\mathbf{t}) = 0,$$

since θ_2 has no atoms.

Finally, suppose $2 \in I_1$. Then $\theta_2 = \lambda_2$, normalized Lebesgue measure on $\mathbb{T}^2$. Since $\partial(U_2 \times U_2')$ is a square in $\mathbb{T}^2$, it satisfies $\theta_2(\partial(U_2 \times U_2')) = 0$. Using the same argument as in the case $i = 1$, it follows that $\theta_H(\partial(V_2)) = 0$. Thus W is metrically regular.

It remains to show that W is aperiodic. Suppose to the contrary that $W + h = W$ for some $h \neq 0$. In our identification of H with $\varepsilon(H)$, let us write $h = (h_1, h_2, h_3, \ldots)$ where $h_i = \chi_i(h)$. Since $h \neq 0$, there is a smallest i so that $h_i \neq 0$. If $i = 1$, then $U_1 + h_1 \neq U_1$ so $W + h \neq W$. Thus we assume $i > 1$.

If $i \in I_1$ then $\chi_i(H) = \mathbb{T}$ so $U_i' \cap \chi_i(H) = (c_i, d_i)$ and $(U_i' + h_i) \cap \chi_i(H) = (c_i + h_i, d_i + h_i)$. We have $(c_i + h_i, d_i + h_i) \setminus [c_i + d_i] \neq \varnothing$ since $d_i - c_i < \pi$ and $h_i \neq 0$. Choose $g \in H$ so that $\chi_i(g) \in (c_i + h_i, d_i + h_i) \setminus [c_i + d_i]$. In other words,

$$\exists g \in H \text{ such that } \chi_i(g) \in (U_i' + h_i) \setminus \overline{U_i'}. \tag{6–3}$$

Similarly, if $i \in I_2$ then $U_i' \cap \chi_i(H) = \{e_i\}$ and $(U_i' + h_i) \cap \chi_i(H) = \{e_i + h_i\}$. Note that $e_i + h_i \neq e_i$ and let $g \in H$ be such that $\chi_i(g) = e_i + h_i$. In this case we also have (6–3).

But (6–3) implies $g \in (W + h) \setminus W$, so $W + h \neq W$. Thus W is aperiodic. $\qquad\square$

7. The diffraction spectrum

In this final part of the paper, we describe how the dynamical spectrum of a model set Λ is related to its *diffraction spectrum*. To this end, given a LCA group G let us fix a van Hove sequence $\mathcal{A} = \{A_n\}$.

We first assume only that $\Lambda \subseteq X_{r,R}$ is locally finite and define a measure on G associated with Λ by

$$\Delta_\Lambda = \sum_{g \in \Lambda} \delta_g, \tag{7–1}$$

where δ_g denotes unit point measure at g.

For $K \subseteq G$ compact, we define a new measure on G by

$$\Delta_{\Lambda,K}^{(2)} = \frac{(\chi_K \Delta_\Lambda) * (\chi_K \Delta_{-\Lambda})}{\theta(K)},$$

where $*$ denotes convolution, and θ denotes Haar measure on G. An *autocorrelation* $\Delta_\Lambda^{(2)}$ of Δ_Λ is defined to be any weak-* topology limit point of the sequence $\{\Delta_{\Lambda,A_n}^{(2)}\}$. In a slight abuse of terminology we call $\Delta_\Lambda^{(2)}$ an autocorrelation of Λ.

It is easy to see that an autocorrelation is a discrete, positive measure on G that is supported on $\Lambda - \Lambda$. In particular, it has the form

$$\Delta_\Lambda^{(2)} = \sum_{g \in \Lambda - \Lambda} f(g) \delta_g,$$

for some $f(g) \geq 0$.

As one can also show, $\Delta_\Lambda^{(2)}$ is *positive definite* (see [23]) on G. Thus by Bochner's Theorem there is a unique positive measure γ on $\widehat{G}$ so that $\Delta_\Lambda^{(2)} = \widehat{\gamma}$. Here $\widehat{\gamma}$ denotes the *Fourier transform* of γ. We will write $\widehat{\Delta}_\Lambda := \gamma$ and call $\widehat{\Delta}_\Lambda$ a *diffraction spectrum* of Λ. The motivation for this definition is essentially physical.

In general, neither $\Delta_\Lambda^{(2)}$ nor $\widehat{\Delta}_\Lambda$ is unique. However, we say that Λ has a *unique autocorrelation* if there is just one autocorrelation measure $\Delta_\Lambda^{(2)}$. In this case there is also a unique diffraction spectrum $\widehat{\Delta}_\Lambda$ on G.

PROPOSITION 7.1 (SCHLOTTMAN, [29]). *Suppose Λ is a Delone set with finite local complexity such that $X(\Lambda)$ is uniquely ergodic. Then Λ has a unique autocorrelation and, in particular, a unique diffraction spectrum $\widehat{\Delta}_\Lambda$.*

Let $(\widehat{\Delta}_\Lambda)_d$ be the discrete part of $\widehat{\Delta}_\Lambda$ (in the sense of the Lebesgue Decomposition Theorem) and define $\widehat{\Lambda} := \operatorname{supp}(\widehat{\Delta}_\Lambda)$. We call $\widehat{\Lambda}$ the *diffraction point spectrum* of Λ. We say Λ has *pure point diffraction spectrum* if $\widehat{\Delta}_\Lambda = (\widehat{\Delta}_\Lambda)_d$. A Delone set with pure point diffraction spectrum that is nor periodic is called a *quasicrystal*.

PROPOSITION 7.2 (SCHLOTTMAN, [29], MOODY, [17]). *Suppose Λ is a Delone set with finite local complexity, such that the translation action T on $X(\Lambda)$ is uniquely ergodic. Let σ_T be the dynamical spectral measure of T. Then $\widehat{\Delta}_\Lambda << \sigma_T$ (in the Radon–Nikodým sense), so that in particular, $\widehat{\Lambda} \subseteq \Sigma_T$. If T has dynamical pure point spectrum, then Λ has pure point diffraction spectrum.*

It is easy to construct examples showing that, in general, $\widehat{\Lambda} \neq \Sigma_T$.

References

[1] Michael Baake, Robert V. Moody, and Martin Schlottmann, *Limit-(quasi)periodic point sets as quasicrystals with p-adic internal spaces*, J. Phys. A **31** (1998), no. 27, 5755–5765.

[2] Arno Berger, Stefan Siegmund, and Yingfei Yi, *On almost automorphic dynamics in symbolic lattices*, Ergodic Theory Dynam. Systems **24** (2004), no. 3, 677–696.

[3] Nicolas Bourbaki, *General topology, I*, Addison–Wesley, Reading, MA, 1966.

[4] N. G. de Bruijn, *Algebraic theory of Penrose's nonperiodic tilings of the plane, I, II*, Nederl. Akad. Wetensch. Indag. Math. **43** (1981), no. 1, 39–52, 53–66.

[5] Alan Forrest, John Hunton, and Johannes Kellendonk, *Topological invariants for projection method patterns*, Mem. Amer. Math. Soc. **159**, American Mathematical Society, Providence, RI, 2002.

[6] Walter H. Gottschalk, *Orbit-closure decompositions and almost periodic properties*, Bull. Amer. Math. Soc. **50** (1944), 915–919.

[7] Walter H. Gottschalk and Gustav A. Hedlund, *Topological dynamics*, American Mathematical Society Colloquium Publications **36**, American Mathematical Society, Providence, RI, 1955.

[8] Paul R. Halmos, and J. von Neumann, *Operator methods in classical mechanics, II*, Ann. of Math. (2) **43** (1942), 332–350.

[9] Edwin Hewitt and Kenneth A. Ross, *Abstract harmonic analysis, I: Structure of topological groups. Integration theory, group representations*, Die Grundlehren der Mathematischen Wissenschaften **115**, Academic Press, New York, 1963.

[10] A. Hof, *Uniform distribution and the projection method*, Quasicrystals and discrete geometry (Toronto, ON, 1995), 201–206, Fields Inst. Monogr. **10**, American Mathematical Society, Providence, RI, 1998.

[11] Jeffrey C. Lagarias, *Geometric models for quasicrystals, I: Delone sets of finite type*, Discrete Comput. Geom. **21** (1999), no. 2, 161–191.

[12] Thang T. Q. Le, *Local rules for quasiperiodic tilings*, The mathematics of long-range aperiodic order (Waterloo, ON, 1995), 331–366, Kluwer, Dordrecht, 1997.

[13] Nelson G. Markley, *Characteristic sequences*, Z. Wahrscheinlichkeitstheorie und Verw. Gebiete **30** (1974), 321–342.

[14] Nelson G. Markley and Michael E. Paul, *Almost automorphic symbolic minimal sets without unique ergodicity*, Israel J. Math. **34** (1979), no. 3, 259–272 (1980).

[15] Yves Meyer, *Algebraic numbers and harmonic analysis*, North-Holland Mathematical Library **2**, North-Holland, Amsterdam, 1972,

[16] Robert V. Moody, *Meyer sets and their duals*, The mathematics of long-range aperiodic order (Waterloo, ON, 1995), 403–441, NATO Adv. Sci. Inst. Ser. C Math. Phys. Sci. **489**, Kluwer, Dordrecht, 1997.

[17] Robert V. Moody, *Uniform distribution in model sets*, Canad. Math. Bull. **45** (2002), no. 1, 123–130.

[18] Robert V. Moody and Nicolae Strungaru, *Point sets and dynamical systems in the autocorrelation topology*, Can. Math. Bull. **47** (2004), no. 1, 82–99.

[19] Sidney A. Morris, *Pontryagin duality and the structure of locally compact abelian groups*, London Mathematical Society Lecture Notes Series **29**, Cambridge University Press, Cambridge, 1977.

[20] John C. Oxtoby, *Ergodic sets*, Bull. Amer. Math. Soc. **58** (1952), 116–136.

[21] Michael E. Paul, *Construction of almost automorphic symbolic minimal flows*, General Topology and Appl. **6** (1976), no. 1, 45–56.

[22] Michael E. Paul, *Almost automorphic symbolic minimal flows*, Ph.D. thesis, Wesleyan University, October 1973.

[23] Hans Reiter and Jan D. Stegeman, *Classical harmonic analysis and locally compact groups*, 2nd ed., London Mathematical Society Monographs New Series **22**, Clarendon Press, Oxford, 2000.

[24] E. Arthur Robinson, Jr., *Symbolic dynamics and tilings of* $\mathbb{R}^d$, Symbolic dynamics and its applications, 81–119, Proc. Sympos. Appl. Math. **60**, American Mathematical Society, Providence, RI, 2004.

[25] E. Arthur Robinson, Jr., *The dynamical properties of Penrose tilings*, Trans. Amer. Math. Soc. **348** (1996), no. 11, 4447–4464.

[26] E. Arthur Robinson, Jr., *On the table and the chair*, Indag. Math. (N.S.) **10** (1999), no. 4, 581–599.

[27] Daniel J. Rudolph, *Markov tilings of* $\mathbb{R}^n$ *and representations of* $\mathbb{R}^n$ *actions*, Measure and measurable dynamics (Rochester, NY, 1987), 271–290, American Mathematical Society, Providence, RI, 1989.

[28] Martin Schlottmann, *Cut-and-project sets in locally compact abelian groups*, Quasicrystals and discrete geometry (Toronto, ON, 1995), 247–264, Fields Inst. Monogr. **10**, American Mathematical Society, Providence, RI, 1998.

[29] Martin Schlottmann, *Generalized model sets and dynamical systems*, Directions in mathematical quasicrystals, 143–159, CRM Monogr. Ser. **13**, American Mathematical Society, Providence, RI, 2000.

[30] Boris Solomyak, *Dynamics of self-similar tilings*, Ergodic Theory Dynam. Systems **17** (1997), no. 3, 695–738.

[31] William A. Veech,. *Almost automorphic functions on groups.* Amer. J. Math. **87** (1965), 719–751.

[32] Susan Williams, *Toeplitz minimal flows which are not uniquely ergodic*, Z. Wahrscheinlichkeitstheorie Verw. Gebiete **67** (1984), no. 1, 95–107.

E. ARTHUR ROBINSON, JR.
DEPARTMENT OF MATHEMATICS
GEORGE WASHINGTON UNIVERSITY
WASHINGTON, DC 20052
UNITED STATES
robinson@gwu.edu

Recent Progress in Dynamics
MSRI Publications
Volume **54**, 2007

Problems in dynamical systems
and related topics

BORIS HASSELBLATT

CONTENTS

1. Introduction

At the Clay Mathematics Institute/Mathematical Sciences Research Institute Workshop on "Recent Progress in Dynamics" in September–October 2004 the speakers and participants were asked to state open problems in their field of research, and much of this problem list resulted from these contributions. Thanks are due, therefore, to the Clay Mathematics Institute and the the Mathematical Sciences Research Institute for generously supporting and hosting this workshop, and to the speakers, who graciously responded to the suggestion that open problems be stated whenever possible, and who in many cases kindly corrected or expanded the renditions here of the problems that they had posed. It is my hope that this list will contribute to the impact that the workshop has already had. It was helpful to this endeavor and is a service to the community that most lectures from the workshop can be viewed as streaming video at http://www.msri.org/calendar/workshops/WorkshopInfo/267/show_workshop.

In this list, almost all sections are based on an original version written by myself about the problems as presented by the proposer in a talk during the workshop. The proposer is identified by the attribution "(presented by…)" in the section heading. Where the proposer undertook significant modification of this original version, the section became attributed to the proposer (without "presented by"). Section 8 and Section 13 were contributed by their authors without any preliminary draft by myself and were only slightly edited by me, and Section 11 is based in good part on the questions raised by Keith Burns in his talk but was written collaboratively. Finally, the collection of problems I describe in Section 20 was not as such presented at the workshop, but includes problems familiar to many participants.

2. Smooth realization of measure-preserving maps (Anatole Katok)

QUESTION 2.1. *Given an ergodic measure-preserving transformation T of a Lebesgue space X with probability measure μ, under which conditions is there a diffeomorphism f of a compact manifold M that preserves a smooth volume v for which (f, v) is measurably isomorphic to (T, μ)? In particular, is there any T with finite μ-entropy for which there is no such f?*

Put differently, is finiteness of entropy (which, as shown first by Kushnirenko, holds for diffeomorphisms of manifolds with respect to any invariant Borel probability measure, see *e.g.*, [82, Corollary 3.2.10]) the only restriction imposed on smooth models of measure-preserving transformations?

A potentially useful method is that of Anosov and Katok [2] (see also [34] for a modern exposition) which provides nonstandard smooth realizations of certain dynamical systems. An important pertinent result is due to Pesin: For a smooth dynamical system on a surface with positive entropy, weak mixing implies Bernoulli [120]. Thus there are restrictions on smooth realizations on particular types of manifolds.

It is expected that there are indeed restrictions on realizability other than finiteness of entropy, so long as one considers smooth measures. (Lind and Thouvenot [98] showed that every finite-entropy measure-preserving transformation can be realized as an automorphism of the 2-torus with respect to a suitable invariant Borel probability measure.) Here the picture may be different for infinite versus finite smoothness. In order to establish the existence of such restrictions one needs to construct some suitable invariants. Again, on one hand one may look at specific manifolds or dimensions, such as in Pesin's aforementioned result for maps with positive entropy. For zero entropy an interesting observation is Herman's "Last Geometric Theorem" [33]:

THEOREM 2.2. *An area-preserving C^∞ diffeomorphism f of the disk that has Diophantine rotation number on the boundary has a collection of invariant circles accumulating on the boundary.*

The Anosov–Katok construction provides examples of nonstandard realization of rotations with Liouvillian rotation numbers. In particular, given *any* Liouvillian rotation number ρ, Fayad, Saprykina and Windsor ([36], using the methods of [35]) constructed an area-preserving C^∞ diffeomorphism of the disc that acts as the rotation by ρ on the boundary and is measurably isomorphic to it.

It should be mentioned here in passing that no nonstandard smooth realizations of Diophantine rotations are known:

QUESTION 2.3. *Given an ergodic Diophantine rotation, is there an ergodic volume-preserving diffeomorphism on a manifold of dimension greater than 1 that is measurably conjugate to the rotation?*

Herman's Theorem 2.2 suggests that this would be very hard to achieve on a disk.

There are several systems where existing methods might help decide whether a nonstandard smooth realization exists, such as Gaussian systems, some interval exchanges, and maybe the horocycle flow on the modular surface.

3. Coexistence of KAM circles and positive entropy in area-preserving twist maps (presented by Anatole Katok)

Consider the standard (twist) map

$$f_\lambda(x, y) := (x + y, y + \lambda \sin 2\pi(x + y))$$

of the cylinder (or annulus) $C := S^1 \times \mathbb{R}$, which preserves area.

QUESTION 3.1. *Is the measure-theoretic entropy* $h_{\text{area}}(f_\lambda)$ *positive (with respect to area as the invariant Borel probability measure)*

(1) *for small* $\lambda > 0$?

(2) *for some* $\lambda > 0$ *if the problem is considered instead on* $\mathbb{T}^2 = S^1 \times S^1$ *(to provide an invariant Borel probability measure)?*

Positive entropy implies the existence of ergodic components of positive area by a theorem of Pesin [120]. It is generally believed that the answers should be positive.

As to the first part of this question, the KAM theorem is clearly a pertinent issue: For small $\lambda > 0$ a large portion of the area of the cylinder is the union of invariant circles. Nevertheless, the complement consists of regions of instability that give rise to positive *topological* entropy due to heteroclinic tangles associated with hyperbolic periodic points. (In higher dimension the invariant tori don't even separate these regions of complicated dynamics.) The horseshoes due to these tangles have zero measure, however, and everything one can prove using estimates of hyperbolicity type is necessarily confined to sets of measure zero. To establish positive measure-theoretic entropy, by contrast, requires control on a set of positive measure. In particular, one must ensure that the invariant circles of all scales do not fill a set of full measure. This set is easily seen to be closed, and it has a Cantor structure. Unfortunately the boundary circles are *not* among those obtained by the KAM theorem (this was apparently first observed by Herman), and they are generally believed to be nonsmooth, which suggests that proving this to be the boundary of the hyperbolic domain will be difficult indeed; no imaginable technique can be expected to serve the purpose.

This illustrates a fundamental problem: Just as Kolmogorov discovered the essential tools for describing complicated dynamics, the KAM theorem established, as illustrated in this essential example, that the applicability of these tools even to mechanical problems faces fundamental limitations.

As to the second part of the question, it is known that near $\lambda = 0.98 \cdot 2\pi$ the last KAM circles disappear, so one might hope for the problem to become tractable. However the elliptic periodic points don't disappear at that stage. There is a plausible scheme to make all elliptic points disappear for certain large parameter values which circumvents the global constraints of index theory by creating orientation-reversed hyperbolic points and is inspired by Jakobson's parameter-exclusion method for 1-dimensional maps[70]. This is aimed at finding parameters for which useful estimates can be carried out.

A "realistic" variant of this problem might be to consider random perturbations of this system. This is not devoid of difficulties, but might be more tractable.

4. Orbit growth in polygonal billiards (Anatole Katok)

Consider the billiard system in a triangle or, more generally, a polygon $P \subset \mathbb{R}^2$. This is an area-preserving dynamical system. The challenge is to understand the global complexity of such a system. For example, let $S(T)$ be the number of orbits of length at most T that begin and end in vertices.

PROBLEM 4.1. *Find upper and lower bounds for $S(T)$.*

QUESTION 4.2. *Is there a periodic orbit for every choice of P?*

This is open even for most obtuse triangles; R. Schwartz has shown however that if the maximal angle in a triangle is less than $100°$ then periodic orbits always exist [129].

PROBLEM 4.3. *Find conditions for ergodicity of the billiard flow with respect to Liouville measure (area). In particular, is the billiard flow ergodic for almost every P?*

Boshernitzan and Katok observed that based on the work of Kerckhoff–Masur–Smillie [88], a Baire category argument produces a dense G_δ of ergodic polygons. Vorobets [134] improved this by giving an explicit sufficient condition for ergodicity in terms of the speed of simultaneous approximation of all angles mod π by rationals. Existence of even a single ergodic example with Diophantine angles remains an open and probably very hard question

Katok [80] showed that $T^{-1} \log S(T) \to 0$, which is far from effective. For *rational* polygons Masur [105] showed that there are positive constants C_1 and C_2 such that $C_1 T^2 \leq S(T) \leq C_2 T^2$. For some examples, (*e.g.*, those leading to *Veech surfaces* [67]) existence of quadratic multiplicative asymptotics has been shown and even the constant calculated.

Any effective subexponential estimate (such as $e^{-T^{3/4}}$, say) for arbitrary polygons would be a major advance.

CONJECTURE 4.4. $\lim_{T \to \infty} S(T)/T^{2+\varepsilon} = 0$ *for every polygon and every* $\varepsilon > 0$, *but* $S(T)/T^2$ *is often irregular and unbounded.*

It should be said that understanding orbit growth in measure-theoretic terms with respect to the Liouville measure is not a difficult matter; one can calculate slow entropy and gets a quadratic growth rate. Indeed, Mañé observed that the number of connections of length up to T between two randomly chosen boundary points is on average quadratic in T, *i.e.*, statistically one sees quadratic orbit growth. Accordingly, any deviation from quadratic orbit growth would be connected to different invariant measures.

The basic problem is the lack of structure here, except for rational polygons where one can represent the problem in terms of a Riemann surface with a quadratic differential and then bring tools of Teichmüller theory to bear. For irrational polygons one could try to associate a Riemann surface of infinite genus in an analogous fashion, but this sacrifices recurrence. There are some border-line cases where one can use recurrence in dynamical systems that preserve an infinite measure.

The basic problem related to this circle of questions is that in these parabolic systems dynamical complexity arises from a combination of stretching and cutting. The stretching is well understood for polygonal billiards, and produces quadratic growth (geometrically a shear), but usually produces no periodic orbits; the interesting phenomena arise from cutting, which is poorly understood beyond the fact that growth is subexponential.

5. Flat surfaces and polygonal billiards (presented by Anton Zorich)

This topic is closely related with the previous one. In fact, by unfolding, rational billiards produce flat surfaces of a special kind. Powerful methods based on the study of the Teichmüller geodesic flow on various strata of quadratic differentials usually are not directly applicable to billiards.

QUESTION 5.1. *Is the geodesic flow on a generic flat sphere with 3 singularities ergodic?*

QUESTION 5.2. *Is there a closed regular geodesic?*

Two copies of a triangle with boundaries identified gives such a space, so this problem is related to polygonal billiards. For some rational triangles the initial direction is preserved and thus provides a first integral.

QUESTION 5.3. *Is there a precise quadratic asymptotic for the growth of closed geodesics on every genus 2 flat surface?*

There is much recent progress (such as the classification by Calta and McMullen of Veech surfaces in $H(2)$ – genus 2 with a single conical singularity [20; 106]) to put this question into reach.

6. Symbolic extensions (Michael Boyle and Sheldon Newhouse)

Briefly, the effort to understand possible symbolic dynamics for a general topological dynamical system leads to the Downarowicz theory of "entropy structure", a master entropy invariant which provides a refined and precise structure for describing the emergence of chaos on refining scales. This leads to problems of the compatibility of entropy structure with varying degrees of smoothness.

In the remainder of this section all homeomorphisms are assumed to have finite topological entropy and to be defined on compact metrizable spaces. If g is the restriction of some full shift on a finite alphabet to a closed shift-invariant subsystem Y, then (Y, g) is said to be a *subshift*.

DEFINITION 6.1. Given a homeomorphism f of a compact metrizable space X with finite topological entropy, a *symbolic extension* of (X, f) is a continuous surjection $\varphi : Y \to X$ such that $f \circ \varphi = \varphi \circ g$ and (Y, g) is a subshift.

Given φ as above, we may also refer to the subshift (Y, g) as a symbolic extension of (X, f). A coding of a hyperbolic dynamical system by a topological Markov shift provides the classical example. In general, the subshift (Y, g) is required to be a subsystem of some full shift on a *finite* set of symbols, but it need not be a Markov shift and its topological entropy (though finite) need not equal that of f.

DEFINITION 6.2. The *topological symbolic extension entropy* of f is

$$\mathbf{h}_{\mathrm{sex}}(f) := \inf\{\mathbf{h}_{\mathrm{top}}(g)\},$$

where the inf is over all symbolic extensions of f. (If there is no symbolic extension of (X, f), then $\mathbf{h}_{\mathrm{sex}}(f) = \infty$.) The *topological residual entropy* of f is $\mathbf{h}_{\mathrm{res}}(f) := \mathbf{h}_{\mathrm{sex}}(f) - \mathbf{h}_{\mathrm{top}}(f)$.

If $\mathbf{h}_{\mathrm{top}}(f) = 0$, then $\mathbf{h}_{\mathrm{sex}}(f) = 0$; otherwise, the residual and topological entropies are independent, as follows.

THEOREM 6.3 ([14; 27]). *If $0 < \alpha < \infty$ and $0 \leq \beta \leq \infty$ then there is a homeomorphism f with $h_{top}(f) = \alpha$ and $h_{res}(f) = \beta$.*

Let $\mathcal{M}_f$ be the space of f-invariant Borel probabilities, and let h denote the entropy function on $\mathcal{M}_f$, $h(\mu) = h(\mu, f)$. The key to a good entropy theory for symbolic extensions [12] is to study the extensions in terms of $\mathcal{M}_f$ (as begun in [27]).

DEFINITION 6.4. Let $\varphi : (Y, S) \to (X, f)$ be a symbolic extension of (X, f). The extension entropy function of φ is the function

$$h_{\mathrm{ext}}^{\varphi} : \mathcal{M}_f \to [0, +\infty)$$

$$\mu \mapsto \sup\{h(\nu, S) : \nu \in \mathcal{M}_g, \varphi\nu = \mu\}.$$

For a given $\mu \in \mathcal{M}_f$ and a given symbolic extension φ, the number $h_{\mathrm{ext}}^{\varphi}(\mu)$ measures the information in the symbolic system used to encode the trajectories in the support of μ. Define the symbolic extension entropy function of f,

$$h_{\mathrm{sex}}^{f} : \mathcal{M}_f \to [0, \infty)$$

$$\mu \mapsto \inf_{\varphi} h_{\mathrm{ext}}^{\varphi}(\mu),$$

where the inf is over all symbolic extensions φ of (X, f). (If there is no symbolic extension of (X, f), define $h_{\text{sex}}^f := \infty$.) The function h_{sex}^f is capturing for all μ in $\mathcal{M}_f$ the lower bound on the information required in any finite encoding of the system (*i.e.*, any symbolic extension of (X, f)) to describe the trajectories supported by μ. This function is a highly refined quantitative reflection of the emergence of chaos (entropy) in the system (X, f), as it reflects "where" the chaos arises (on the supports of which measures) and also "how" (as the scale of resolution on which the system is examined refines to zero). There is a more thorough elaboration of this intuition in [28]; in any case, the final justification of the claim is the full theory of entropy structure [12; 28].

To make this precise we follow the path of [27; 12; 28] and study an allowed sequence of functions $h_n : \mathcal{M}_f \to [0, \infty)$ which increase to h. An allowed sequence determines the *entropy structure* of (X, f). There are many choices of allowed sequence for (h_n), studied in [28]; here is one concrete and crucial (though not completely general) example, which reflects the intuition of "refining scales". Suppose X admits a refining sequence of finite partitions $\mathcal{P}_n$, with diameters of partition elements going to zero uniformly in n, and such that the boundary of P has μ measure zero, for all μ in $\mathcal{M}_f$, for all n, and for all P in $\mathcal{P}_n$. (Such partitions exist for example if X is finite dimensional with zero-dimensional periodic point set [93] or if (X, f) has an infinite minimal factor [101; 99].) Set $h_n(\mu) = h_n(\mu, f, \mathcal{P}_n)$. Then (h_n) defines the entropy structure.

In [12], one general construction of the entropy structure is given, and the collection of all the functions h_{ext}^φ is given a useful functional analytic characterization in terms of the entropy structure. Together with [29], this reduced many problems involving symbolic extensions to problems of pure functional analysis on a metrizable Choquet simplex. For example, it became possible [12] to show the following

- There is a homeomorphism f with $h_{\text{res}}(f) < \infty$, but with no symbolic extension (Y, g) such that $h_{\text{top}}(g) = h_{\text{top}}(f) + h_{\text{res}}(f)$.
- The function h_{sex}^f is upper semicontinuous and its maximum need not be achieved at any ergodic measure.
- The topological symbolic extension entropy of f is the maximum value achieved by its symbolic extension entropy function.

Another outcome was an inductive characterization of the function h_{sex}^f from the given sequence h_n. Define the tail sequence $\tau_n := h - h_n$, which decreases to zero. For ordinals α, β, define recursively

- $u_0 \equiv 0$
- $u_{\alpha+1} = \lim_k \widetilde{(u_\alpha + \tau_k)}$
- $u_\beta =$ the u.s.c. envelope of $\sup\{u_\alpha : \alpha < \beta\}$, if β is a limit ordinal.

With these definitions, there is the following theorem.

THEOREM 6.5 [12; 28]. $u_\alpha = u_{\alpha+1} \iff u_\alpha + h^f = h^f_{\text{sex}}$, and such an α exists among countable ordinals (even if $h_{\text{sex}} \equiv \infty$).

The convergence above can be transfinite [12], and this indicates the subtlety of the emergence of complexity on ever smaller scales. However the characterization is also of practical use for constructing examples.

Downarowicz unified the whole theory with an appropriate notion of equivalence. Following [28], declare two nondecreasing sequences of nonnegative functions (h_n) and (h'_n) to be *uniformly equivalent* if for every integer n and $\varepsilon > 0$ there exists N such that $h_N > h'_n - \varepsilon$ and $h'_N > h_n - \varepsilon$. Now, let (h_n) be a sequence defining the entropy structure in [12] (given by a complicated general construction from [12]). Let (h'_n) be another nondecreasing sequence of nonnegative functions on $\mathcal{M}_f$. Then by definition, (h'_n) also defines the entropy structure if and only if it is uniformly equivalent to the reference sequence (h_n). Thus the entropy structure for a system (X, f) is a certain uniform equivalence class of sequences of functions on $\mathcal{M}_f$. The many approaches to defining entropy lead to many candidate sequences (h_n), and Downarowicz examined them [28]. With few exceptions, the approaches yield sequences in the same uniform equivalence class as the reference sequence (and most of these sequences are considerably more simple to define then the reference sequence). A sequence uniformly equivalent to the reference sequence determines all the same entropy invariants (*e.g.*, the topological entropy, the entropy function on $\mathcal{M}_f$, h^f_{sex}, and the transfinite order of accumulation in Theorem 6.5), by application of the same functional analytic characterizations as apply to derive the invariants from the reference sequence. Because so many sequences lead to the same encompassing collection of entropy invariants, it makes sense to refer to the entire equivalence class of these sequences as the *entropy structure* of the system.

Viewing the entropy structure as fundamental, one asks which structures can occur. At the level of topological dynamics there is a complete answer, due to Downarowicz and Serafin.

THEOREM 6.6 [29]. *The following are equivalent.*

(1) (g_n) *is a nondecreasing sequence of functions on a metrizable Choquet simplex C, beginning with $g_0 \equiv 0$ and converging to a bounded function g, and with $g_{n+1} - g_n$ upper semicontinuous for all n.*

(2) *There is a homeomorphism f of a compact metrizable space, with entropy structure given by a sequence (h_n), such that there exists an affine homeomorphism from $\mathcal{M}_f$ to C which takes (h_n) to a sequence uniformly equivalent to (g_n).*

Given (g_n) above, Downarowicz and Serafin actually construct a model f on the Cantor set such that the affine homeomorphism takes (h_n) to (g_n). Moreover, f can be made minimal.

More generally one asks what entropy structures are compatible with what degrees of smoothness.

QUESTION 6.7. *Let X be a compact Riemannian manifold and $1 \leq r \leq \infty$. What entropy structures are possible for C^r diffeomorphisms on X?*

Precisely, Question 6.7 asks the following: given (g_n) a nondecreasing sequence of nonnegative upper semicontinuous functions on a metrizable Choquet simplex C, and converging to a bounded function g, does there exist a C^r diffeomorphism f on X, with entropy structure given by a sequence (h_n), such that there exists an affine homeomorphism from $\mathcal{M}_f$ to C which takes (h_n) to a sequence uniformly equivalent to (g_n)?

Question 6.7 is more a program for the decades than one problem. We move to particular (still very difficult) problems within this program.

First, we isolate the one good distinguished class in the entropy structure theory: this is the Misiurewicz class of *asymptotically h-expansive* systems [109]. It turns out that (X, f) is asymptotically h-expansive if and only if its entropy structure is given by a sequence (h_n) which converges to h uniformly [28], if and only if it has a principal symbolic extension in the sense of Ledrappier (the factor map preserves the entropy of every invariant measure) [14; 12]. Buzzi showed a C^∞ system (X, f) is *asymptotically h-expansive* [19].

QUESTION 6.8. *Which asymptotically h-expansive entropy structures occur for some C^∞ diffeomorphism on some (or a given) compact manifold X?*

Note that the Newhouse conjecture (Conjecture 7.1) implies severe constraints to realization if X is a surface.

At the 1991 Yale conference for Roy Adler, Boyle presented the first examples of systems with finite entropy but with no symbolic extension (these were constructed in response to a 1988 question of Joe Auslander). This provoked a question from A. Katok.

QUESTION 6.9 (KATOK, 1991). *Are there smooth finite entropy examples with no symbolic extension?*

We have seen that there are no bad C^∞ examples. For lesser smoothness, Downarowicz and Newhouse showed that the situation is quite different.

THEOREM 6.10 [30]. *A generic area-preserving C^1 diffeomorphism of a surface is either Anosov or has no symbolic extension. If $1 < r < \infty$ and $\dim(M) > 1$ then there are residual subsets $\mathfrak{R}$ of open sets in $\mathrm{Diff}^r(M)$ such that $h_{\mathrm{res}}(f) > 0$—and hence f has no principal symbolic extension—for every $f \in \mathfrak{R}$.*

The first result implies that a generic area-preserving C^1 surface diffeomorphism that is not Anosov is not topologically conjugate to any C^∞ diffeomorphism; this includes all diffeomorphisms on surfaces other than $\mathbb{T}^2$. The difficult proof of [30] merges the detailed entropy theory of symbolic extensions with genericity arguments for persistent homoclinic tangencies. Concrete examples of C^r maps $1 \le r < \infty$ with positive residual entropy, based on old examples of Misiurewicz [107; 108], are given in [13]. The most important open problem currently is the following.

QUESTION 6.11. *Suppose f is a C^r diffeomorphism of a compact Riemannian manifold, with $1 < r < \infty$. Is it possible for f to have infinite residual entropy?*

The arguments of [30] led Downarowicz and Newhouse to the following more specific version of this problem.

QUESTION 6.12. *Let M be a compact manifold, $f : M \to M$ a C^r map with $r > 1$. Is it necessarily true that*

$$h_{\text{sex}}(f) \le h_{top}(f) + \frac{\dim M \log \text{Lip}(f)}{r-1}?$$

The right-hand side here is effectively an iterated Yomdin-type defect. Yomdin proved that the defect in upper semicontinuity given by local volume growth is $\dim M \log \text{Lip}(f)/r$. In the constructions one tends to carry out in this field, one iterates the procedure that gives this estimate and divides again by r each time. The right-hand side above is the sum of the resulting geometric series. For $r = 1$ and $r = \infty$ this right-hand side agrees with the known results: C^1-maps may not have a symbolic extension at all, and C^∞ maps have a principal symbolic extension. The question is related to the sense that maps of intermediate regularity should have symbolic extensions, and the entropies of these should not be too much larger than that of a map and by a margin that is less for maps with higher regularity.

7. Measures of maximal entropy (presented by Sheldon Newhouse)

CONJECTURE 7.1 (NEWHOUSE). *Let M be a compact surface and $f : M \to M$ a C^∞ diffeomorphism with $h_{top}(f) > 0$. Then there are at most finitely many measures of maximal entropy.*

Evidence for this conjecture can be found in many places. Franz Hofbauer essentially proved the analogous fact for piecewise monotone maps of the interval.

There are a countable number of homoclinic closures, and all ergodic measures of sufficiently high entropy are supported on these.

The product of an Anosov diffeomorphism of $\mathbb{T}^2$ with the identity on the circle shows that in higher dimension such a claim can only hold with some additional hypotheses.

8. Properties of the measure-theoretic entropy of Sinai–Ruelle–Bowen measures of hyperbolic attractors (contributed by Miaohua Jiang)

Let $\mathrm{Diff}^{1+\alpha}(M)$ be the collection of all $C^{1+\alpha}$-diffeomorphisms on a compact smooth Riemannian manifold M. Assume that a map $f_0 \in \mathrm{Diff}^{1+\alpha}(M)$ is transitive and has a hyperbolic attractor Λ as its nonwandering set. By structural stability, any $g \in \mathrm{Diff}^{1+\alpha}(M)$ in a sufficiently small C^1-neighborhood of f_0 is topologically conjugate to f_0 on the attractor and its nonwandering set is also a hyperbolic attractor. We denote this neighborhood of f_0 by $U_\varepsilon^{C^1}(f_0)$. Let $U(f_0)$ be the collection of those diffeomorphisms in $\mathrm{Diff}^{1+\alpha}(M)$ that can be connected with f_0 by a finite chain of such neighborhoods, *i.e.*,

$$U(f_0) = \left\{ g \mid \exists n \in \mathbb{N} \; \forall i = 1, 2, \ldots, n \; \exists f_i \in \mathrm{Diff}^{1+\alpha}(M), \; \varepsilon_i > 0 : \right.$$
$$\left. g \in U_{\varepsilon_n}^{C^1}(f_n) \text{ and } U_{\varepsilon_{i-1}}^{C^1}(f_{i-1}) \cap U_{\varepsilon_i}^{C^1}(f_i) \neq \varnothing \text{ for } i = 1, 2, \ldots, n \right\}.$$

The set $U(f_0)$ is an open set of $\mathrm{Diff}^{1+\alpha}(M)$. Any map f in $U(f_0)$ possesses a hyperbolic attractor and there exists an SRB measure μ_f for f. Any two maps in $U(f_0)$ are conjugate by a Hölder continuous map that is not necessarily close to the identity. Each map in $U(f_0)$ also has the same topological entropy. However, the measure-theoretic entropy $h_{\mu_f}(f)$ of $f \in U(f_0)$ with respect to its SRB measure μ_f can vary. It was shown by David Ruelle that the dependence of μ_f on the map f is differentiable when the maps are C^4.

QUESTION 8.1. *Is* $\inf_{f \in U(f_0)} h_{\mu_f}(f) = 0$? (Added in proof: Miaohua Jiang, Huyi Hu and Yunping Jiang have answered the question in the affirmative.)

QUESTION 8.2. *Does this functional have a local minimum?*

For expanding maps on the circle, the infimum being zero was confirmed by Mark Pollicott. The problems were raised during conversations between Miaohua Jiang and Dmitry Dolgopyat.

9. Sinai–Ruelle–Bowen measures and natural measures (presented by Michał Misiurewicz)

DEFINITION 9.1. Let X be a compact metric space, $f : X \to X$ a continuous map, $\mathcal{M}$ the space of all probability measures on X and $f_* : \mathcal{M} \to \mathcal{M}$, $(f_*(\mu))\varphi :=$ $\mu(\varphi \circ f)$, where $\mu(\varphi) := \int_X \varphi \, d\mu$ for $\varphi : X \to \mathbb{R}$. Given a "reference" measure

m on X for which $f_*(m) \ll m$ and $A_n(\mu) := \sum_{k=0}^{n-1} f_*^k(\mu)/n$, a Sinai–Ruelle–Bowen measure for f is a measure m_f such that there is an open $U \subset X$ with $m(U) > 0$ such that $\lim_{n \to \infty} A_n(\delta_x) = m_f$ for m-a.e. $x \in U$ [135; 136]. A *natural measure* is a measure m_f for which there is an open set U with $m(U) > 0$ such that $\lim_{n \to \infty} A_n(\mu) = m_f$ for every $\mu \in \mathcal{M}$ with $\mu(U) = 1$ and $\mu \ll m$.

THEOREM 9.2. *A Sinai–Ruelle–Bowen measure is natural.*

PROOF. Integrate $A_n(\delta_x)$ (which tends to m_f as $n \to \infty$) over x with respect to m. □

The converse does not hold.

THEOREM 9.3 [110; 111; 71]. *If g is an expanding algebraic endomorphism or an algebraic Anosov automorphism of a torus $\mathbb{T}^d$ then there exists $f : \mathbb{T}^d \to \mathbb{T}^d$ that is topologically conjugate to g and such that*

(1) $f_*(m) \ll m$ *for* $m = $ *Lebesgue measure,*
(2) $\lim_{n \to \infty} A_n(m) = m_f$,
(3) m_f *has maximal entropy,*
(4) $\{A_n(\delta_x) \mid n \in \mathbb{N}\}$ *is dense in the space of f-invariant Borel probability measures for m-a.e. $x \in \mathbb{T}^d$.*

This particular situation is impossible for smooth (even C^1) f, which motivates

QUESTION 9.4. *With Lebesgue measure as the reference measure, are there smooth dynamical systems for which the ergodic natural measure is not a Sinai–Ruelle–Bowen measure?*

There is an example of piecewise continuous, piecewise smooth interval map for which the natural measure is the average of two delta-measures at fixed points and an SRB measure does not exist [10, page 391].

One could say that in these examples the conjugacy sends Lebesgue measure to one that is completely unrelated to Lebesgue measure.

10. Billiards (Domokos Szasz)

Consider a dispersing billiard on the two-dimensional torus with a finite horizon (*i.e.*, assume that the length of orbit segments between impacts with scatterers is bounded). The Lorentz process is the $\mathbb{Z}^2$-cover of this billiard. (In other words, it is a billiard on $\mathbb{R}^2$ with periodically arranged convex scatterers.) The phase space of the billiard can, of course, be embedded isomorphically into the phase space of the Lorentz process, into its cell 0, say. Assume that the initial phase point of the Lorentz process is selected in cell 0 according to the Liouville measure.

It is known that, for the billiard dynamics, correlations of Hölder functions decay exponentially fast (cf. [136]). As a consequence, for Hölder functions the central limit theorem holds, implying that for the corresponding Lorentz process the typical displacement of orbits increases as the square root of the number of collisions. However, for periodic trajectories of the billiard the displacement is either bounded or is ballistic (*i.e.*, it grows linearly with the number of collisions). According to a construction of [16] (for more details and further references see [132]) there do exist ballistic orbits.

QUESTION 10.1. *How large is the set of ballistic orbits? Could one give a lower bound for its Hausdorff dimension?*

This is a geometric question because it is not a matter of studying typical behavior. When one aims at constructing ballistic orbits different from those arising from periodic ones the problem is that there are "shadows" of the scatterers, which makes this situation different from geodesic flows in negative curvature because it introduces an analog of positive curvature, and there is no good geometric picture here.

11. Stable ergodicity (with Keith Burns)

DEFINITION 11.1. An embedding f is said to be *partially hyperbolic* on Λ if there exists a Riemannian metric for which there are continuous positive functions λ_i, μ_i, $i = 1, 2, 3$ on M such that

$$0 < \lambda_1 \le \mu_1 < \lambda_2 \le \mu_2 < \lambda_3 \le \mu_3 \text{ with } \mu_1 < 1 < \lambda_3$$

and an invariant splitting

$$T_x M = E^s(x) \oplus E^c(x) \oplus E^u(x), \quad d_x f E^\tau(x) = E^\tau(f(x)), \ \tau = s, c, u$$

into pairwise orthogonal subspaces $E^s(x)$, $E^c(x)$ and $E^u(x)$ such that

$$\lambda_1 \le |||d_x f \restriction E^s(x)||| \le \|d_x f \restriction E^s(x)\| \le \mu_1,$$
$$\lambda_2 \le |||d_x f \restriction E^c(x)||| \le \|d_x f \restriction E^c(x)\| \le \mu_2,$$
$$\lambda_3 \le |||d_x f \restriction E^u(x)||| \le \|d_x f \restriction E^u(x)\| \le \mu_3,$$

where $|||A||| := \min\{\|Av\| \mid \|v\| = 1\}$. In this case we set $E^{cs} := E^c \oplus E^s$ and $E^{cu} := E^c \oplus E^u$.

REMARK. Each subbundle E^τ for $\tau = u, s, c, cu, cs$ is Hölder continuous.

Denote the set of C^2 partially hyperbolic diffeomorphisms of a compact manifold M by $\mathrm{PHD}^2(M)$ and the set of volume-preserving such by $\mathrm{PHD}^2_{\mathrm{vol}}(M)$.

CONJECTURE 11.2 (PUGH–SHUB). *The set of diffeomorphisms that are volume-ergodic contains a C^2-dense and C^1-open subset of $\mathrm{PHD}^2_{\mathrm{vol}}(M)$.*

Since the usual method available for establishing ergodicity from hyperbolicity is the Hopf argument, it is natural to consider us-paths, that is, paths obtained by concatenating finitely many segments each of which lies entirely in a stable or an unstable leaf. The property of being joined by a us-path is obviously an equivalence relation on points of M. If there is just one equivalence class, in other words if any two points are joined by a us-path, we say that the diffeomorphism has the *accessibility property*. One also wants to consider this property modulo sets of measure 0, which leads to the *essential accessibility property*, which says that a measurable set which is a union of equivalence classes must have zero or full measure.

This suggests approaching the above conjecture via the two following ones:

CONJECTURE 11.3. *$PHD^2_{\mathrm{vol}}(M)$ and $PHD^2(M)$ contain subsets consisting of diffeomorphisms with the accessibility property that are both C^2-dense and C^1-open.*

CONJECTURE 11.4. *Essential accessibility implies ergodicity in $PHD^2_{\mathrm{vol}}(M)$.*

Pertinent known results are:

THEOREM 11.5 [26]. Conjecture 11.3 *is true if C^2 dense is weakened to C^1 dense.*

THEOREM 11.6 [114; 65; 17]. Conjecture 11.3 *is true for diffeomorphisms with 1-dimensional center.*

Removing the assumption of 1-dimensional center bundle will require substantially new ideas.

Results towards Conjecture 11.4 are the classical ones by Hopf [66], Anosov and Anosov–Sinai [1; 3] as well as those by Grayson, Pugh and Shub [52; 123], Pugh and Shub [124; 125] and the most refined version due to Burns and Wilkinson [18].

DEFINITION 11.7. We say that f is *center-bunched* if $\mu_1 < \lambda_2/\mu_2$ and $\lambda_3^{-1} < \lambda_2/\mu_2$.

This holds automatically whenever the center bundle is 1-dimensional.

THEOREM 11.8 [18]. *An (essentially) accessible, center-bunched, partially hyperbolic diffeomorphism is ergodic (and, in fact, has the K-property).*

With this in mind one can rephrase Conjecture 11.4 as follows:

QUESTION 11.9. *Can one dispense with the center-bunching hypothesis in* Theorem 11.8?

This would require a substantially new insight. The present techniques crucially require center bunching, even though it has been weakened significantly from its earliest formulations.

Maybe a different approach is needed:

QUESTION 11.10. *Can one show that accessibility implies ergodicity of the stable and unstable foliations in the sense that sets saturated by stable leaves and sets saturated by unstable leaves must have either zero or full measure?*

It is not known whether a diffeomorphism that satisfies the hypotheses of Theorem 11.8 must be Bernoulli.

QUESTION 11.11. *Are systems as in* Theorem 11.8 *Bernoulli?*

The answer is expected to be negative, but the known examples of K-systems that are not Bernoulli are not of this type. It may be possible that a study of early smooth examples by Katok may be instructive. They are not partially hyperbolic but might be sufficiently "soft" to be useful here.

12. Mixing in Anosov flows (Michael Field)

Let Λ be a basic set for the Axiom A flow Φ and $\mathcal{P}$ denote the periodic spectrum of $\Phi \upharpoonright \Lambda$ (set of prime periods of periodic orbits). Bowen showed that $\mathcal{P}$ is an invariant of mixing. The analyticity and extension properties of the ζ-function ζ_Φ of Φ are (obviously) determined by $\mathcal{P}$ (for the definition of ζ_Φ we assume the measure of maximal entropy on Λ). In view of the close relation between exponential mixing of Φ and extension properties of ζ_Φ [122], we ask

QUESTION 12.1. *Is the periodic spectrum an invariant of exponential mixing?*

(For conditions on $\mathcal{P}$ related to rapid mixing, see [41, Theorem 1.7].)

Let x be a homoclinic point for the periodic orbit Γ. In [41] a definition is given of 'good asymptotics' for the pair (Γ, x). Without going into detail, the definition involves precise asymptotic estimates for a sequence of periodic orbits which converge to the Φ-orbit of x. Typically, good asymptotics is an open condition in the C^2-topology. If there exists (Γ, x) with good asymptotics then Φ is (rapidly) mixing [41].

DEFINITION 12.2. We say Φ has *very good asymptotics* if every pair (Γ, x) has good asymptotics in the sense of [41].

This is a generic condition on Axiom A flows.

QUESTION 12.3. *Does very good asymptotics imply exponential mixing?*

A weaker (but perhaps more tractable) version of this question is

QUESTION 12.4. *Does very good asymptotics imply analytic extension of the ζ-function?*

A flow Φ is C^r-stably mixing if there exists a C^r-open neighborhood of Φ consisting of mixing flows. It was shown in [41] that if $r \geq 2$ then a C^r-Axiom A flow can be C^r-approximated by a C^2-stably mixing Axiom A flow (in fact, by a C^2-stably rapid mixing flow). If the flow is Anosov or an attractor one may approximate by a C^1-stably mixing flow.

QUESTION 12.5. *If the dimension of the basic set is at least two, can one always approximate by a C^1-stably mixing flow?*

(This is really a question about the local geometry of the basic set. For example, it suffices to know that $W^{uu}(x) \cap \Lambda$ is locally path-connected for all $x \in \Lambda$. This condition is automatically satisfied for attractors.)

For results and background related to the following question see [40].

PROBLEM 12.6. *Suppose the dimension of the basic set is one (suspension of a subshift of finite type). Show that if Φ is C^1-stably mixing, then Φ cannot be C^r, $r > 1$.*

Of course, it is interesting here to find examples where $C^{1+\alpha}$-stable mixing of Φ implies that Φ cannot be more regular than $C^{1+\alpha}$.

Although the results in [41] show that every Anosov flow can be approximated by a C^1-stably mixing Anosov flow, there remains the

CONJECTURE 12.7 (PLANTE [121]). *For transitive Anosov flows, mixing is equivalent to stable mixing.*

As Plante showed, the conjecture amounts to showing that if the strong foliations are integrable then they cannot have dense leaves.

13. The structure of hyperbolic sets (contributed by Todd Fisher)

As stated in Section 16 below, there are a number of fundamental questions about the structure of Anosov diffeomorphisms. It is then not surprising that there are a number of problems concerning the structure of general hyperbolic sets.

A question posed by Bonatti concerns the topology of hyperbolic attractors.

On surfaces a hyperbolic attractor can be either the entire manifold (Anosov case) or a 1-dimensional lamination ("Plykin attractors").

On 3-dimensional manifolds there are many kinds of hyperbolic attractors: Let A be a hyperbolic attractor of a diffeomorphism f on a compact 3-manifold M. The following cases are known to exist.

(1) If the unstable manifold of the points $x \in A$ are bidimensional, then A is either the torus T^3 (Anosov case), or a bidimensional lamination.

(2) If the unstable manifolds of the points $x \in A$ are 1-dimensional, then the attractor can be

   (i) a 1-dimensional lamination which is transversely Cantor ("Williams attractors") or

   (ii) an invariant topological 2-torus $\mathbb{T}^2$, and the restriction of f to this torus is conjugate to an Anosov diffeomorphism (however, the torus $\mathbb{T}^2$ can be fractal with Hausdorff dimension strictly bigger than 2).

QUESTION 13.1. *Is there some other possibility? For example, is it possible to get an attractor such that the transversal structure of the unstable lamination is a Sierpinski carpet?*

THEOREM 13.2 [42]. *If M is a compact surface and Λ is a nontrivial mixing hyperbolic attractor for a diffeomorphism f of M, and Λ is hyperbolic for a diffeomorphism g of M, then Λ is either a nontrivial mixing hyperbolic attractor or a nontrivial mixing hyperbolic repeller for g.*

It is also shown in [42] by counterexample that this does not hold in higher dimensions for general attractors. However, if we add some assumptions on the attractor or weaken the conclusion we have the following problems.

QUESTION 13.3. *Suppose M is a compact smooth boundaryless manifold of dimension n and Λ is a mixing hyperbolic attractor for f with $\dim(\mathbb{E}^u) = n - 1$ and hyperbolic for a diffeomorphism g. Does this imply that Λ is a mixing hyperbolic attractor or repeller for g?*

QUESTION 13.4. *Suppose Λ is a locally maximal hyperbolic set for a diffeomorphism f and hyperbolic for a diffeomorphism g. Does this imply that Λ is locally maximal for g? or that Λ is contained in a locally maximal hyperbolic set for g?*

Related to Question 13.4 we note that in [43] it is shown that on any manifold, of dimension greater than one, there is an open set of diffeomorphisms containing a hyperbolic set that is not contained in a locally maximal one. Furthermore, it is shown if the dimension of the manifold is at least four that there is an open set of diffeomorphisms containing a transitive hyperbolic set that is not contained in a locally maximal one.

QUESTION 13.5. *Suppose M is a compact surface and $\Lambda \subset M$ is a transitive hyperolic set for a diffemorphism f of M. If Λ is transitive, then is Λ contained in a locally maximal hyperbolic set?*

Inspired by Hilbert's famous address in 1900, Arnold requested various mathematicians to provide great problems for the twenty-first century. Smale gave his list in [130]. Smale's Problem 12 deals with the centralizer of a "typical" diffeomorphism. For $f \in \mathrm{Diff}^r(M)$ (the set of C^r diffeomorphisms from M to M) the centralizer of f is

$$C(f) = \{g \in \mathrm{Diff}^r(M) \mid fg = gf\}.$$

Let $r \geq 1$, M be a smooth, connected, compact, boundaryless manifold, and

$$T = \{f \in \mathrm{Diff}^r(M) \mid C(f) \text{ is trivial}\}.$$

Smale asks whether T is dense in $\mathrm{Diff}^r(M)$. Smale also asks if there is a subset of T that is open and dense in $\mathrm{Diff}^r(M)$. Smale states: "I find this problem interesting in that it gives some focus in the dark realm beyond hyperbolicity where even the problems are hard to pose clearly." [130]

Even though Smale states that the problem of studying the centralizer gives focus on nonhyperbolic behavior, unfortunately even the hyperbolic case, in general, remains open. However, a number of people have partial results to Smale's question for Axiom A diffeomorphisms.

Palis and Yoccoz [115] have shown that there is an open and dense set of C^∞ Axiom A diffeomorphisms with the strong transversality property and containing a periodic sink that have a trivial centralizer. Togawa [133] has shown that on any manifold there is a C^1 residual set among C^1 Axiom A diffeomorphisms with a trivial centralizer.

QUESTION 13.6. *For any manifold and any $r \geq 1$ is there an open and dense set U contained in the set of C^r Axiom A diffeomorphisms such that any $f \in U$ has a trivial centralizer.*

14. The dynamics of geodesic flows (presented by Gerhard Knieper)

CONJECTURE 14.1. *For any compact manifold the geodesic flow of a generic Riemannian metric has positive topological entropy.*

This holds for surfaces. Specifically, for tori this is achieved using methods of Hedlund, Birkhoff and others to construct a horseshoe, and for higher genus this is a consequence of the exponential growth forced by entropy. Consequently only the sphere requires substantial work. For the sphere Contreras and Paternain [21] showed this in the C^2-topology (for metrics) using dominated splitting and Knieper and Weiss [92] proved this in the C^∞ topology using global Poincaré sections (pushed from the well-known case of positive curvature using work of Hofer and Wysocki in symplectic topology) and the theory of prime ends as applied by Mather. A consequence (via a theorem of Katok) is that generically there is a horseshoe and hence exponential growth of closed geodesics.

QUESTION 14.2. *Can one make similar statements for Liouville entropy?*

QUESTION 14.3. *Is there a metric of positive curvature whose geodesic flow has positive Liouville entropy?*

The underlying question is whether there is a mechanism for the generation of hyperbolicity on a large set from positive curvature.

If a manifold of nonpositive curvature has rank 1 (*i.e.*, every geodesic is hyperbolic), then the unit tangent bundle splits into two sets that are invariant under the geodesic flow, the regular set, which is open and dense, and the singular set.

QUESTION 14.4. *Does the singular set have zero Liouville measure?*

An affirmative answer would imply ergodicity of the geodesic flow. For analytic metrics on surfaces the singular set is a finite union of closed geodesics. In higher dimensions there are examples where the singular set carries positive topological entropy.

Irreducible nonpositively curved manifolds of higher rank are locally symmetric spaces by the rank rigidity theorem. In this case closed geodesics are equidistributed. For $\varepsilon > 0$ let $P_\varepsilon(M)$ be a maximal set of ε-separated closed geodesics and $P_\varepsilon(t) := \{c \in P_\varepsilon(M) \mid \ell(c) \leq t\}$. By a result of Spatzier there is an $\varepsilon > 0$ such that $\lim_{t\to\infty}(1/t)\log \operatorname{card} P_\varepsilon(t) = h_{\text{top}}(\varphi^t)$. This implies that closed geodesics are equidistributed with respect to the measure of maximal entropy.

QUESTION 14.5. *Can one replace "ε-separated" by "nonhomotopic"?*

This is likely but unknown.

15. Averaging (Yuri Kifer)

The basic idea in averaging is to start from an "ideal" (unperturbed) system

$$\frac{dX(t)}{dt} = 0, \qquad X(0) = 0,$$

$$\frac{dY(t)}{dt} = b(x, Y(t)), \quad Y(0) = y,$$

which gives rise to the flow $\varphi_0^t : \mathbb{R}^d \times M \to \mathbb{R}^d \times M$, $(x, y) \mapsto (x, F_x^t(y))$. Integrable Hamiltonian systems are of this type. One then perturbs the system by adding a slow motion in the first coordinate:

$$\frac{dX^\varepsilon(t)}{dt} = \varepsilon B(X^\varepsilon(t), Y^\varepsilon(t)), \quad X^\varepsilon(0) = x,$$

$$\frac{dY^\varepsilon(t)}{dt} = b(X^\varepsilon(t), Y^\varepsilon(t)), \qquad Y^\varepsilon(0) = y.$$

We write $X^\varepsilon = X^\varepsilon_{x,y}$ and $Y^\varepsilon = Y^\varepsilon_{x,y}$. This results in a flow

$$\varphi^t_\varepsilon(x, y) = (X^\varepsilon_{x,y}(t), Y^\varepsilon_{x,y}(t)),$$

with X representing the slow motion. The question is whether on a time scale of t/ε the slow motion can be approximated by solving the averaged equation where B is replaced by $\bar{B}$ which is obtained from the former by averaging it along the fast motion (see [89] and [90]).

$$\frac{\bar{X}^\varepsilon(t)}{dt} = \varepsilon\bar{B}(X^\varepsilon(t)) \tag{15–1}$$

In discrete time the "ideal" unperturbed system is of the form

$$\varphi_0(x, y) = (x, F_x(y)),$$

with $x \in \mathbb{R}^d$, $F_x : M \to M$. The perturbed system is

$$\varphi_\varepsilon(x, y) = (x + \varepsilon\Psi(x, y), F_x(y)),$$

and we can bring this into a form analogous to the one for continuous time by writing it as difference equations:

$$X^\varepsilon(n + 1) = X^\varepsilon(n) + \varepsilon\Psi(X^\varepsilon(n), Y^\varepsilon(n)),$$
$$Y^\varepsilon(n + 1) = F_{X^\varepsilon(n)}(Y^\varepsilon(n)).$$

It is natural to rescale time to t/ε, and a basic averaging result (Artstein and Vigodner [4]) is that (in a sense of differential inclusion) any limit point of

$$\{X^\varepsilon(t/\varepsilon) \mid \varepsilon > 0, \ t \in [0, T]\}$$

is a solution of

$$\frac{dZ^0(t)}{dt} = B_{\mu_z}(Z^0(t))$$

with $Z^0(0) = x$, where μ_z is F_z-invariant and $B_\mu(x) := \int B(x, y) \, d\mu(y)$. Heuristically one instead uses the following averaging principle. Suppose the limit $\bar{B}(x) := \lim_{t\to\infty} \int_0^t B(x, F^s_x(y)) \, ds/t$ exists for "most" (x, y) and "almost" does not depend on y. Then try to approximate $X^\varepsilon(t)$ in some sense over time intervals of order $1/\varepsilon$ by the averaged motion $\bar{X}$ given by (15–1) with $\bar{X}^\varepsilon(0) = x$. (This goes back to Clairaut, Lagrange, Laplace, Fatou, Krylov–Bogolyubov, Anosov, Arnold, Neishtadt, Kasuga and others, and there are also stochastic versions.) One then would like to know whether

$$\lim_{\varepsilon\to 0} \sup_{0\le t\le T/\varepsilon} |X^\varepsilon(t) - \bar{X}^\varepsilon(t)| = 0,$$

and in which sense this happens. Next one can inquire about the error

$$X^\varepsilon(t) - \bar{X}^\varepsilon(t).$$

New results assume that the fast motion is chaotic, typically the second factor is assumed to be hyperbolic. One can then hope for approximation in measure, and there are theorems to that effect in 3 main cases: The fast motion is independent of the slow one (this is the easy case, and one gets a.e.-convergence), when the fast motion preserves a smooth measure that is ergodic for a.e. x (this is due to Anosov and covers the Hamiltonian situation), and much more recently, when the fast motion is an Axiom-A flow (depending C^2 on x as a parameter) in a neighborhood of an attractor endowed with Sinai–Ruelle–Bowen measure.

One may in the latter case ask whether there is a.e.-convergence rather than in measure. For instance, this is not true in general in the case of perturbations of integrable Hamiltonian systems. In the latter case we may have no convergence for any fixed initial condition from a large open set (see Neishtadt's example in [90]). This is related to the question of whether there are resonances and whether these affect convergence and how. To make this concrete, consider the discrete-time system

$$X^\varepsilon(n+1) = X^\varepsilon(n) + \varepsilon \sin 2\pi Y^\varepsilon(n),$$

$$Y^\varepsilon(n+1) = 2Y^\varepsilon(n) + X^\varepsilon(n) \pmod 1.$$

QUESTION 15.1. *Is it the case that*

$$\lim_{\varepsilon \to 0} \sup_{0 \le n \le T/\varepsilon} |X^\varepsilon(n) - X^\varepsilon(0)| = 0$$

for Lebesgue-almost every (x, y)? (Added in proof: Kifer and Bakhtin now seem to have a negative answer to this.)

It is known by a large-deviations argument (see [90]) that given $\delta > 0$ the measure of the set of (x, y) for which $\sup_{0 \le n \le T/\varepsilon} |X^\varepsilon(n) - X^\varepsilon(0)| > \delta$ is at most $e^{-C/\varepsilon}$ for some $C > 0$.

When one considers the rescaled averaged motion $\bar{Z}(t) = \bar{X}^\varepsilon(t/\varepsilon)$ (averaged with respect to Sinai–Ruelle–Bowen measure) an *adiabatic invariant* is an invariant function, *i.e.*, a function H such that $H(\bar{Z}(t)) = H(\bar{Z}(0))$.

CONJECTURE 15.2. *On* (M, vol), $H(X^\varepsilon(t/\varepsilon^2))$ *converges weakly to a diffusion, assuming that the fast motion is hyperbolic.*

It may be simpler to start with expanding fast motions.

16. Classifying Anosov diffeomorphisms and actions (presented by Anatole Katok and Ralf Spatzier)

The classical question on which these questions are based is whether one can classify all Anosov diffeomorphisms. This has been done up to topological conjugacy on tori and nilmanifolds and for codimension-1 Anosov diffeomorphisms [48; 102; 113]. The central ingredient is the fundamental observation by Franks that if an Anosov diffeomorphism on a torus acts on the fundamental group in the same way as a hyperbolic automorphism then there is a conjugacy. Manning proved that any Anosov map of the torus is indeed of this type and extended the result to nilmanifolds.

QUESTION 16.1. *Is every Anosov diffeomorphism of a compact manifold M topologically conjugate to a finite factor of an automorphism of a nilmanifold N/Γ?*

If there are indeed other examples, then there is currently a lack of imagination regarding the possibilities for Anosov diffeomorphisms. In the framework of the proofs mentioned above the central assumption is that the universal cover is $\mathbb{R}^n$ and the map is globally a product, but there is no a priori reason that this should be so. Indeed, for Anosov flows the situation is quite different, and there are many unconventional examples of these, beginning with one due to Franks and Williams that is not transitive [49].

In higher rank there are plausible mechanisms to rule out topologically exotic discrete Anosov actions.

Perturbation of periodic points shows that one cannot expect better than topological classification of Anosov diffeomorphisms, and an example by Farrell and Jones suggest a different reason: There is an Anosov diffeomorphism on an exotic torus.

One may ask about characterizations of algebraic Anosov actions up to C^∞ conjugacy.

THEOREM 16.2 [8]. *An Anosov diffeomorphism with C^∞ Anosov splitting that preserves an affine connection (e.g., is symplectic) is C^∞ conjugate to an algebraic one.*

CONJECTURE 16.3. *Instead of preservation of an affine connection this can be done assuming preservation of some sensible higher-order geometric structure, e.g., a Gromov-rigid structure.*

QUESTION 16.4. *Does preservation of an affine connection alone suffice?*

QUESTION 16.5. *Does smooth splitting alone suffice?*

This might be possible. A natural approach would be to construct invariant structures on the stable and unstable foliations and glue these together to a global invariant structure. The problem is that in some of the standard nilpotent examples the natural structure is not of the type one gets this way.

A different and more recent result is the following:

THEOREM 16.6 [76; 32]. *A uniformly quasiconformal Anosov diffeomorphism is C^∞ conjugate to an algebraic one.*

CONJECTURE 16.7 (KATOK). *An Anosov diffeomorphism whose measure of maximal entropy is smooth is smoothly conjugate to an algebraic one.*

Anosov flows are much more flexible, there are many examples that make a classification seem unlikely [49; 59]. There are some characterizations of algebraic flows. These provide some results similar to the above, as well as analogous problems.

The situation is rather different for algebraic actions so long as they are irreducible (*e.g.*, not products of Ansov diffeomorphisms). These actions are usually hard to perturb. Katok and Spatzier showed that such actions with semisimple linear part are rigid [85], and Damjanovic and Katok pushed this to partially hyperbolic actions on tori [23].

CONJECTURE 16.8. *Any C^∞ Anosov $\mathbb{R}^k$- (or $\mathbb{Z}^k$-) action for $k \geq 2$ on a compact manifold without rank 1 factors is algebraic.*

No counterexamples are known even with lower regularity than C^∞.

We next quote a pertinent result by Federico Rodríguez Hertz:

THEOREM 16.9 [64]. *A $\mathbb{Z}^2$ action on $\mathbb{T}^3$ with at least one Anosov element and whose induced action on homology has only real eigenvalues (one less than 1 and 2 bigger than 1) is C^∞ conjugate to an algebraic one.*

In fact, more generally, let Γ be a subgroup of $\mathrm{GL}(N, \mathbb{Z})$, the group of $N \times N$ matrices with integer entries and determinant ± 1, and say that the standard action of Γ on $\mathbb{T}^N$ is globally rigid if any Anosov action of Γ on $\mathbb{T}^N$ which induces the standard action in homology is smoothly conjugate to it.

THEOREM 16.10 [64]. *Let $A \in GL(N, \mathbb{Z})$, be a matrix whose characteristic polynomial is irreducible over $\mathbb{Z}$. Assume also that the centralizer $Z(A)$ of A in $GL(N, \mathbb{Z})$ has rank at least 2. Then the associated action of any finite index subgroup of $Z(A)$ on $\mathbb{T}^N$ is globally rigid.*

The assumption on the rank of the centralizer is hardly restrictive. Due to the Dirichlet unit theorem, in the above case, $Z(A)$ is a finite extension of $\mathbb{Z}^{r+c-1}$ where r is the number of real eigenvalues and c is the number of pairs of complex eigenvalues. So, $r + 2c = N$, and $Z(A)$ has rank 1 only if $N = 2$ or if $N = 3$ and A has a complex eigenvalue or if $N = 4$ and A has only complex eigenvalues.

In [64] Hertz also states:

QUESTION 16.11. *Consider $N \geq 3$, $A \in GL(N, \mathbb{Z})$ such that $Z(A)$ is "big enough". Under which assumptions is the standard action of every finite-index subgroup Γ on $\mathbb{T}^N$ globally rigid?*

Another rigidity result is due to Kalinin and Spatzier:

THEOREM 16.12 [77]. *If M is a compact manifold with a Cartan $\mathbb{R}^k$-action such that $k \geq 3$, there is a dense set of Anosov elements and every 1-parameter subgroup is topologically transitive (hence there are no rank-1 factors) then this action is C^∞ conjugate to an algebraic one, indeed a homogeneous one (i.e., the left action of $\mathbb{R}^k$ embedded in a group G on G/Γ for a cocompact discrete subgroup Γ).*

(Homogeneous Anosov actions of this type are not classified because it is not known how to classify suspensions of Anosov $\mathbb{Z}^k$-actions on nilmanifolds.)

QUESTION 16.13. *Does the Kalinin–Spatzier result hold for $k = 2$?*

QUESTION 16.14. *Does the Kalinin–Spatzier result hold assuming only the existence of an Anosov element?*

QUESTION 16.15. *Does the Kalinin–Spatzier result hold without the transitivity assumption (and maybe even without excluding rank-1 factors)?*

17. Invariant measures for hyperbolic actions of higher-rank abelian groups (Anatole Katok)

The basic example is Furstenberg's "×2 × 3"-example of the $\mathbb{N}^2$-action on S^1 generated by $E_i : x \mapsto ix$ (mod 1) for $i = 1, 2$. For a single of these transformations there are plenty of invariant measures, but for $\{E_i \mid i \in \mathbb{N}\}$ the only jointly invariant measures are easily seen to be Lebesgue measure and the Dirac mass at 0. The same holds if one takes a polynomial $P(\cdot)$ with integer coefficients and considers $\{E_{P(n)} \mid n \in \mathbb{N}\}$. Furstenberg asked whether Lebesgue measure is the only nonatomic invariant Borel probability measure for E_2 and E_3.

The second example is $M = \mathrm{SL}(n, \mathbb{R})/\Gamma$ for $n \geq 3$ and a lattice $\Gamma \subset \mathrm{SL}(n, \mathbb{R})$. The *Weyl chamber flow* is the action of the set D of positive diagonal elements on M by left translation. (D is isomorphic to $\mathbb{R}^{n-1}$).

PROBLEM 17.1. *Find all invariant measures for these two examples.*

Rudolph [127] showed in 1990 that a measure invariant under both E_2 and E_3 for which one of E_2 and E_3 has positive entropy is Lebesgue measure. Geometric methods which form the basis of most of the work up to now were introduced in [87]. In 2003 Einsiedler, Katok and Lindenstrauss [31] proved the analogous result for the Weyl chamber flow (assuming positive entropy for one element of the action). See[100] for a survey of the this rapidly developing subject at a recent (but not present) date.

Fundamentally the issues for the case of positive entropy are reasonably well understood although (possibly formidable) technical problems remain. However, even simple questions remain in the general case. Here is an example.

QUESTION. 17.2. *Given an Anosov diffeomorphism and a generic ergodic invariant measure* (i.e., *neither Lebesgue measure nor an atomic one), is there a diffeomorphism that preserves this measure and that is not a power of the Anosov diffeomorphism itself?*

Indeed, the zero-entropy case is entirely open, and experts differ on the expected outcome. One can take a geometric or Fourier-analytic approach. The difficulty with the latter one is that even though one has a natural dual available, measures don't behave well with respect to passing to the dual. At the level of invariant distributions there is little difference between rank 1 and higher rank whereas the wealth of invariant measures is quite different between these two situations. The geometric approach produced the results for positive entropy, but in the zero-entropy situation conditional measures on stable and unstable leaves are atomic.

Thus, we either lack the imagination to come up with novel invariant measures or the structure to rule these out.

As to the reason for concentrating on abelian actions, this simply provides the first test case for understanding *hyperbolic* actions in this respect. Recently, substantial progress was achieved beyond the algebraic or uniformly hyperbolic cases [75; 84]. The paradigm here is that positive entropy hyperbolic invariant measures are forced to be absolutely continuous if the rank of the action is sufficiently high, *e.g.*, for $\mathbb{Z}^k$ actions on $k + 1$-dimensional manifolds for $k \geq 2$.

For *unipotent* actions, by contrast, the Ratner rigidity theory is fairly comprehensive, but here the paradigm is in essence unique ergodicity, which is quite different from the hyperbolic situation.

18. Rigidity of higher-rank abelian actions (presented by Danijela Damjanovic)

Consider actions of $A = \mathbb{Z}^k$ or $A = \mathbb{R}^k$ on a compact manifold, where k is at least 2. One class of such actions consists of $\mathbb{Z}^k$-actions on $\mathbb{T}^N$ by toral automorphisms; we say that these are *genuinely of higher rank* if there is a subgroup isomorphic to $\mathbb{Z}^2$ that acts by ergodic automorphisms. Another class consists of the action by the diagonal $A = \mathbb{R}^{n-1}$ on $M = \mathrm{SL}(n, \mathbb{R})/\Gamma$, or actions by a generic hyperplane $\mathbb{R}^d$ for $2 \leq d < n - 1$. This is a partially hyperbolic action whose neutral direction is the neutral direction for the full Cartan. For the first of these cases Katok and Damjanovic have proved rigidity [24], and this raises the following

QUESTION 18.1. *Can the KAM-methods of Damjanovic and Katok be used to establish rigidity in actions of the second type?*

The second situation provides much more geometric structure than the first one, and this can be put to use. Methods of Katok and Spatzier [86] reduce the problem to studying perturbations in the neutral direction, each of which is given by a cocycle over the perturbed action. Therefore the same objective can be achieved by answering the following:

QUESTION 18.2. *Can one show cocycle rigidity for the perturbed actions?*

Katok and Kononenko [83] have established Hölder cocycle stability for partially hyperbolic diffeomorphisms with the accessibility property that could be put to use here. Progress has been achieved recently by first introducing a new method for proving cocycle rigidity [23] which uses results and methods from algebraic K-theory and then developing this method further so that it applies to cocycle over perturbations and hence produces local rigidity (up to small standard perturbations within the full Cartan subgroup) for the actions of the second kind [25].

19. Local rigidity of actions (presented by David Fisher)

For more background, references, more information on the questions raised in this section as well as other interesting questions in this area, the reader should refer to the survey by Fisher in this volume.

DEFINITION 19.1. A homomorphism $i : \Gamma \to D$ from a finitely generated group Γ to a topological group D is said to be *locally rigid* if any i' sufficiently close to i in the compact-open topology is conjugate to i by a small element of D.

This is the case for the inclusion of an irreducible cocompact lattice in a semisimple Lie group with no compact or 3-dimensional factors (Calabi–Vesentini, Selberg, Weil).

A basic question posed by Zimmer around 1985 is whether one can do anything of interest if the topological group is the diffeomorphism group of a compact manifold and Γ is a lattice in a group G that has no rank-1 factors (for example, $G = SL(n, \mathbb{R})$, $G = SL(n, \mathbb{Z})$ for $n \geq 3$). (Katok and collaborators have studied $\Gamma = \mathbb{Z}^d$ for $d \geq 2$ with this in view.) Benveniste showed that every isometric action of a cocompact group is locally rigid in $\mathrm{Diff}^\infty(M)$, and shortly thereafter Margulis and Fisher showed that any isometric action of a group Γ is locally rigid if the group has proprty (T) of Kazhdan, *i.e.,* $H^1(\Gamma, \pi) = 0$ for every unitary representation π.

A theorem of Kazhdan asserts that there are cocompact lattices in $SU(1, n)$ that admit nontrivial homomorphisms to $\mathbb{Z}$, and as an easy consequence any action of these with connected centralizer has deformations. These deformations are not very interesting, so one can ask.

QUESTION 19.2. *Are these the only deformations?*

There are cocompact lattices in $SO(1, n)$ with embeddings in $SO(n + 1)$, where the resulting action on S^n has an infinite-dimensional deformation space. The known deformations do not preserve volume:

QUESTION 19.3. *Are there volume-preserving deformations in this situation?*

If Γ is an irreducible lattice in a semisimple Lie group G that has no compact factors and is contained in a Lie group H with a cocompact lattice Λ then Γ acts on H/Λ. In many cases these actions are not locally rigid, such as for lattices in $SU(1, n)$ and in $SO(1, n)$.

CONJECTURE 19.4. *An action of a lattice Γ is locally rigid if there is a map from G onto a group G_1 that is locally isomorphic to either $SO(1, n)$ or $SU(1, n)$ and acts on a space X in such a way that there is a factorization of $H\Lambda$ to X that intertwines the actions of Γ on H/Λ and X (the latter induced by $\Gamma \subset G_1$.*

This is even open for Anosov actions. A simple linear example of these would be the natural action of $SL(2, \mathbb{Z}\sqrt{2})$ on $\mathbb{T}^4$ obtained by the linear action on $\mathbb{R}^4$ of the 2 Galois-conjugate embeddings, which have an invariant lattice. In the case of Anosov actions this should be an approachable question. Crossing the preceding example with the identity on a circle should be much harder.

20. Smooth and geometric rigidity

CONJECTURE 20.1. *A compact negatively curved Riemannian manifold with $C^{1+\text{zygmund}}$ horospheric foliations is locally symmetric.*

It is believed that smooth rigidity of systems with smooth invariant foliations should hold with low regularity. Yet this remains an open issue. There is some evidence that this is a hard question. For example, investigations of the Anosov obstruction to C^2 foliations [61] made clear that its vanishing does not have immediate helpful consequences. And the basic bootstrap [60] does not start at C^2.

The invariant subbundles E^u and E^s, called the *unstable* and *stable* bundles, are always *Hölder continuous*. For E^u this means that there exist $0 < \alpha \leq 1$ and $C, \delta > 0$ such that $d_G(E^u(p), E^u(q)) \leq C d_M(p, q)^\alpha$ whenever $d_M(p, q) \leq \delta$, where d_G is an appropriate metric on subbundles of TM. We say that E^u is C^α or α-Hölder; in case $\alpha = 1$ we say E^u is C^{Lip} or *Lipschitz continuous*. A continuous function $f : U \to \mathbb{R}$ on an open set $U \subset \mathbb{R}$ is said to be *Zygmund-regular* if there is $K > 0$ such that $|f(x + h) + f(x - h) - 2f(x)| \leq K|h|$ for all $x \in U$ and sufficiently small h. To specify a value of K we may refer to a function as being K-Zygmund. The function is said to be *"little Zygmund"* or C^{zygmund} if $|f(x + h) + f(x - h) - 2f(x)| = o(|h|)$. Zygmund regularity implies modulus of continuity $O(|x \log |x||)$ and hence H-Hölder continuity for all $H < 1$ [139, Theorem (3·4)]. It follows from Lipschitz continuity and hence from differentiability. Being "little Zygmund" implies having modulus of continuity $o(|x \log |x||)$. For $r \in \mathbb{N}$ denote by $C^{r,\omega}$ the space of maps whose r-th derivatives have modulus of continuity ω. For $r > 0$ let $C^r = C^{\lfloor r \rfloor, O(x^{r - \lfloor r \rfloor})}$.

For the dependence of the leaves on the base point several slightly different definitions are possible. The canonical definition is via the highest possible regularity of lamination charts. One may also look into the transverse regularity of k-jets. Alternatively, one can examine the holonomy semigroup, *i.e.*, for pairs of nearby smooth transversals to the lamination one considers the locally defined map between them that is obtained by "following the leaves". By transversality this is well-defined, and for smooth transversals one can discuss the regularity of these maps, which turns out to be largely independent of the transversals chosen. We adopt this notion here and refer to it as the regularity of holonomies or (transverse) regularity of the lamination. There is little difference between these definitions in our context. Following the discussion in [126] we can summarize the relation as follows:

THEOREM 20.2 ([126, Theorem 6.1]). *If $r \in \mathbb{R} \cup \{\infty\}$, $r \notin \mathbb{N} \setminus \{1\}$ then a foliation with uniformly C^r leaves and holonomies has C^r foliation charts.*

However, if $r \in \mathbb{N} \setminus \{1\}$, a foliation with uniformly C^r leaves and holonomies need not have C^r foliation charts. The problem are mixed partials. Without assuming uniform regularity the above statements can fail drastically: There is a foliation with uniformly C^∞ leaves and with (nonuniformly) C^∞ holonomies that does not have a C^1 foliation chart [126, Figure 9]. In our context the regularity is always uniform, so the above result implies that one can define regularity equally well via holonomies or foliation charts. The essential ingredient for Theorem 20.2 is

THEOREM 20.3 ([74]). *Let M be a C^∞ manifold, F^u, F^s continuous transverse foliations with uniformly smooth leaves, $n \in \mathbb{N}_0$, $\alpha > 0$, $f : M \to \mathbb{R}$ uniformly $C^{n+\alpha}$ on leaves of F^u and F^s. Then f is $C^{n+\alpha}$.*

This leads to the following observation.

THEOREM 20.4. *If $r \in \mathbb{R} \cup \{\infty\}$, $r \notin \mathbb{N} \setminus \{1\}$ and the stable and unstable foliations have uniformly C^r holonomies, then there are C^r bifoliation charts, i.e., charts that straighten both foliations simultaneously.*

PROOF. By hypothesis every point p has a neighborhood U on which the inverse $[x, y] \mapsto (x, y) \in W_\varepsilon^u(p) \times W_\varepsilon^s(p)$ of the local product structure map is uniformly C^r in either entry. By Theorem 20.3 it is C^r. □

There is a connection between the regularity of the subbundles and that of the lamination: For any $r \in \mathbb{N} \cup \{\infty\}$ and $\alpha \in [0, 1)$ or "$\alpha = $Lip" a foliation tangent to a $C^{r+\alpha}$ subbundle is itself $C^{r+\alpha}$ [126, Table 1]. (The reverse implication holds only for $r = \infty$ because leaves tangent to a C^r subbundle are C^{r+1}.)

The invariant subbundles are always Hölder continuous. It should be noted, however, that for $\alpha < 1$ the α-Hölder condition on subbundles does not imply any regularity of the foliations. Indeed, without a Lipschitz condition even a one-dimensional subbundle may not be uniquely integrable, so already continuity of the foliation cannot be obtained this way. On the other hand, there turns out to be a converse connection:

THEOREM 20.5 [63]. *If the holonomies are α-Hölder and individual leaves are C^∞ then the subbundles are β-Hölder for every $\beta < \alpha$.*

There are variants of this for leaves of finite smoothness and almost-everywhere Hölder conditions. Furthermore, whenever bunching-type information gives a particular degree of regularity for the subbundles, one can usually get the same regularity for the holonomies, and vice versa.

CONJECTURE 20.6. *If both invariant foliations of an Anosov system are C^2 then they are both C^∞.*

Bolder variants of this would replace C^2 by $C^{1+\text{Lip}}$, $C^{1+\text{BV}}$, $C^{1+\text{zygmund}}$ ("little Zygmund") or $C^{1+o(x|\log x|)}$, but the C^2 version would be spectacular enough, even in the symplectic case.

Note that such rigidity results can only be expected assuming high regularity of both foliations simultaneously because [62] gives a sufficient condition for one foliation to be C^2 that holds for an open set of dynamical systems.

To prove such results it may be necessary to restrict to the geometric context, where there are extra ingredients that might help. The leaves are spheres, and they are "tied together" by the sphere at infinity (ideal boundary) of the universal cover. An important result by Hamenstädt [58] should help substantially as well:

THEOREM 20.7. *If the horospheric foliations are C^2 then the topological and Liouville entropies of the geodesic flow coincide.*

If the Katok entropy conjecture were known this would finish the problem.

Thus, the following problem remains: By exploiting geometric information show that if the horospheric foliations are C^2 and the topological and Liouville entropies of the geodesic flow coincide then the horospheric foliations are C^∞

(or C^k for sufficiently large k to invoke the bootstrap [60]). This leads to a geometric counterpart of Conjecture 20.6.

CONJECTURE 20.8. *A compact negatively curved Riemannian manifold with C^2 horospheric foliations is locally symmetric.*

By [7; 9] this follows from Conjecture 20.6. A related proposal is the following: Give an alternate proof of the Hamenstädt result by showing that if the horospheric foliations are C^2 then the Jacobian cocycle is cohomologous to a constant (which implies coincidence of Bowen–Margulis measure and Liouville measure, *i.e.*, coincidence of topological and Liouville entropy). The reason that this route is interesting to explore is that it provides a motivation to return to the Anosov cocycle and investigate whether it is at all connected with the Jacobian cocycle in subtle ways.

As noted above, smoothness of invariant structures associated with a hyperbolic dynamical system is necessary for smooth conjugacy to an algebraic model. There are several important instances where such conditions are sufficient.

Smoothness of the invariant foliations of a hyperbolic dynamical system has turned out to be sufficient for smooth conjugacy to an algebraic model in the symplectic case. For geodesic flows even more can be said. Open questions concern the precise amount of smoothness needed and possible conclusions in the absence of symplecticity.

Smoothness of the invariant foliations. The most basic result in this direction is implicit: The proof by Avez [5] that an area-preserving Anosov diffeomorphism of $\mathbb{T}^2$ is topologically conjugate to an automorphism actually gives a conjugacy as smooth as the invariant foliations. The definitive result in this setting is worth giving here, because it is suggestive of the work yet to be done in higher dimension.

THEOREM 20.9 [68]. *Let f be a C^∞ area-preserving Anosov diffeomorphism of $\mathbb{T}^2$. Then the invariant subbundles are differentiable and their first derivatives satisfy the Zygmund condition [139, Section II.3, (3·1)] and hence have modulus of continuity $O(x|\log x|)$ [139, Theorem (3·4)]. There is a cocycle, the Anosov cocycle, which is a coboundary if and only if these derivatives have modulus of continuity $o(x|\log x|)$ or, equivalently, satisfy a "little Zygmund" condition. In this case, or if the derivatives have bounded variation [54], the invariant foliations are C^∞ and f is C^∞ conjugate to an automorphism.*

Note the sharp divide between the general and the smoothly rigid situation. Indeed, the constant defining $O(x|\log x|)$ is nonzero a.e., except when the Anosov cocycle is trivial. Therefore this is the finest possible dichotomy.

To obtain C^∞ foliations it is actually shown first that triviality of the Anosov cocycle implies C^3 subbundles, and a separate argument then yields C^∞ foliations.

Following Guysinsky one can explain the Anosov cocycle using local normal forms. For a smooth area-preserving Anosov diffeomorphism on $\mathbb{T}^2$ deLatte [94] showed that one can find local smooth coordinate systems around each point that depend continuously (actually C^1) on the point and bring the diffeomorphism f into the *Moser normal form* [112]

$$f(x, y) = \begin{pmatrix} \lambda_p^{-1} x / \varphi_p(xy) \\ \lambda_p y \varphi_p(xy) \end{pmatrix},$$

where (x, y) are in local coordinates around a point p and the expression on the right is in coordinates around $f(p)$. The terms involving φ_p that depend on the product xy correspond to the natural resonance $\lambda_p \lambda_p^{-1} = 1$ that arises from area-preservation (actually from the family of resonances $\lambda_p = \lambda_p^{n+1} \lambda_p^{-n}$). The function φ_p is as smooth as f, and $\varphi_p(0) = 1$. Now we suppress the (continuous) dependence of λ and φ on p. Note that for a point $(0, y)$ we have

$$Df = \begin{pmatrix} \lambda^{-1}xy(1/\varphi)'(xy) + \lambda^{-1}/\varphi(xy) & \lambda^{-1}x^2(1/\varphi)'(xy) \\ \lambda y^2 \varphi'(xy) & \lambda xy \varphi'(xy) + \lambda \varphi(xy) \end{pmatrix}$$
$$= \begin{pmatrix} \lambda^{-1} & 0 \\ \lambda y^2 \varphi'(0) & \lambda \end{pmatrix}.$$

In these local coordinates the unstable direction at a point $(0, y)$ on the stable leaf of p is spanned by a vector $(1, a(y))$. Since this subbundle is invariant under Df and since $f(0, y) = (0, \lambda y)$, the coordinate representation of Df from above gives $a(\lambda y) = \lambda^2 y^2 \varphi'(0) + \lambda^2 a(y)$. If the unstable subbundle is C^2 then differentating this relation twice with respect to x at 0 gives $\lambda^2 a''(0) = 2\lambda^2 \varphi'(0) + \lambda^2 a''(0)$, *i.e.*, $\varphi'(0) = 0$. This means that the Anosov obstruction is $\varphi'(0)$, where φ arises from the nonstationary Moser-deLatte normal form. (Thus this is also the obstruction to C^1 linearization.)

Hurder and Katok verify that $A(p) := \varphi_p'(0)$ is a cocycle and show that it is nonzero a.e. unless it is null-cohomologous. (Guysinsky's result that $C^{1+BV} \Rightarrow C^\infty$ follows because bounded variation implies differentiability almost everywhere.)

The work by Hurder and Katok is actually carried out for the weak subbundles of volume-preserving Anosov flows on three-manifolds. In this situation analogous issues arise relative to the strong subbundles. These can be worked out with closely related techniques:

THEOREM 20.10 [46]. *Let M be a 3-manifold, $\varphi : \mathbb{R} \times M \to M$ a C^k volume-preserving Anosov flow. Then $E^u \oplus E^s$ is Zygmund-regular, and there is an* *obstruction to higher regularity that can be described geometrically as the curvature of the image of a transversal under a return map. This obstruction defines the cohomology class of a cocycle, and the following are equivalent:*

(1) *The longitudinal KAM-cocycle is a coboundary.*
(2) $E^u \oplus E^s$ *is "little Zygmund".*
(3) $E^u \oplus E^s$ *is Lipschitz.*
(4) $E^u \oplus E^s \in C^{k-1}$.
(5) φ *is a suspension or contact flow.*

PROBLEM 20.11. *Extend this result to higher dimension.*

The complications in higher dimension are due in large part to the simple fact that when the invariant foliations are not one-dimensional there may be different contraction and expansion rates at any given point. Therefore a first step in working on this problem would be to assume uniform quasiconformality in stable and unstable directions. This has strong structural implications in itself, though (Theorem 16.6, [32; 76; 128]).

Different contraction and expansion rates are responsible already for the fact that in higher dimension the transverse regularity is usually lower than in the two-dimensional case. Note that the results there never assert higher regularity for both foliations than in the two-dimensional area-preserving case. If the obstruction vanishes that was used to show optimality of those results, then the regularity "jumps" up a little, and a further obstruction, associated with different contraction and expansion rates, may prohibit regularity $C^{1+O(x|\log x|)}$. Only when all those finitely many obstructions vanish can we have $C^{1+O(x|\log x|)}$. These obstructions are best described in normal form [55], as is the Anosov cocycle.

To give a sample we show that a "1-2-resonance" produces an obstruction to C^1 foliations. To work with the simplest possible situation consider a 3-dimensional Anosov diffeomorphism f with fixed point p such that the eigenvalues $0 < \lambda < \mu < 1 < \eta < \infty$ of Df_p satisfy $\mu = \lambda\eta$. (This is a variant of the 1-2-resonance $\lambda_1 = \lambda_2^2$ for a symplectic system.) Up to higher-order terms that might arise from higher resonances the normal form of f at p is $f(x, y, z) = (\eta x, \mu y + axz, \lambda z)$. Representing E^u along the z-axis by $(1, v_1(z), v_2(z))$ gives

$$Df_{(0,0,z)}(1, v_1(z), v_2(z)) = (\eta, az + \mu v_1(z), \lambda v_2(z)),$$

which rescales to $(1, az/\eta + \mu v_1(z)/\eta, \lambda v_2(z)/\eta)$. Invariance of E^u therefore yields

$$v_1(\lambda z) = az/\eta + \mu v_1(z)/\eta.$$

Differentiating twice with respect to z gives $\lambda v_1'(0) = a/\eta + (\mu/\eta)v_1'(0)$, which implies $a = 0$ since $\lambda = \mu/\eta$. Thus the resonance term a in the normal form is an obstruction to C^1 Anosov splitting. (One can verify this without using normal forms, but the calculation is somewhat longer.) By the way, the work of Kanai mentioned in the next subsection made a rather stringent curvature pinching assumption to rule out a number of low resonances. The refinements by Feres and Katok that led to an almost complete proof of Theorem 20.17 centered on a careful study of the resonances that might arise without such pinching assumptions. This was delicate work because the issue are not only resonances at periodic points, but "almost resonances" between Lyapunov exponents. The papers [39; 37] contain an impressive development of these ideas.

While there is an analog of the Anosov cocycle in higher dimension, its vanishing is known to be necessary only for C^2 foliations [61] and is not known to lead to higher regularity of the invariant foliations. Thus it has not yielded any effective application, and the central portion of the above approach falls apart.

The bootstrap to C^∞ subbundles works in full generality, even without area-preservation, although it usually starts at regularity higher than C^3 (see [60; 47]). In other words, once the invariant foliations have a sufficiently high degree of regularity, they are always C^∞.

Smooth rigidity. The main issue in higher dimension is to conclude from smoothness of the invariant foliations that there is a smooth conjugacy to an algebraic model, and to identify the right algebraic model in the first place.

A result that appeared after systematic development of the continuous time situation (see also [7, Theorem 3]) will serve to illustrate this:

THEOREM 20.12 ([8]). *Let M be a C^∞ manifold with an C^∞ affine connection ∇, $f : M \to M$ a topologically transitive Anosov diffeomorphism preserving ∇ with E^u, $E^s \in C^\infty$. Then f is C^∞ conjugate to an automorphism of an infranilmanifold. The invariant connection hypothesis can be replaced by invariance of a smooth symplectic form.*

Note the absence of a topological hypothesis. (There is a finite-smoothness sharpening of this result [38] that does not use the powerful theorem of Gromov central to the proof by Benoist and Labourie.)

Now we turn to the continuous time case, where these developments are most significant.

The history begins with the work of Ghys [51], who classified volume-preserving Anosov flows on 3-manifolds with smooth invariant foliations into three types: suspensions of hyperbolic automorphisms of the torus, geodesic flows on surfaces of constant negative curvature (up to finite coverings), and a new type of flow that differs from the old ones by a special time change. If the flow is known to be geodesic then the smooth conjugacy to the constant curvature geodesic flow preserves topological and measure-theoretic entropies, and hence by entropy rigidity (see page 310 and [79]) the original metric is constantly curved. The work towards classification of flows with smooth invariant foliations has followed this model closely. Before describing this, let us mention in passing the secondary issue of reducing the regularity at which the classification becomes possible. In the situation of Ghys one can use an analysis of 3-dimensional volume-preserving Anosov flows and a result entirely analogous to Theorem 20.9 [68] to conclude:

THEOREM 20.13 [51; 68]. *A negatively curved metric on a compact surface is hyperbolic if its horocycle foliations are* $C^{1+o(x|\log x|)}$.

In higher dimension the seminal work is due to Kanai [78]. He was the first to implement the following strategy: If one assumes that the invariant foliations are smooth then one can study Lie bracket relations between the stable and unstable subbundles. The interaction between these and the dynamics can be used to build an invariant connection (named after him now [91]) and to show that it is flat, which in turn is used to build a Lie algebra structure that is identifiable as a standard model.

He obtained the following result:

THEOREM 20.14 [78]. *The geodesic flow of a strictly 9/4-pinched negatively curved Riemannian metric on a compact manifold is smoothly conjugate to the geodesic flow of a hyperbolic manifold if the invariant foliations are* C^∞.

Two groups picked up this lead, with the primary aim of removing the pinching hypothesis, which in particular rules out nonconstantly curved locally symmetric spaces as models. It also emerged that the main import of the assumptions is dynamical rather than geometric, and that therefore one should look for theorems about flows more general than geodesic ones.

Feres and Katok [39; 37] built on Kanai's idea by refining his arguments with intricate analyses of resonance cases for Lyapunov exponents to cover most of the ground in terms of the admissible algebraic models.

THEOREM 20.15 [37]. *Consider a compact Riemannian manifold* M *of negative sectional curvature. Suppose the horospheric foliations are smooth. If the metric is 1/4-pinched or* M *has odd dimension then the geodesic flow is smoothly conjugate to that of a hyperbolic manifold. If the dimension is 2 (mod 4) then the geodesic flow is smoothly conjugate to that of a quotient of complex hyperbolic space.*

Some of the results proved along the way to this conclusion did not assume that the flow under consideration is geodesic. The refinements over Kanai's work were, in the case of the first hypothesis, a more delicate argument for vanishing of the curvature of the Kanai connection. Under the second hypothesis Feres shows that if the Kanai connection is not flat then the invariant subbundles split further (resonance considerations enter here), and a connection associated with this further splitting must be locally homogeneous.

Roughly simultaneously the complete result about smooth conjugacy was obtained by Benoist, Foulon and Labourie [7]. Not only does it include all geodesic flows, but it requires only a contact structure, which turned out to require substantial additional work. This makes it a proper counterpart of the three-dimensional result of Ghys:

THEOREM 20.16 [7]. *Suppose Φ is a contact Anosov flow on a compact manifold of dimension greater than 3, with C^∞ Anosov splitting. Then there is an essentially unique time change and a finite cover on which the flow is C^∞ conjugate to the geodesic flow of a negatively curved manifold.*

What enables the authors to give a monolithic proof (as opposed to covering the various classes of symmetric spaces one by one) is a rigidity result by Gromov [53; 6; 138]. This is the place where substantial regularity is needed, and on an m-dimensional manifold one can replace C^∞ in hypothesis and conclusion by C^k with $k \geq m^2 + m + 2$. This theorem is invoked in the first major step of the proof, to produce a homogeneous structure: The diffeomorphisms of the universal cover that respect the splitting and the flow form a Lie group that acts transitively. (Gromov's theorem produces this structure on an open dense set, and the Kanai connection is used to extend it.) Step two determines the structure of this group and its Lie algebra, and step three develops the dynamics of the group and relates it to the expected algebraic model.

The Feres–Katok approach needs a slightly different minimal regularity. In fact, if one adds the a posteriori redundant assumption of (nonstrict) $\frac{1}{4}$-pinching (or merely strict $\frac{4}{25}$-pinching) then C^5 horospheric foliations always force rigidity [60].

We note an amplified version for the case of geodesic flows in which the conjugacy conclusion for geodesic flows is replaced by isometry of the metrics due to a more recent rigidity result by Besson, Courtois and Gallot, Theorem 20.29.

THEOREM 20.17. *If the horospheric foliations of a negatively curved compact Riemannian manifold are C^∞ then the metric is locally symmetric (up to isometry).*

The above result subsumes several classification steps. First of all, one obtains an orbit equivalence, which implies coincidence of the Lyapunov cocycles (periodic data). But furthermore, the original result in [7] directly arrives at a smooth conjugacy, which means that periods of periodic orbits are preserved as well.

This is an extra collection of moduli for the continuous time case. Finally, in the case of geodesic flows, there is, in addition, the Besson–Courtois–Gallot Theorem 20.29, which gives the isometry.

While the regularity of the invariant subbundles is usually substantially lower than in the two-dimensional case, it is widely believed that the minimal regularity for such smooth rigidity results should be C^2 or even $C^{1+\mathrm{Lip}}$, *i.e.*, quite close to that in Theorem 20.9. Indeed, these foliations are hardly ever $C^{1+\mathrm{Lip}}$:

THEOREM 20.18. *For an open dense set of* symplectic *Anosov systems the regularity predicted by computing B^u only from periodic points is not exceeded (i.e., if the rates compare badly at a single periodic point then the regularity is correspondingly low—at that periodic point) [62]. An open dense set of Riemannian metrics do not have $C^{1+\mathrm{Lip}}$ horospheric foliations [62].*

Furthermore, for any $\varepsilon > 0$ there is an open set of symplectic Anosov diffeomorphisms for which the subbundles and holonomies are C^ε at most on a (Lebesgue) null set [63].

If the invariant subbundles are C^2 then the Liouville measure coincides with the Bowen–Margulis measure of maximal entropy [58; 91]. (For Finsler metrics this is false [117].) According to the Katok Entropy Rigidity Conjecture (page 310), this should imply that the manifold is locally symmetric. Optimists might suspect that rigidity already appears from $C^{1+o(x|\log x|)}$ or $C^{1+\mathrm{zygmund}}$ on, but there is no evidence to that effect (save for Theorem 20.18).

Another result of Ursula Hamenstädt is worth remarking on here. It says that for contact Anosov flows with C^1 invariant foliations fixing the time parametrization fixes all other moduli of smooth conjugacy.

THEOREM 20.19 [57]. *If two conjugate (not just orbit equivalent) Anosov flows both have C^1 Anosov splitting and preserve a C^2 contact form then the conjugacy is C^2.*

The C^1 assumption on the splitting is not vacuous, but not stringent either, being satisfied by an open set of systems. Note that the conjugacy preserves both Lebesgue and Bowen–Margulis measure. If one keeps in mind that smooth conjugacy has been established mainly with one side being algebraic, this result is striking in its generality.

Inasmuch as they refer to flows, the hypotheses of the preceding rigidity results do not distinguish between the regularity of the strong versus weak invariant foliations. The reason is that for geodesic flows strong and weak foliations have the same regularity due to the invariant contact structure: The strong subbundles are obtained from the weak ones by intersecting with the kernel of the smooth canonical contact form.

Plante [121] showed that the strong foliations may persistently fail to be C^1, namely when the asymptotic cycle of volume measure is nonzero. Even though the latter is not the case for (noncontact) perturbations of geodesic flows, these flows may still fail to have C^1 strong foliations (see [116; 11], where the contact form is "twisted" by an extra "magnetic force term", which does not produce a nontrivial asymptotic cycle).

Entropy rigidity. A different rigidity conjecture was put forward by Katok in a paper that proved it for surfaces [79].

The result that prompted the conjecture is

THEOREM 20.20 [79]. *For the geodesic flow of a unit-area Riemannian metric without focal points on a surface of negative Euler characteristic E the Liouville and topological entropies lie on either side of $\sqrt{-2\pi E}$, with equality (on either side) only for constantly curved metrics.*

CONJECTURE 20.21 [79, p. 347]. *Liouville measure has maximal entropy only for locally symmetric metrics, i.e., only in these cases do the topological and Liouville entropies agree.*

This says that equivalence of Bowen–Margulis and Lebesgue measure only occurs for locally symmetric spaces. This conjecture has engendered an enormous amount of activity and remains unresolved. The exact nature of the results in [79] suggests some variants of this conjecture, however, that have been adressed more successfully.

THEOREM 20.22 [44]. *The conjecture holds locally along one-parameter perturbations of constantly curved metrics, but in dimension 3 it is no longer the case that a hyperbolic metric (with unit volume) maximizes Liouville entropy.*

The Katok entropy rigidity conjecture cannot take quite so neat a form as it does for surfaces. Foulon notes that for flows in dimension three it extends beyond the realm of geodesic flows:

THEOREM 20.23 [45]. *A smooth contact Anosov flow on a three-manifold whose topological and Liouville entropies coincide is, up to finite covers, conjugate to the geodesic flow of a constantly curved compact surface.*

CONJECTURE 20.24 (FOULON). *Three-dimensional C^∞ Anosov flows for which Bowen–Margulis and Lebesgue measure are equivalent must be C^∞ conjugate to either a suspension of a toral automorphism or the geodesic flow of a compact hyperbolic surface.*

That a metric is locally symmetric has been proved under a stronger but suggestive hypothesis [97]. Consider the universal cover M of the manifold in question and for each $x \in M$ define a measure λ_x on the sphere at infinity by projecting the Lebesgue measure on the sphere $S_x M$ along geodesics starting at x (Lebesgue or visibility measure). Use a construction of the (Bowen–)Margulis measure [104] to define measures ν_x on the sphere at infinity [56].

THEOREM 20.25. *If there is a constant a such that $\lambda_x = a\nu_x$ for all x then M is symmetric.*

PROOF. By [96; 137] it is asymptotically harmonic, and by [47] and Theorem 20.29 below it is symmetric.   □

In fact, one can also define a *harmonic* measure η_x at infinity for every $x \in M$ by using Brownian motion.

THEOREM 20.26 [95; 81]. *In the case of surfaces the harmonic measure class coincides with the Lebesgue class only when the curvature is constant.*

CONJECTURE 20.27 (THE "SULLIVAN CONJECTURE", [131, p. 724]). *In higher dimension the coincidence of the harmonic and visibility measure classes happens only for locally symmetric spaces.*

THEOREM 20.28. *If any two of these three measures here defined are proportional for every x then M is symmetric.*

PROOF. This again follows from [96; 137; 47; 9].   □

The goal can be restated as the requirement to relax the hypothesis from proportionality to mutual absolute continuity [97].

Coming from rather a different direction, Besson, Courtois and Gallot found themselves addressing a related issue by showing that topological entropy is minimized only by locally symmetric metrics. Strictly speaking, their result concerns the volume growth entropy h of a compact Riemannian manifold, which is the exponential growth rate of the volume of a ball in the universal cover as a function of the radius. This is a lower bound for the topological entropy of the geodesic flow with equality if the sectional curvature is nonpositive [103] (in fact, when there are no conjugate points [50]).

THEOREM 20.29 ([9]). *Let X, Y be compact oriented connected n-dimensional manifolds, $f : Y \to X$ continuous of nonzero degree. If g_0 is a negatively curved locally symmetric metric on X then every metric g on Y satisfies $h^n(Y, g) \operatorname{Vol}(Y, g) \geq |\deg(f)| h^n(X, g_0) \operatorname{Vol}(X, g_0)$ and for $n \geq 3$ equality occurs iff (Y, g) is locally symmetric (of the same type as (X, g_0)) and f is homotopic to a homothetic covering $(Y, g) \to (X, g_0)$. In particular, locally symmetric spaces minimize entropy when the volume is prescribed.*

A version of this result holds for nonpositively curved locally symmetric spaces of rank 1, and one may ask whether their method extends to higher rank.

A complementary result, about leaving the realm of geodesic flows, is contained in the work [118] of the brothers Paternain: "Twisting" any Anosov geodesic flow (by adding a "magnetic" term to the Hamiltonian) strictly decreases topological entropy.

21. Quantitative symplectic geometry (Helmut Hofer)

Denote by Symp^{2n} the category of $2n$-dimensional symplectic manifolds with embeddings serving as the morphisms. This carries the action of $(0, \infty)$ by rescaling: $\alpha^*(M, \omega) = (M, \alpha\omega)$. Consider a subcategory $\mathfrak{C}$ that is invariant under this action and $[0, \infty]$ with the standard ordering (on which one has the same action). We do not require it to be a full subcategory.

DEFINITION 21.1. A *(generalized) symplectic capacity* for $\mathfrak{C}$ is an equivariant functor $c : \mathfrak{C} \to [0, \infty]$ with the property that $c((M, \omega)) > 0$ if $M \neq \varnothing$. For $1 \leq d \leq n$ a *d-capacity* is a capacity such that $0 < c(B^{2d} \times \mathbb{R}^{2n-2d}) < \infty$ and $c(B^{2d-2} \times \mathbb{R}^{2n-2d+2}) = \infty$, where B^d denotes the open unit ball in $\mathbb{R}^d$.

An example of a n-capacity is

$$c(M, \omega) := \left(\int_M \omega^n \right)^{1/n}.$$

Let $B^{2n}(a)$ denote the ball of radius $(a/\pi)^{1/2}$ and define $Z^{2n}(a) = B^2(a) \times \mathbb{R}^{2n-2}$ We put $B^{2d} = B^{2d}(1)$ and similarly $Z^{2d} = Z^{2d}(1)$. Gromov's non-squeezing result implies the existence of 1-capacities. If $B^{2n}(a)$ symplectically embeds into $Z^{2n}(b)$ then $a \leq b$. Therefore one can take

$$c_{B^{2n}}(M, \omega) := \sup\{a \mid (B^{2n}, a\omega) \text{ symplectically embeds into } (M, \omega)\}$$

or

$$c^{Z^{2n}}(M, \omega) := \inf\{a \mid (M, \omega) \text{ symplectically embeds into } (Z^{2n}, a\omega)\}.$$

There are many 1-capacities one can construct from Floer theory, Gromov–Witten theory, symplectic field theory or contact homology, and many questions in symplectic geometry can be answered by constructing such a functor. No example of a d-capacity for $1 < d < n$ is known. Therefore it is important to ask

QUESTION 21.2. *Are there d-capacities for d other than 1 and n in dimension $2n$, where $n \geq 3$?*

This is a fundamental question about the nature of symplectic geometry. Clearly the n-capacities are volume-related invariants and 1-capacities are invariants of a 2-dimensional kind related to 2-dimensional cross-sections. The essence of Question 21.2 is captured by the next question concerned with dimension six.

QUESTION 21.3. *Is there an $\varepsilon > 0$ such that for every $r > 0$ there is a symplectic embedding of $B^2(\varepsilon) \times B^4(r)$ into $B^4(1) \times \mathbb{R}^2$?*

If the answer is "no" it means that in dimension six a 2-capacity exists, and in this case it is very likely that a proof has to be based on some "new symplectic technology".

The next question is exploring the problem if in some sense the technology to deal with symplectic geometry in dimension four is complete. Given a positive quadratic form Q define $E_Q := \{Q < 1\} \subset \mathbb{R}^{2n}$ to be the associated ellipsoid. Then there is a unique $a \in \Sigma := \{a \in (0, \infty)^n \mid a_1 \leq a_2 \leq \ldots \leq a_n\}$ such that E is by a linear symplectic map the same as the ellipsoid $E(a) := \{x = (z_1, \ldots, z_n) \in \mathbb{R}^{2n} \mid \sum |z_i|^2/a_i < 1\}$.

On Σ define a "linear" partial ordering $\leq_l$ by $a \leq_l b : \iff$ there exists a linear symplectic map T such that $T(E(a)) \subset E(b)$. By some linear algebra this order structure is the same as requiring $a_i \leq b_i$ for all i. Define a "nonlinear" partial ordering $\leq_{nl}$ by $a \leq_{nl} b : \iff$ there is a symplectic embedding of $E(a)$ into $E(b)$. It is a nontrivial result (due to Ekeland and Hofer) that on the set of points "between" $(1, \ldots, 1)$ and $(2, \ldots, 2)$ these two orderings are the same, but this fails on any larger set (Lalonde and McDuff for $n = 2$, Schlenk in general).

Consider capacities on ellipsoids and for $a \in \Sigma$ order the numbers $\{ja_i \mid j \in \mathbb{N}, i = 1, \ldots, n\}$ by size with multiplicities and denote this sequence by $c_k(a)$ (If $a = (1, 5)$ we get 1 2 3 4 5 5 6 7 8 9 10 10....). These are capacities for each k (and are due to Ekeland and Hofer).

QUESTION 21.4. *Is $a \leq_{nl} b$ in $\mathbb{R}^4$ equivalent to $c_k(a) \leq c_k(b)$ for all k and $a_1 a_2 \leq b_1 b_2$ (this is the volume constraint)?*

If the answer is indeed "yes" a proof can be expected to be very hard. Particular cases of this question are:

QUESTION 21.5. *Is $(1, 8) \leq_{nl} (3, 3)$? Is $(1, 4) \leq_{nl} (2, 2)$?*

Schlenk can symplectically embed

$$E(1, 8) \to E(3.612, 3.612) \text{ and } E(1, 4) \to E(2.692, 2.692).$$

Observe that $E(1, 4)$ and $E(2, 2)$ have the same volume, so an embedding will be very tight.

In general, consider a symplectic category $\mathfrak{C}$ and denote the collection of capacities on it by S. Then one can generate new ones. Consider a function $f : [0, \infty]^n \to [0, \infty]$ with $f(1, \ldots, 1) > 0$ that is positively 1-homogeneous ($f(ta) = tf(a)$) and monotone (if $a_i \leq b_i$ then $f(a) \leq f(b)$). Then, given capacities $c_1, \ldots, c_n \in S$ we get a new capacity $f(c_1, \ldots, c_n)$. Also, if $c_k \to c$ as $k \to \infty$ and $c(B^{2n}) > 0$ then c is a capacity as well. The following is essentially a rephrasing of Question 21.4.

QUESTION 21.6. *Do the c_k defined above together with $(\int \omega^2)^{1/2}$ generate S in this way?*

22. Hilbert's sixteenth problem (presented by Yulij Ilyashenko)

QUESTION 22.1 (HILBERT'S 16TH PROBLEM). *What can be said about the number and location of the limit cycles of a polynomial ordinary differential equation in the plane?*

This has been among the most persistent in Hilbert's list, and therefore even simplified versions make for substantial problems:

QUESTION 22.2 (HILBERT'S 16TH PROBLEM FOR QUADRATIC POLYNOMI-ALS). *What can be said about the number and location of the limit cycles of an ordinary differential equation in the plane whose right-hand side is a quadratic polynomial vector field?*

This question remains unresolved as well. There are partial results by Ilyashenko and Llibre of the following type. For a Zariski-open set of quadratic vector fields one can define a numerical characteristic of each vector field and then bound the number of limit cycles in terms of this parameter.

Numerous related problems may be found in the survey [69]

23. Foliations (presented by Steven Hurder)

Consider a compact manifold M with a foliation F.

QUESTION 23.1. *Can a leaf L in a minimal set $Z \subset M$ of the foliation be deformed? Or can the minimal set be deformed?*

Reeb showed that if there is a compact leaf with trivial holonomy (*i.e.*, only the identity) then it has a foliated neighborhood that is a product, *i.e.*, the situation is far from rigid. On the other hand, results by Stowe show that if there is enough cohomology data then one cannot move the leaf.

QUESTION 23.2. *If Z is a minimal set in M, are all leaves in Z diffeomorphic up to covers?*

Duminy proved in 1982 that if the foliation is C^2 with codimension 1 leaves and Z is exceptional (*i.e.*, neither M nor a single leaf) then all leaves in Z have Cantor ends.

24. "Fat" self-similar sets (Mark Pollicott)

A *similarity* on $\mathbb{R}^d$ is a map $T : \mathbb{R}^d \to \mathbb{R}^d$ for which $\|T(x, y) - T(u, v)\| = r\|(x, y) - (u, v)\|$, where $0 < r < 1$ and $\|\cdot\|$ is the Euclidean norm. Given similarities $T_1, \ldots, T_n : \mathbb{R}^d \to \mathbb{R}^d$ a set Λ is said to be *self-similar* if $\Lambda = \bigcup_{i=1}^{n} T_i(\Lambda)$. One may ask how "big" such sets can be, for example, how close to d the Hausdorff dimension can be, whether they can have positive Lebesgue measure or open interior.

In the case that $d = 2$, there are examples of self-similar sets with empty interior and positive Lebesgue measure (this is due to Csörnyei, Jordan, Pollicott, Preiss and Solomyak [22] and answers a question of Peres and Solomyak [119]). The construction uses 10 contractions (all by a factor of 3), but there is some latitude in how the similarities are chosen, and a different construction accomplishes the same result using 6 similarities.

QUESTION 24.1. *Can one find examples using fewer similarities?*

It is interesting to note that there are apparently no analogous results when $d = 1$.

QUESTION 24.2. *Are the examples of self similar sets with positive measure but empty interior in $\mathbb{R}$?*

Easier results are obtained from Sierpinski triangles. If $1/2 < \lambda < 1$ the similarities

$$T_0(x, y) = (\lambda x, \lambda y) + (0, 0)$$

$$T_1(x, y) = (\lambda x, \lambda y) + (1/2, 0)$$

$$T_2(x, y) = (\lambda x, \lambda y) + (0, 1/2)$$

produce "fat" Sierpinski triangles Λ_λ (the case $\lambda = 1/2$ gives the standard Sierpinski triangle).

It is easy to check that when $0 < \lambda < 1/2$ one obtains a Cantor set with $\dim_H(\Lambda_\lambda) = -\log 3 / \log \lambda$.

THEOREM 24.3 (JORDAN [72]).

$$\dim_H(\Lambda_\lambda) = -\frac{\log 3}{\log \lambda} \quad for\ a.e.\ \lambda \in [1/2, (4/3)^{1/3}].$$

There is a dense set $D \subset [1/2, 1/\sqrt{3}]$ such that $\dim_H(\Lambda_\lambda) < -\log 3 / \log \lambda$ for $\lambda \in D$. One of the most interesting remaining questions is the following.

QUESTION 24.4. *How large is the exceptional set D? Is it uncountable? Does it have nonzero Hausdorff dimension?*

The next result was proved in [73] by Jordan and Pollicott, and in [15] by Broomhead, Montaldi and Sidorov. Let m be the d-dimensional Lebesgue measure.

THEOREM 24.5. $m(\Lambda_\lambda) > 0$ for a.e. $\lambda \in [0.585\ldots, 0.647\ldots]$ and $int(\Lambda_\lambda) \neq \varnothing$ for $\lambda > .647\ldots$.

This suggests two natural questions.

QUESTION 24.6. *What is the largest value of λ such that Λ_λ has empty interior?*

Sidorov conjectures that the correct value is the reciprocal of the golden ratio.

QUESTION 24.7. *Is there any $\lambda \in [0.585\ldots, 0.647\ldots]$ for which $int(\Lambda_\lambda) = \varnothing$ and $m(\Lambda_\lambda) > 0$?*

References

[1] Dmitri Viktorovich Anosov: *Geodesic flows on Riemann manifolds with negative curvature*, Proc. Steklov Inst. Math. **90** (1967); *Roughness of geodesic flows on compact Riemannian manifolds of negative curvature*, Dokl. Akad. Nauk SSSR **145** (1962), 707–709; *Ergodic properties of geodesic flows on closed Riemannian manifolds of negative curvature*, Dokl. Akad. Nauk SSSR **151** (1963), 1250–1252.

[2] Dmitri Viktorovich Anosov and Anatole B. Katok: *New examples in smooth ergodic theory: Ergodic diffeomorphisms*, Trans. Moscow Math. Soc. **23** (1970), 1–35.

[3] Dmitriĭ Viktorovich Anosov and Yakov Sinai: *Some smooth ergodic systems*, Russian Math. Surveys **22** (1967), no. 5, 103–167.

[4] Zvi Artstein and Alexander Vigodner: *Singularly perturbed ordinary differential equations with dynamic limits*, Proc. Roy. Soc. Edinburgh Sect. A **126** (1996), 541–569.

[5] André Avez: *Anosov diffeomorphisms*, Proceedings of the Topological Dynamics Symposium (Fort Collins, CO, 1967), 17–51, W. Gottschalk and J. Auslander, eds., Benjamin, New York, 1968.

[6] Yves Benoist: *Orbites des structures rigides (d'après M. Gromov)*, Integrable systems and foliations/Feuilletages at systèmes intégrables (Montpellier, 1995), 1–17, Progr. Math. **145**, Birkhäuser, Boston, 1997.

[7] Yves Benoist, Patrick Foulon, and François Labourie: *Flots d'Anosov à distributions stable et instable différentiables,* J. Amer. Math. Soc. **5** (1992), no. 1, 33–74.

[8] Yves Benoist and François Labourie: *Sur les difféomorphismes d'Anosov affins à feuilletages stable et instables différentiables*, Invent. Math. **111** (1993), no. 2, 285–308.

[9] Gérard Besson, Gilles Courtois, and Sylvestre Gallot: *Entropies et rigidités des espaces localement symétriques de courbure strictement négative*, Geom. Funct. Anal. **5** (1995), no. 5, 731–799; *Minimal entropy and Mostow's rigidity theorems*, Ergodic Theory Dynam. Systems **16** (1996), no. 4, 623–649.

[10] M. Blank and L. Bunimovich: *Multicomponent dynamical systems: SRB measures and phase transitions*, Nonlinearity **16** (2003), no. 1, 387–401.

[11] Jeffrey Boland: *On the regularity of the Anosov splitting and conjugacy types for magnetic field flows*, unpublished.

[12] M. Boyle and T. Downarowicz: *The entropy theory of symbolic extensions*, Invent. Math. **156** (2004), 119–161.

[13] M. Boyle and T. Downarowicz: *Symbolic extension entropy: C^r examples, products and flows*, Discrete Contin. Dynam. Systems **16** (2006), no. 2, 329–341.

[14] M. Boyle, D. Fiebig and U. Fiebig: *Residual entropy, conditional entropy and subshift covers*, Forum Math. **14** (2002), 713–757.

[15] D. Broomhead, J. Montaldi, and N. Sidorov: *Golden gaskets: Variations on the Sierpiński sieve*, Nonlinearity **17** (2004), 1455–1480.

[16] Leonid A. Bunimovich, Nicolai I. Chernov, and Yakov G. Sinai: *Statistical properties of two-dimensional hyperbolic billiards*, Russian Math. Surveys **46** (1991), 47–106.

[17] Keith Burns, Federico Rodríguez Hertz, Jana Rodríguez Hertz, Anna Talitskaya, and Raúl Ures: *Density of accessibility for partially hyperbolic diffeomorphisms with one dimensional center*, in preparation.

[18] Keith Burns and Amie Wilkinson: *On the ergodicity of partially hyperbolic systems*, preprint, 2005.

[19] J. Buzzi: *Intrinsic ergodicity of smooth interval maps*, Israel J. Math. **100** (1997), 125–161.

[20] Kariane Calta: *Veech surfaces and complete periodicity in genus two*, J. Amer. Math. Soc. **17** (2004), no. 4, 871–908.

[21] Gonzalo Contreras and Gabriel Paternain: *Genericity of geodesic flows with positive topological entropy on S^2*, J. Differential Geom. **61** (2002), no. 1, 1–49.

[22] M. Csörnyei, T. Jordan, M. Pollicott, D. Preiss, and B. Solomyak: *Positive measure self-similar sets without interior*, Ergodic Theory Dynam. Systems **26** (2006), no. 3, 755–758.

[23] Danijela Damjanovic and Anatole Katok: *Periodic cycle functionals and cocycle rigidity for certain partially hyperbolic $\mathbb{R}^k$-actions*, Discrete Contin. Dynam. Systems **13** (2005), 985–1005.

[24] Danijela Damjanovic and Anatole Katok: *Local rigidity of partially hyperbolic actions of $\mathbb{Z}^k$ and $\mathbb{R}^k$, $k \geq 2$, I: KAM method and actions on the torus*, preprint. http://www.math.psu.edu/katok_a/papers.html/

[25] Danijela Damjanovic and Anatole Katok: *Local rigidity of partially hyperbolic actions, II: Restrictions of Weyl chamber flows on $\mathrm{SL}(n, \mathbb{R})/\Gamma$*, preprint. http://www.math.psu.edu/katok_a/papers.html/

[26] D. Dolgopyat and A. Wilkinson: *Stable accessibility is C^1 dense*, Geometric methods in dynamics, II, Astérisque **287** (2003), xvii, 33–60.

[27] T. Downarowicz: *Entropy of a symbolic extension of a totally disconnected dynamical system*, Ergodic Theory Dynam. Systems **21** (2001), 1051–1070.

[28] T. Downarowicz: *Entropy Structure*, J. Anal. Math. **96** (2005), 57–116.

[29] T. Downarowicz and J. Serafin: *Possible entropy functions*, Israel J. Math. **135** (2003), 221–251.

[30] T. Downarowicz and S. Newhouse: *Symbolic extensions in smooth dynamical systems*, Invent. Math. **160** (2005), no. 3, 453–499.

[31] Manfred Einsiedler, Elon Lindenstrauss, and Anatole Katok: *Invariant measures and the set of exceptions to Littlewood's conjecture*, Ann. of Math. (2) **164** (2006), no. 2, 513–560.

[32] Yong Fang: *Smooth rigidity of uniformly quasiconformal Anosov flows*. Ergodic Theory Dynam. Systems **24** (2004), no. 6, 1937–1959.

[33] Bassam Fayad and Raphaël Krikorian: manuscript in preparation.

[34] Bassam Fayad and Anatole B. Katok: *Constructions in elliptic dynamics*, Herman memorial issue, Ergodic Theory Dynam. Systems **24** (2004), no. 5, 1477–1520.

[35] Bassam Fayad and M. Saprykina: *Weak mixing disc and annulus diffeomorphisms with arbitrary Liouville rotation number on the boundary*, Ann. Sci. École Norm. Sup. (4) **38** (2005), no. 3, 339–364.

[36] Bassam Fayad, M. Saprykina, and Alistair Windsor: manuscript in preparation.

[37] Renato Feres: *Geodesic flows on manifolds of negative curvature with smooth horospheric foliations*, Ergodic Theory Dynam. Systems **11** (1991), no. 4, 653–686.

[38] Renato Feres: *The invariant connection of a $\frac{1}{2}$-pinched Anosov diffeomorphism and rigidity*, Pacific J. Math. **171** (1995), no. 1, 139–155.

[39] Renato Feres and Anatole Katok: *Invariant tensor fields of dynamical systems with pinched Lyapunov exponents and rigidity of geodesic flows*, Ergodic Theory Dynam. Systems **9** (1989), no. 3, 427–432; *Anosov flows with smooth foliations and rigidity of geodesic flows on three-dimensional manifolds of negative curvature*, Ergodic Theory Dynam. Systems **10** (1990), no. 4, 657–670.

[40] M. Field, I. Melbourne, and A. Török: *Stable ergodicity for smooth compact Lie group extensions of hyperbolic basic sets*, Ergodic Theory Dynam. Systems **25** (2005), 517–551.

[41] M. J. Field, I. Melbourne, and A. Török: *Stability of mixing and rapid mixing for hyperbolic flows*, Ann. of Math. (2), to appear.

[42] Todd Fisher: *The topology of hyperbolic attractors on compact surfaces*, Ergodic Theory Dynam. Systems **26** (2006), no. 5, 1511–1520.

[43] Todd Fisher: *Hyperbolic sets that are not locally maximal*, Ergodic Theory Dynam. Systems **26** (2006), no. 5, 1491–1509.

[44] Livio Flaminio: *Local entropy rigidity for hyperbolic manifolds*, Comm. Anal. Geom. **3** (1995), no. 3–4, 555–596.

[45] Patrick Foulon: *Entropy rigidity of Anosov flows in dimension three*, Ergodic Theory Dynam. Systems **21** (2001), no. 4, 1101–1112.

PROBLEMS IN DYNAMICAL SYSTEMS   319

[46] Patrick Foulon and Boris Hasselblatt: *Zygmund strong foliations*, Israel J. Math. **138** (2003), 157–169.

[47] Patrick Foulon and François Labourie: *Sur les variétés compactes asymptotiquement harmoniques*, Invent. Math. **109** (1992), no. 1, 97–111.

[48] John Franks: *Anosov diffeomorphisms*, Global analysis, 61–93, Proc. Sympos. Pure Math. **14**, American Mathematical Society, Providence, RI, 1970.

[49] John Franks and Robert Williams: *Anomalous Anosov flows*, Global theory of dynamical systems, 158–174, Zbigniew Nitecki and R. Clark Robinson, eds., Lecture Notes in Mathematics **819**, Springer, Berlin, 1980.

[50] A. Freire and Ricardo Mañé: *On the entropy of the geodesic flow in manifolds without conjugate points*, Invent. Math. **69** (1982), no. 3, 375–392.

[51] Etienne Ghys: *Déformations de flots d'Anosov et de groupes Fuchsiens*, Ann. Inst. Fourier (Grenoble) **42** (1992) no. 1–2, 209–247.

[52] Matthew Grayson, Charles C. Pugh, and Michael Shub: *Stably ergodic diffeomorphisms*, Ann. of Math. (2) **140** (1994), no. 2, 295–329.

[53] Michael Gromov: *Rigid transformations groups*, Géometrie differentielle (Paris, 1986), 65–139, Daniel Bernard and Yvonne Choquet-Bruhat, eds., Travaux en Cours **33**, Hermann, Paris, 1988,

[54] Misha Guysinsky: *Smoothness of holonomy maps derived from unstable foliation*, Smooth ergodic theory and its applications (Seattle, WA, 1999), 785–790, Proc. Sympos. Pure Math. **69**, American Mathematical Society, Providence, RI, 2001.

[55] Misha Guysinsky and Anatole Katok: *Normal forms and invariant geometric structures for dynamical systems with invariant contracting foliations*, Math. Res. Lett. **5** (1998), no. 1–2, 149–163.

[56] Ursula Hamenstädt: *A new description of the Bowen–Margulis measure*, Ergodic Theory Dynam. Systems **9** (1989), no. 3, 455–464; Boris Hasselblatt: *A new construction of the Margulis measure for Anosov flows*, Ergodic Theory Dynam. Systems **9** (1989), no. 3, 465–468.

[57] Ursula Hamenstädt: *Regularity of time-preserving conjugacies for contact Anosov flows with C^1 Anosov splitting*, Ergodic Theory Dynam. Systems **13** (1993), no. 1, 65–72.

[58] Ursula Hamenstädt: *Invariant two-forms for geodesic flows*, Math. Ann. **301** (1995), no. 4, 677–698.

[59] Michael Handel and William Thurston: *Anosov flows on new three manifolds*, Invent. Math. **59** (1980), no. 2, 95–103.

[60] Boris Hasselblatt: *Bootstrapping regularity of the Anosov splitting*, Proc. Amer. Math. Soc. **115**, (1992), no. 3, 817–819.

[61] Boris Hasselblatt: *Anosov obstructions in higher dimension*, Internat. J. Math. **4** (1993), no. 3, 395–407.

[62] Boris Hasselblatt: *Regularity of the Anosov splitting and of horospheric foliations*, Ergodic Theory Dynam. Systems **14** (1994), no. 4, 645–666.

[63] Boris Hasselblatt and Amie Wilkinson: *Prevalence of non-Lipschitz Anosov foliations*, Ergodic Theory Dynam. Systems **19** (1999), no. 3, 643–656.

[64] Federico Rodríguez Hertz: *Global rigidity of certain abelian actions by toral automorphisms*, in preparation.

[65] Federico Rodríguez Hertz, Jana Rodríguez Hertz, and Raúl Ures: *Acessibility and stable ergodicity for partially hyperbolic diffeomorphisms with 1D center bundle*, preprint, 2005.

[66] Eberhard Hopf: *Statistik der geodätischen Linien in Mannigfaltigkeiten negativer Krümmung*, Berichte über die Verhandlungen der Sächsischen Akademie der Wissenschaften zu Leipzig, Mathematisch-Physikalische Klasse **91** (1939), 261–304.

[67] Pascal Hubert and T. A. Schmidt: *An introduction to Veech surfaces*, Handbook of dynamical systems, 1B, 501–526, Elsevier, Amsterdam, 2006.

[68] Steven Hurder and Anatole Katok: *Differentiability, rigidity, and Godbillon–Vey classes for Anosov flows*, Publ. Math. IHES **72** (1990), 5–61.

[69] Y. Ilyashenko: Centennial history of Hilbert's 16th problem, Bull. Amer. Math. Soc. **39**, (2002), no. 3, 301–354.

[70] Michael V. Jakobson: *Absolutely continuous invariant measures for one-parameter families of one-dimensional maps*, Comm. Math. Phys. **81** (1981), no. 1, 39–88.

[71] Esa Järvenpää and Tapani Tolonen: *Relations between natural and observable measures*, Nonlinearity **18** (2005), no. 2, 897–912.

[72] Thomas Jordan: *Dimension of fat Sierpiński gaskets*, Real Anal. Exchange **31** (2005/06), no. 1, 97–110.

[73] Thomas Jordan and Mark Pollicott: *Properties of measures supported on fat Sierpinski carpets*, Ergodic Theory Dynam. Systems **26** (2006), no. 3, 739–754.

[74] Jean-Lin Journé: *A regularity lemma for functions of several variables*, Rev. Mat. Iberoamericana **4** (1988), no. 2, 187–193.

[75] Boris Kalinin and Anatole B. Katok: *Measure rigidity beyond uniform hyperbolicity: Invariant measures for Cartan actions on tori*, J. Mod. Dyn. **1** (2007), no. 1, 123–146. http://www.math.psu.edu/katok_a/papers.html/

[76] Boris Kalinin and Victoria Sadovskaya: *On local and global rigidity of quasiconformal Anosov diffeomorphisms*, J. Inst. Math. Jussieu **2** (2003), no. 4, 567–582.

[77] Boris Kalinin and Ralf Spatzier: *Rigidity of the measurable structure for algebraic actions of higher-rank abelian groups*, Ergodic Theory Dynam. Systems **25** (2005), no. 1, 175–200.

[78] Masahiko Kanai: *Geodesic flows of negatively curved manifolds with smooth stable and unstable foliations*, Ergodic Theory Dynam. Systems **8** (1988), no. 2, 215–239.

[79] Anatole B. Katok: *Entropy and closed geodesics*, Ergodic Theory Dynam. Systems **2** (1982), no. 2, 339–367.

[80] Anatole B. Katok: *The growth rate for the number of singular and periodic orbits for a polygonal billiard*, Comm. Math. Phys. **111** (1987), 151–160.

[81] Anatole B. Katok: *Four applications of conformal equivalence to geometry and dynamics*, Charles Conley memorial issue, Ergodic Theory Dynam. Systems **8*** (1988), 139–152.

[82] Anatole B. Katok and Boris Hasselblatt: *Introduction to the modern theory of dynamical systems*, Cambridge University Press, Cambridge, 1995.

[83] Anatole B. Katok and Alexey Kononenko: *Cocycle stability for partially hyperbolic systems*, Math. Res. Lett. **3** (1996), 191–210.

[84] Anatole Katok and Federico Rodríguez Hertz: *Uniqueness of large invariant measures for Z^k actions with Cartan homotopy data*, J. Mod. Dyn. **1** (2007), no. 2, 287–300. http://www.math.psu.edu/katok_a/papers.html/

[85] Anatole B. Katok and Ralf Jürgen Spatzier: *First cohomology of Anosov actions of higher rank abelian groups and applications to rigidity*, Publ. Math. IHES **79** (1994), 131–156.

[86] Anatole B. Katok and Ralf Jürgen Spatzier: *Subelliptic estimates of polynomial differential operators and applications to rigidity of abelian actions*, Math. Res. Lett. **1** (1994), 193–202.

[87] Anatole B. Katok and Ralf Jürgen Spatzier: *Invariant measures for higher rank hyperbolic abelian actions*, Ergodic Theory Dynam. Systems **16** (1996), 751–778.

[88] Steven Kerckhoff, Howard Masur, and John Smillie: *Ergodicity of billiard flows and quadratic differentials*. Ann. of Math. (2) **124** (1986), no. 2, 293–311; *A rational billiard flow is uniquely ergodic in almost every direction*, Bull. Amer. Math. Soc. (N. S.) **13** (1985), no. 2, 141–142.

[89] Yuri Kifer: *Averaging principle for fully coupled dynamical systems and large deviations,* Ergodic Theory Dynam. Systems **24** (2004), 847–871.

[90] Yuri Kifer: *Some recent advances in averaging*, Modern dynamical systems and applications, 385–403, Cambridge University Press, Cambridge, 2004.

[91] Gerhard Knieper: *Hyperbolic dynamics and Riemannian geometry*. Handbook of dynamical systems, 1A, 453–545, Boris Hasselblatt and Anatole Katok, eds., Elsevier, Amsterdam, 2002.

[92] Gerhard Knieper and Howard Weiss: C^∞ *genericity of positive topological entropy for geodesic flows on S^2*, J. Differential Geom. **62** (2002), no. 1, 127–141.

[93] J. Kulesza: *Zero-dimensional covers of finite-dimensional dynamical systems*, Ergodic Theory Dynam. Systems **15** (1995), 939–950.

[94] David DeLatte: *Nonstationary normal forms and cocycle invariants*, Random Comput. Dynam. **1** (1992/93), no. 2, 229–259; *On normal forms in Hamiltonian*

dynamics, a new approach to some convergence questions, Ergodic Theory Dynam. Systems **15** (1995), no. 1, 49–66.

[95] François Ledrappier: *Propriété de Poisson et courbure négative*, C. R. Acad. Sci. Paris Sér. I Math. **305** (1987), no. 5, 191–194.

[96] François Ledrappier: *Harmonic measures and Bowen–Margulis measures*, Israel J. Math. **71** (1990), no. 3, 275–287.

[97] François Ledrappier: *Applications of dynamics to compact manifolds of negative curvature*, Proceedings of the International Congress of Mathematicians, 1, 2 (Zürich, 1994), 1195–1202, Birkhäuser, Basel, 1995.

[98] Douglas Lind and Jean-Paul Thouvenot: *Measure-preserving homeomorphisms of the torus represent all finite entropy ergodic transformations*, Math. Systems Theory **11** (1977/78), no. 3, 275–282.

[99] E. Lindenstrauss: *Mean dimension, small entropy factors and an embedding theorem*, Publ. Math. IHES **89** (1999), 227–262.

[100] Elon Lindenstrauss: *Rigidity of multiparameter actions*, Probability in mathematics, Israel J. Math. **149** (2005), 199–226.

[101] E. Lindenstrauss and B. Weiss: *Mean topological dimension*, Israel J. Math **115** (2000), 1–24.

[102] Anthony Manning: *There are no new Anosov diffeomorphisms on tori*, Amer. J. Math. **96** (1974), 422–429.

[103] Anthony Manning: *Topological entropy for geodesic flows*, Ann. of Math. (2) **110** (1979), no. 3, 567–573.

[104] Grigoriĭ A. Margulis: *Certain measures that are connected with $\mathcal{Y}$-flows on compact manifolds* (in Russian), Funkcional. Anal. i Priložen. **4** (1970), no. 1, 62–76; English transl. in Functional Anal. Appl. **4** (1970), 55–67.

[105] Howard Masur: *the growth rate of trajectories of a quadratic differential*, Ergodic Theory Dynam. Systems **10** (1990), 151–176.

[106] Curtis T. McMullen: *Dynamics of* SL(2, ℝ) *over moduli spaces in genus two*, preprint.

[107] M. Misiurewicz: *On non-continuity of topological entropy*, Bull. Acad. Polon. Sci. Sér. Sci. Math. Astronom. Phys. **19** (1971), 319–320.

[108] M. Misiurewicz: *Diffeomorphism without any measure with maximal entropy*, Bull. Acad. Polon. Sci. Sér. Sci. Math. Astronom. Phys. **21** (1973), 903–910.

[109] M. Misiurewicz: *Topological conditional entropy*, Studia Math. **55** (1976), 175–200.

[110] Michał Misiurewicz: *Ergodic natural measures*, Algebraic and topological dynamics, 1–6, S. Kolyada, Y. Manin, and T. Ward, eds., Contemp. Math. **385**, American Mathematical Society, Providence, RI, 2005.

[111] M. Misiurewicz and A. Zdunik: *Convergence of images of certain measures,* Statistical physics and dynamical systems (Köszeg, 1984), 203–219, Progr. Phys. **10**, Birkhäuser, Boston, 1985.

[112] Jürgen K. Moser: *The analytic invariants of an area-preserving mapping near a hyperbolic fixed point,* Comm. Pure Appl. Math. **9** (1956), 673–692.

[113] Sheldon Newhouse: *On codimension one Anosov diffeomorphisms,* Amer. J. Math. **92** (1970), 761–770.

[114] Viorel Niţică and Andrei Török: *An open dense set of stably ergodic diffeomorphisms in a neighborhood of a non-ergodic one,* Topology **40** (2001), no. 2, 259–278.

[115] Jacob Palis and Jean-Christophe Yoccoz: *Rigidity of centralizers of diffeomorphisms.* Ann. Sci. École Norm. Sup. (4) **22** (1998), no. 1, 81–98.

[116] Gabriel Pedro Paternain: *On the regularity of the Anosov splitting for twisted geodesic flows,* Math. Res. Lett. **4** (1997), no. 6, 871–888.

[117] Gabriel Pedro Paternain: *On two noteworthy deformations of negatively curved Riemannian metrics,* Discrete Contin. Dynam. Systems **5** (1999), no. 3, 639–650.

[118] Gabriel Pedro Paternain and Miguel Paternain: *Anosov geodesic flows and twisted symplectic structures,* International Conference on Dynamical Systems (Montevideo, 1995), 132–145, Longman, Harlow, 1996; *First derivative of topological entropy for Anosov geodesic flows in the presence of magnetic fields,* Nonlinearity **10** (1997), no. 1, 121–131.

[119] Y. Peres and B. Solomyak: *Problems on self-similar and self-affine maps: An update,* Progr. Probab. **46** (2000), 95–106.

[120] Yakov B. Pesin: *Characteristic Ljapunov exponents, and smooth ergodic theory* (in Russian), Uspehi Mat. Nauk **32** (1977), no. 4 (196), 55–112, 287; English translation in Russian Math. Surveys **32** (1977), no. 4, 55–114.

[121] Joseph F. Plante: *Anosov flows,* Amer. J. Math. **94** (1972), 729–754.

[122] M. Pollicott: *On the rate of mixing of Axiom A flows,* Invent. Math. **81** (1985), 427–447.

[123] Charles C. Pugh and Michael Shub: *Stable ergodicity and partial hyperbolicity,* International Conference on Dynamical Systems (Montevideo, 1995), 182–187, Longman, Harlow, 1996; *Stably ergodic dynamical systems and partial hyperbolicity,* J. Complexity **13** (1997), no. 1, 125–179.

[124] Charles C. Pugh and Michael Shub: *Stably ergodic dynamical systems and partial hyperbolicity,* J. Complexity **13** (1997), no. 1, 125–179.

[125] Charles C. Pugh and Michael Shub: *Stable ergodicity and julienne quasi-conformality,* J. Eur. Math. Soc. (JEMS) **2** (2000), no. 1, 1–52.

[126] Charles C. Pugh, Michael Shub and Amie Wilkinson: *Hölder foliations,* Duke Math. J. **86** (1997), no. 3, 517–546.

[127] Daniel Rudolph: *×2 and ×3 invariant measures and entropy,* Ergodic Theory Dynam. Systems **10** (1990), no. 2, 395–406.

[128] Victoria Sadovskaya: *On uniformly quasiconformal Anosov systems*, Math. Res. Lett. **12** (2005), 425–441.

[129] Richard Evan Schwartz: *Obtuse triangular billiards, II: 100 Degrees worth of periodic billiard trajectories*, preprint, 2005 http://www.math.brown.edu/-res/Papers/deg100.pdf. (See also *Obtuse triangular billiards, I: Near the (2, 3, 6) triangle*, Experiment. Math. **15** (2006), no. 2, 161–182.)

[130] Steven Smale: *Mathematical problems for the next century*, Mathematics: Frontiers and perspectives, 271–294, American Mathematical Society, Providence, RI, 2000.

[131] Dennis Sullivan: *The Dirichlet problem at infinity for a negatively curved manifold*, J. Differential Geom. **18** (1983), no. 4, 723–732.

[132] Domokos Szász and Tamás Varjú: *Local limit theorem for the Lorentz process and its recurrence in the plane*, Ergodic Theory Dynam. Systems. **24** (2004), 257–278.

[133] Yoshio Togawa: *Centralizers of C^1-diffeomorphisms*, Proc. Amer. Math. Soc. **71** (1978), no. 2, 289–293.

[134] Yaroslav B. Vorobets: *Ergodicity of billiards in polygons* (in Russian) Mat. Sb. **188** (1997), no. 3, 65–112; English translation in Sb. Math. **188** (1997), no. 3, 389–434.

[135] Lai-Sang Young: *Ergodic theory of differentiable dynamical systems*, Real and complex dynamical systems: Proceedings of the NATO Advanced Study Institute (Hillerød, 1993), 293–336, Bodil Branner and Paul Hjorth, eds., NATO Adv. Sci. Inst. Ser. C Math. Phys. Sci. **464**, Kluwer, Dordrecht, 1995.

[136] Lai Sang Young: *Statistical properties of dynamical systems with some hyperbolicity*, Ann. of Math. (2) **147** (1998), 558–650.

[137] Cheng Bo Yue: *On the Sullivan conjecture*, Random Comput. Dynam. **1** (1992), no. 2, 131–145.

[138] Abdelghani Zeghib: *On Gromov's theory of rigid transformation groups: A dual approach*, Ergodic Theory Dynam. Systems **20** (2000), no. 3, 935–946; *Sur les groupes de transformations rigides: théorème de l'orbite dense-ouverte*, pp. 169–188 in Rigidité, groupe fondamental et dynamique, Panor. Synthèses **13**, Soc. Math. France, Paris, 2002.

[139] Antoni Szczepan Zygmund: *Trigonometric series*, Cambridge University Press, Cambridge, 1959 (and 1968, 1979, 1988), revised version of *Trigonometrical series*, Monografje Matematyczne **5**, Z subwencji Funduszu kultury narodowej, Warszawa, 1935.

BORIS HASSELBLATT
DEPARTMENT OF MATHEMATICS
TUFTS UNIVERSITY
MEDFORD, MA 02155
Boris.Hasselblatt@Tufts.edu